URBAN DEVELOPMENT FOR THE 21ST CENTURY

Managing Resources and Creating Infrastructure

URBAN DEVELOPMENT FOR THE 21ST CENTURY

Managing Resources and Creating Infrastructure

Edited by

Kimberly Etingoff

Apple Academic Press Inc.
3333 Mistwell Crescent
Oakville, ON L6L 0A2
Canada

Apple Academic Press Inc.
9 Spinnaker Way
Waretown, NJ 08758
USA

First issued in paperback 2021

Exclusive worldwide distribution by CRC Press, a member of Taylor & Francis Group

ISBN 13: 978-1-77463-578-0 (pbk)
ISBN 13: 978-1-77188-257-6 (hbk)

Library and Archives Canada Cataloguing in Publication

Urban development for the 21st century: managing resources and creating infrastructure / edited by Kimberly Etingoff.

Includes bibliographical references and index.
ISBN 978-1-77188-257-6 (bound)
1. City planning. 2. City planning--Environmental aspects. 3. Sustainable living.
I. Etingoff, Kim, editor

HT166.U65 2015 307.1'216 C2015-902545-1

Library of Congress Cataloging-in-Publication Data

Urban development for the 21st century: managing resources and creating infrastructure / editor, Kimberly Etingoff. -- 1st ed.

pages cm
Includes bibliographical references and index.
ISBN 978-1-77188-257-6 (alk. paper)
1. City planning. 2. Sustainable urban development. I. Etingoff, Kim, editor.

HT165.5.U73 2015 307.1'216--dc23 2015013370

Apple Academic Press also publishes its books in a variety of electronic formats. Some content that appears in print may not be available in electronic format. For information about Apple Academic Press products, visit our website at **www.appleacademicpress.com** and the CRC Press website at **www.crcpress.com**

About the Editor

KIMBERLY ETINGOFF

Kim Etingoff has a master's degree in urban and environmental policy and planning from Tufts University. Her recent experience includes researching a report on food resiliency within the city of Boston with Initiative for a Competitive Inner City. She worked in partnership with Dudley Street Neighborhood Initiative and Alternatives for Community and Environment to support a community food-planning process based in a Boston neighborhood, which was oriented toward creating a vehicle for community action around urban food issues, providing extensive background research to ground the resident-led planning process. She has worked in the Boston Mayor's Office of New Urban Mechanics, and has also coordinated and developed programs in urban agriculture and nutrition education. In addition, she has many years of experience researching, writing, and editing educational and academic books on environmental and food issues.

Contents

Acknowledgment and How to Cite

The editor and publisher thank each of the authors who contributed to this book. The chapters in this book were previously published in various places in various formats. To cite the work contained in this book and to view the individual permissions, please refer to the citation at the beginning of each chapter. Each chapter was carefully selected by the editor; the result is a book that looks at how to develop and design sustainable cities from a variety of perspectives. The chapters included are broken into three sections, which describe the following topics:

- Cities can present challenges in the form of unsustainable resource use. Chapter 1covers some strategies already in existence that can capitalize on the opportunities cities present to reduce resource use while maintaining a high standard of living.
- The authors of Chapter 2 present a survey-based emissions accounting methodology to determine profiles for case study households, in order to provide better context for emissions-reducing policies.
- Chapter 3 explores the relationship between population segments' environmental attitudes and urban resource consumption, finding little correlation. It then explores reasons and policy implications for the gap between environmental awareness and actual behavior.
- Chapter 4 seeks to understand the content of waste streams in urban Ghana, and explores opportunities to reconceptualize waste as resources to divert it from landfills and create opportunity.
- The authors of Chapter 5 focus on problems associated with and challenges to sanitation in a rapidly urbanizing region in Rwanda, finding cost as a major barrier.
- Urban transportation systems contribute to many unsustainable outcomes, but several potential solutions exist. Chapter 6 considers Bus Rapid Transit systems as city-shaping and sustainable opportunities to improve cities.
- Using human mobility data sets, the authors of Chapter 7 consider how to integrate personal and system constraints to create personalized, more efficient routes that contribute to an overall decrease in traffic.

- Chapter 8 presents an analytical GIS-based tool to assess transportation accessibility in terms of monetary cost and distance, finding that under current case study conditions car trips are the overall least costly mode of transportation.
- The authors of Chapter 9 measure and analyze the overall emissions from low-rise lifestyles, comparing them to emissions from high-rise apartment lifestyles, and consider the conditions under which each contribute to more emissions.
- Chapter 10 develops methodologies to spatially optimize micro-renewable energy residential development and to integrate residential energy efficiency and other environmental requirements, to be used by land use planners seeking to create sustainable developments.
- Chapter 11 presents a calculator tool to analyze the long-term energy use and ecological impacts of residential buildings and settlements during current use scenarios as well as for projected projects.

List of Contributors

Emanuela Abis
Department of Civil, Environmental Engineering and Architecture, Cagliari University, Cagliari, 09124, Italy

Pamela Abbott
Institute of Policy Analysis and Research, IPAR-Rwanda, Kigali 273, Rwanda

Jane Adogo
School of Law, Faculty of Business, Economics and Law, University of Surrey, Guildford, Surrey GU2 7XH, UK

Sanna Ala-Mantila
Department of Surveying Sciences, Aalto University School of Engineering, P.O. Box 15800, FI-00076 AALTO, Finland

Alex Arenas
Departament d'Enginyeria Informàtica i Matemàtiques, Universitat Rovira i Virgili, Av.da Països Catalans, 26, Tarragona 43007, Spain

Wolfgang Baaske
Studia - Studienzentrum für Internationale Analysen, Schlierbach, 4553, Austria

Xuemei Bai
Fenner School of Environment and Society, Australian National University, Canberra, Australia

Stuart L. Barr
Centre of Earth Systems Engineering, School of Civil Engineering and Geosciences, Newcastle University, Newcastle upon Tyne, NE1 7RU, UK

Robert Cervero
University of California, Berkeley, California, USA

Katrina Charles
Centre for Environmental Strategy, Faculty of Engineering and Physical Sciences, University of Surrey, Guildford GU2 7XH, UK

Jonathan Chenoweth
Centre for Environmental Strategy, Faculty of Engineering and Physical Sciences, University of Surrey, Guildford GU2 7XH, UK

Richard J. Dawson
Centre of Earth Systems Engineering, School of Civil Engineering and Geosciences, Newcastle University, Newcastle upon Tyne, NE1 7RU, UK

Manlio De Domenico
Departament d'Enginyeria Informàtica i Matemàtiques, Universitat Rovira i Virgili, Av.da Països Catalans, 26, Tarragona 43007, Spain

Ling Feng
Key Lab of Urban Environment and Health, Institute of Urban Environment, Chinese Academy of Sciences, Xiamen, China and Xiamen Key Lab of Urban Metabolism, Xiamen, China

Alistair C. Ford
Centre of Earth Systems Engineering, School of Civil Engineering and Geosciences, Newcastle University, Newcastle upon Tyne, NE1 7RU, UK

Marta C. González
Department of Civil and Environmental Engineering, Massachusetts Institute of Technology, 77 Massachusetts Avenue, Cambridge 02139, MA, USA

Jukka Heinonen
Department of Surveying Sciences, Aalto University School of Engineering, P.O. Box 15800, FI-00076 AALTO, Finland

Philip James
Centre of Earth Systems Engineering, School of Civil Engineering and Geosciences, Newcastle University, Newcastle upon Tyne, NE1 7RU, UK

Seppo Junnila
Department of Surveying Sciences, Aalto University School of Engineering, P.O. Box 15800, FI-00076 AALTO, Finland

Karl-Heinz Kettl
Graz University of Technology, Institute of Process and Particle Engineering, Graz, 8010, Austria and Energie Agentur Steiermark GmbH, Graz, 8020, Austria

Bettina Lancaster
Studia - Studienzentrum für Internationale Analysen, Schlierbach, 4553, Austria

Antonio Lima
School of Computer Science, University of Birmingham, Edgbaston, Birmingham B15 2TT, UK

appears to contribute to ever-growing levels of greenhouse gas (GHG) emissions. Meanwhile, in much of Africa and Asia and many nations in Latin America and the Caribbean, urbanization has long outstripped local governments' capacities or willingness to act as can be seen in the high proportion of the urban population living in poor quality, overcrowded, illegal housing lacking provision for water, sanitation, drainage, healthcare and schools. But there is good evidence that urban areas can combine high living standards with relatively low GHG emissions and lower resource demands. Chapter 1, by Satterthwaite, draws on some examples of this and considers what these imply for urban policies in a resource-constrained world. These suggest that cities can allow high living standards to be combined with levels of GHG emissions that are much lower than those that are common in affluent cities today. This can be achieved not with an over-extended optimism on what new technologies can bring but mostly by a wider application of what already has been shown to work.

Devising policies for a low carbon city requires a careful understanding of the characteristics of urban residential lifestyle and consumption. The production-based accounting approach based on top-down statistical data has a limited ability to reflect the total greenhouse gas (GHG) emissions from residential consumption. In Chapter 2, Lin and colleagues present a survey-based GHG emissions accounting methodology for urban residential consumption, and apply it in Xiamen City, a rapidly urbanizing coastal city in southeast China. Based on this, the main influencing factors determining residential GHG emissions at the household and community scale are identified, and the typical profiles of low, medium and high GHG emission households and communities are identified. Up to 70% of household GHG emissions are from regional and national activities that support household consumption including the supply of energy and building materials, while 17% are from urban level basic services and supplies such as sewage treatment and solid waste management, and only 13% are direct emissions from household consumption. Housing area and household size are the two main factors determining GHG emissions from residential consumption at the household scale, while average housing area and building height were the main factors at the community scale. The results show a large disparity in GHG emissions profiles among different households, with high GHG emissions households emitting about five times more than

Introduction

Urban planners around the world are increasingly concerned with creating and maintaining sustainable, healthy cities. Cities are at the forefront of the trend toward sustainable living since they are the site of concentrated population, resource use, and greenhouse gas emissions, yet also have the tools and the resources to address climate change and environmental degradation. Part of the urban planner's challenge is to impact individual behavior on a systemic, urban scale, since sustainable cities are made up of systems that encourage sustainable behavior.

The articles presented in this book cover many aspects of sustainable urban living on an individual yet systematic scale. The first section focuses on how individuals, households, and cities use resources and create greenhouse gas emissions; the concept of resources in this case is expanded to include waste streams. The second section considers the contribution of transportation systems to sustainability in cities, and explores several options for measuring and encouraging sustainable transportation options. In the third section, articles examine the renewable and non-renewable energy demand and use of housing, in an effort to encourage for sustainable housing solutions. Together, the articles provide planners with new options for assessing current contexts and moving cities toward greater sustainable living.

Kimberly Etingoff

With more than half the world's population now living in urban areas and with much of the world still urbanizing, there are concerns that urbanization is a key driver of unsustainable resource demands. Urbanization also

Gernot Stoeglehner
University of Natural Resources and Life Sciences, Department of Landscape, Spatial and Infrastructure Sciences, Institute of Spatial Planning and Rural Development, Peter-Jordan-Straße 82, Vienna, 1190, Austria

Aime Tsinda
Centre for Environmental Strategy, Faculty of Engineering and Physical Sciences, University of Surrey, Guildford GU2 7XH, UK

Christina von Haaren
Department of Environmental Planning, Leibniz University of Hannover, Hannover, 30419, Germany

Jin Wang
Key Lab of Urban Environment and Health, Institute of Urban Environment, Chinese Academy of Sciences, Xiamen, China and Xiamen Key Lab of Urban Metabolism, Xiamen, China

Michael Weiss
University of Natural Resources and Life Sciences, Department of Landscape, Spatial and Infrastructure Sciences, Institute of Spatial Planning and Rural Development, Peter-Jordan-Straße 82, Vienna, 1190, Austria and Environment Agency Austria, Vienna, 1090, Austria

Yunjun Yu
Key Lab of Urban Environment and Health, Institute of Urban Environment, Chinese Academy of Sciences, Xiamen, China and Xiamen Key Lab of Urban Metabolism, Xiamen, China

Tao Lin
Key Lab of Urban Environment and Health, Institute of Urban Environment, Chinese Academy of Sciences, Xiamen, China and Xiamen Key Lab of Urban Metabolism, Xiamen, China

Andrew Lovett
School of Environmental Sciences, University of East Anglia, Norwich, NR4 7TJ, UK

Denny Meyer
Institute for Social Research, Swinburne University of Technology, Melbourne, Victoria 3124, Australia

Hermine Mitter
University of Natural Resources and Life Sciences, Department of Landscape, Spatial and Infrastructure Sciences, Institute of Spatial Planning and Rural Development, Peter-Jordan-Straße 82, Vienna, 1190, Austria

Nora Niemetz
Graz University of Technology, Institute of Process and Particle Engineering, Graz, 8010, Austria and Südwind, Linz, 4020, Austria

Georg Neugebauer
University of Natural Resources and Life Sciences, Department of Landscape, Spatial and Infrastructure Sciences, Institute of Spatial Planning and Rural Development, Peter-Jordan-Straße 82, Vienna, 1190, Austria

Peter Newton
Institute for Social Research, Swinburne University of Technology, Melbourne, Victoria 3124, Australia

Kenan Okurut
Centre for Environmental Strategy, Faculty of Engineering and Physical Sciences, University of Surrey, Guildford GU2 7XH, UK

Martin Oteng-Ababio
Department of Geography and Resource Development, University of Ghana, Legon, Accra 00233, Ghana

Claudia Palmas
Department of Environmental Planning, Leibniz University of Hannover, Hannover, 30419, Germany

Steve Pedley
Robens Centre for Public and Environmental Health, University of Surrey, Guildford GU27XH, UK

David Satterthwaite
International Institute for Environment and Development (IIED), 3 Endsleigh Street, London WC1H 0DD, UK

low GHG emissions households. Emissions from high GHG emissions communities are about twice as high as from low GHG emissions communities. The findings can contribute to better tailored and targeted policies aimed at reducing household GHG emissions, and developing low GHG emissions residential communities in China.

Consumption is a transcending challenge for the 21st century that is stimulating research on multiple pathways required to deliver a more environmentally sustainable future. Chapter 3, by Newton and Meyer, is nested in what is a much larger field of research on sustainable consumption and reports on part of a major Australian Research Council study into the determinants of household resource consumption, based on a survey of 1,250 residents in Melbourne, Australia. Three environmental lifestyle segments are established that represent the spectrum of attitudes, opinions and intentions across the surveyed population: "committed" greens, "material" greens and "enviro-sceptics" (representing respectively 33.5%, 40.3% and 26.3% of the population). Each segment was found to display distinctive socio-demographic attributes, as well as urban geographies. However, few differences were found in relation to each segment's actual consumption of energy, water, housing space, urban travel and domestic appliances. The research findings indicate that in these areas of urban resource consumption—all principal contributors to the ecological footprint of households—there are sets of factors at work that override attitudes, opinions and intentions as indicators of consumer behaviour. Some of these factors are information, organization and finance related and are the focus of much public policy. However, the persistence of well ingrained habits and practices among individuals and households and the lack of norms and values in western societies that explicitly promote environmental conservation among its population, are fundamentally involved in the attitude-action gap and constitute important avenues for future research and action.

Conventional solid waste management protocols and thinking generally tend to assume that waste already exits and therefore needs to be managed. Consequently, most models of solid waste management, especially in the developing countries including Ghana, are simply reactions to the presence of something that needs to be disposed of or discarded. Chapter 4, by Oteng-Ababio, sees this conventional solid waste management philosophy as a potential barrier to an efficient and sustainable management and argues that adopting an integrated systemic approach will both help to control the processes that generate waste (including waste handling and utilization) and enable city managers to minimize waste generation in the first place. This paper uses a project initiated by a community-based organization in Ga Mashie (Accra) to explore the potential of converting household waste into a resource. Adopting a multiple research methodology, the study analyzes the characteristic and composition of waste generated within communities in Accra. The results show that a greater part of the 'waste' is recyclable or potentially recyclable and that a well-coordinated recycling programme will not only ensure a huge reduction of waste volume, but can equally lengthen the life of existing dumpsites and possibly, create wealth and reduce poverty. The paper argues that scaling up the project offers the local authority an opportunity to tap into the innovative strengths embedded in the project, particularly its physical and economic synergies, which may bolster community sustainable development.

Like most cities in developing countries, Kigali is experiencing rapid urbanisation leading to an increase in the urban population and rapid growth in the size and number of informal settlements. More than 60% of the city's population resides in these settlements, where they experience inadequate and poor quality urban services including sanitation. In Chapter 5, Tsinda and colleagues discuss the issues and constraints related to the provision of sustainable sanitation in the informal settlements in Kigali. Two informal settlements (Gatsata and Kimisagara) were selected for the study, which used a mixed method approach for data collection. The research found that residents experienced multiple problems because of poor sanitation and that the main barrier to improved sanitation was cost. Findings from this study can be used by the city authorities in the planning of effective sanitation intervention strategies for communities in informal settlements.

The urban transportation sector's environmental, economic, and social footprint is immense and expanding. Many of the world's most vexing and pressing problems—fossil fuel dependency, global warming, poverty, and social exclusion—are inextricably tied to the transportation sector. Much of the blame for the transportation sector's inordinate environmental footprint lies in the increasing automobile-dependency of cities. Rapid motorization unavoidably shifts future travel from the most sustainable modes—public transport and non-motorized ones (walking and cycling) – to private vehicles. Despite growing concerns over energy futures, climate change, and access for the poor, public transport's market share of trips is expected to erode over the next decade in all world regions if past trends (in how ownership and usage of the private car is priced and public financial resources are spent on transport infrastructure) continue. A paradigm shift is needed in how we think about transportation and its relationship to the city. Chapter 6, by Cervero, argues that the integration of transport infrastructure and urban development must be elevated in importance. In many cities of the Global South, recent Bus Rapid Transit (BRT) investments provide an unprecedented opportunity to do just that. To date, however, BRT systems have failed to leverage compact, mixed-use development due not only to little strategic station-area planning but also factors like siting lines and stations in stagnant urban districts and busy roadway medians. BRT systems are being conceived and designed as mobility investments rather than city-shaping ones. Given that the majority of future urban growth worldwide will be in intermediate-size cities well-suited for BRT investments, the opportunities for making these not only mobility investments but city-shaping investments as well should not be squandered. Transit-oriented development is but one of a number of built forms that hold considerable promise toward placing cities of the Global South on more sustainable mobility and urbanization pathways.

Human mobility in a city represents a fascinating complex system that combines social interactions, daily constraints and random explorations. New collections of data that capture human mobility not only help us to understand their underlying patterns but also to design intelligent systems. Bringing us the opportunity to reduce traffic and to develop other applications that make cities more adaptable to human needs. In Chapter 7, De Domenico and colleagues propose an adaptive routing strategy which accounts for individual constraints to recommend personalized routes and, at the same time, for constraints imposed by the collectivity as a whole. Using big data sets recently released during the Telecom Italia Big Data Challenge, the authors show that their algorithm allows them to reduce the overall traffic in a smart city thanks to synergetic effects, with the participation of individuals in the system, playing a crucial role.

Transport accessibility is an important driver of urban growth and key to the sustainable development of cities. Chapter 8, by Ford and colleagues, presents a simple GIS-based tool developed to allow the rapid analysis of accessibility by different transport modes. Designed to be flexible and use publicly-available data, this tool (built in ArcGIS) uses generalized cost to measure transport costs across networks including monetary and distance components. The utility of the tool is demonstrated on London, UK, showing the differing patterns of accessibility across the city by different modes. It is shown that these patterns can be examined spatially, by accessibility to particular destinations (e.g., employment locations), or as a global measure across a whole city system. A number of future infrastructure scenarios are tested, examining the potential for increasing the use of low-carbon forms of transport. It is shown that private car journeys are still the least cost mode choice in London, but that infrastructure investments can play a part in reducing the cost of more sustainable transport options.

Suburban households living in spacious detached houses and owing multiple cars are often seen as main culprits for negative greenhouse consequences of urban sprawl. Consequently, the effects of sprawl have been mostly studied from the viewpoints of emissions from home energy consumption and private driving. Little attention has been paid to the changes in other consumption. In Chapter 9, Ala-Mantila and colleagues link urban sprawl to the proliferation of semi-detached and detached housing, described as a low-rise lifestyle, at the expense of apartment house living i.e., high-rise lifestyle. The authors analyze differences between the low-rise and the high-rise lifestyles and their environmental effects in the Helsinki Metropolitan Area, taking into account all consumption activities. Environmental effects are assessed by combining greenhouse gas intensities from a consumption-based environmentally-extended input-output (EE I-O) model with expenditure data. Then these carbon footprints are further elucidated with regression analysis. They find that low-rise lifestyles causes approximately 14% more emissions than high-rise lifestyles. However, the relative contributions of emissions from different sources, whether direct or indirect, are almost equal for both. Furthermore, when controlling the level of expenditure, the differences between the two lifestyles unexpectedly disappear and in certain cases are even reversed. The authors believe that their consumption-based approach facilitates the understanding of sprawling lifestyles and offers important insights for sustainable policy-design and urban planning.

In recent years, there has been an increasing interest in using micro-renewable energy sources. However, planning has not yet developed methodological approaches (1) for spatially optimizing residential development according to the different renewable energy potentials and (2) for integrating objectives of optimized energy efficiency with other environmental requirements and concerns. Chapter 10, by Palmas and colleagues, addresses these topics by firstly presenting a new concept for the regional planning. The methodological approach for the evaluation of spatial variations in the available energy potential was based on the combination of existing methods adapted to the local scale and data availability. For assessing the bioenergy potential, a new method was developed. Other environmental criteria for deciding about sustainable locations were identified through a survey of more than 100 expert respondents. This survey involved pairwise comparisons of relevant factors, which were then translated into relative weights using the Analytical Hierarchy Process. Subsequently, these weights were applied to factor maps in a Geographical Information System using a weighted linear combination method. In the test region, the eastern metropolitan area of Cagliari, Sardinia, this analysis resulted in the designation of suitable areas for new settlements and preferred locations for micro-renewable technologies. Based on expert preferences, a number of alternatives for future housing development were identified, which can be integrated in the early stages of land use or development plans. The method proposed can be an effective tool for planners to assess changes and to identify the best solution in terms of sustainable development.

Chapter 11, by Stoeglehner and colleagues, introduces the concepts, methods and the implementation of a calculator for the energetic long term analysis of residential settlement structures (ELAS). The freely available online tool addresses, on the one hand, the complexity of the environmental impacts of buildings and settlements including embodied energy and on the other hand, the mobility of the inhabitants and the necessity to provide ecological and socio-economic valuation on the base of a coherent data set. Regarding the complexity of ecological impacts, housing was represented as a life cycle network, combining the life cycles of energy provision as well as buildings and infrastructure depending on the location and supply structure of the settlement. Comprehensive inventories for these different aspects were included. They were then used to evaluate the whole system of activity linked to buildings and settlements with three different ecological valuation methods and then coupled with a socio-economic appraisal. With the ELAS calculator, a status quo analysis for existing settlements can be carried out, as well as planning alternatives can be assessed which include new developments, the renovation of buildings in the settlement, the change of energy supplies as well as the demolition of settlements with reconstruction on the same or a different site.

PART I

SUSTAINABLE RESOURCE LIFE CYCLES

CHAPTER 1

How Urban Societies Can Adapt to Resource Shortage and Climate Change

DAVID SATTERTHWAITE

1.1 INTRODUCTION

Urban centres and urbanization (the increasing proportion of a population living in urban centres) do not enjoy good reputations from an ecological perspective and are often cited as drivers of unsustainable environmental change [1,2]. Certainly, the much increased pressures on the world's natural resources and ecological systems over the last century or so has been accompanied by a very rapid growth in the world's urban population—from 260 million in 1900 [3] to over 3.4 billion today [4]. The world's population has urbanized rapidly, driven mainly in recent decades by the very large expansion in the world's economy and by changes in its structure. Most new investment, economic value and employment have been in industry and service enterprises and most such enterprises have chosen to locate in urban areas. This helps in explaining why the ratio of rural : urban

How Urban Societies Can Adapt to Resource Shortage and Climate Change. Satterthwaite D. Philosophical Transactions A *373,2038 (2015), DOI: 10.1098/rsta.2010.0350.*

dwellers changed from 7:1 in 1900 to around 1:1 today [5]. The rapid increase in urban population and urban production has been accompanied by an even more rapid increase in the use of fossil fuels (and hence in carbon dioxide emissions) and in most other mineral resources and in freshwater, fish and forestry products. Urbanization is also associated with increasing wealth (at least for a proportion of the growing urban population) and increasing per capita consumption levels. In many nations, urbanization has also been influenced by the locations chosen by multi-national corporations for their production and for the centres where they concentrate their management [6]. Thus, urbanization can be seen as one of the key drivers of high levels of resource use and waste generation that have serious ecological consequences locally (within and around urban centres), regionally (where resource and waste flows from urban centres shift to the wider region) and globally (for instance in regard to climate change and in the reduction in the ozone layer).

However, these local, regional and global ecological consequences are not so much driven by urbanization as by the rapid increase in consumption levels and in the number of people with high-consumption lifestyles—and the (increasingly globalized) production systems that serve and support these. A considerable proportion of this consumption is from middle- and upper-income rural inhabitants and when the consumption and waste generation patterns of rural and urban inhabitants with comparable incomes are compared, urban inhabitants generally have lower consumption and less waste generation. The paper also highlights how not all urbanization is associated with large regional and global ecological impacts—and ends with a discussion of how urbanization can help to delink high standards of living from large local, regional and global ecological impacts. Urbanization brings obvious economic advantages for businesses which explain why private investment concentrates in urban areas but it also brings some environmental advantages that are less obvious—for reducing resource use and greenhouse gas (GHG) emissions, for good environmental health and for building resilience to the likely impacts of climate change. But whether these advantages are realized depends on how urban centres are structured and governed, and, for resource use and waste generation, on the choices made by their wealthier inhabitants.

1.2 URBAN CHARACTERISTICS

Urban centres share some characteristics—a concentration of people and enterprises and their buildings and wastes, infrastructure and usually some public institutions (for instance for schools, police force and for some the seat of local government). They also share a reliance on resources from outside their boundaries (food, usually freshwater and fossil fuels and other mineral resources) and often extra-urban sinks for their wastes. And as urban centres expand, so they transform natural landscapes and land use and land cover—for instance creating more stable sites by filling valleys and swamps and reshaping hills and in the process disrupting local ecologies, changing water flows and increasing impermeable areas [7,8]. The creation and expansion of urban centres also draws in large volumes of materials—for land fill and building materials, food, forestry products and other natural resources and these material flows also change and often transform the areas from which they are drawn. Larger cities often outgrow the capacity of their locality to provide freshwater and many cities have helped decrease such supplies through polluting local sources or mismanaging local aquifers [9].

Urban centres also change the environment of their locality—for instance by reducing rainfall and increasing night-time temperatures [10]. Almost all large urban centres are warmer than their surrounding rural areas, although the size and timing of these differentials are highly variable and influenced by many factors [1]. The urban heat island effect is generally higher in larger urban centres or in particular districts of large urban centres—for instance high-density areas with multi-storey buildings, little open space and little effective natural ventilation. Heat islands have both local and extra-local impacts; locally, they can increase heat stress with particularly serious consequences for vulnerable and working populations [11], pollution (as the higher temperature facilitates more generation of ozone if the precursors are already present) and energy demands for cooling (which when achieved through air conditioning also means more waste heat). Regionally, they usually increase demand for water and globally, they can increase GHG emissions, if a growing use of air conditioning is powered by electricity from thermal power stations.

Thus the energy and material demands of enterprises, residents and institutions in urban areas can alter land use and land cover and freshwater availability locally and often regionally while the disposal of their wastes into water, soils and the atmosphere can impact local to regional and global biogeochemical cycles and climate [12].

The energy and material demands that larger and wealthier urban centres concentrate for natural resources are often increasingly met by imports from beyond their region—including imports from abroad. For instance, around 80 per cent of the food consumed in London is imported from other countries [13]. Many major cities draw on freshwater resources far from their boundaries [14,15]—for instance Mexico City now has to draw some of its water supplies from 150 km away, and pump it up nearly 1000 m [16,17].

But there is such diversity among urban centres in their size, structure, spatial form, economy, wealth, local resource availability and local, regional and global ecological impact that few generalizations are valid. Urban centres vary in size from a few thousand (or in some nations a few hundred) inhabitants to the 21 metropolises that by 2010 had (or were estimated to have) more than 10 million inhabitants [4] and were termed megacities. United Nations (UN) estimates for 2010 suggest that around half the world's urban population lives in urban centres of under 500 000 inhabitants [4] and a considerable proportion of these lives in urban centres with under 20 000 inhabitants (although the proportion living in such urban centres in any nation is much influenced by the criteria used to define an urban centre—for instance if urban centres are all settlements with over 1000 inhabitants or all settlements with over 10 000 inhabitants) [18]. There is no universally accepted definition for an urban centre or for a city or for when an urban centre becomes a city but the term city implies a scale or a political or religious status that would mean that a large section of the world's urban population does not live in cities. Although it is common to see the comment that more than half the world's population lives in cities, this is not correct; they live in urban centres, a high proportion of which are small market towns or service centres that would not be considered to be cities.

Urban centres have economies that vary from the simple—for instance small market towns, mining centres or tourist resorts—to the complexities of large cities and metropolitan regions whose economic base serves local, regional, national and global markets and who sit within large and complex transport systems where a high proportion of the workforce may live outside the city. The differentials between urban centres for all aspects of material and energy use and waste generation are likely to be very large, although the analyses of material and waste flows undertaken to date have given little attention to urban centres in low-income nations. But most small market towns or service centres in low-income nations are likely to have very low levels of use for non-renewable resources as incomes and consumption levels are low so demand for material-intensive capital goods is also low and there is little or no industry. These urban centres draw most of their material needs (food, other natural resources, water) from close by. A high proportion of the urban population in many low-income nations do not have electricity and use biomass fuels [19]. For instance, among the nations that the UN classifies as the Least Developed Countries, close to half the urban population lack access to electricity and more than three-fifths lack access to modern fuels [19]. This also implies very low levels of GHG emissions per person. A proportion of low-income nations has GHG emissions below 0.1 tonnes of carbon dioxide equivalent (CO_2e) per year [20] and it is likely that a proportion of their small urban centres have comparable levels. From a perspective of ecological sustainability, these small urban centres perform extremely well. Even some larger urban centres in low-income nations have GHG emissions that are around 0.1 tonnes CO_2e per person per year while many cities in (for instance) India and Bangladesh have below 1 tonne [21]. But as discussed below, most of the more ecologically sustainable urban centres have low living standards with much of their population facing large preventable health burdens.

Meanwhile, there are many wealthy cities with GHG emissions that are 10–25 tonnes of CO_2e per person per year and so 100–250 times those with 0.1 tonnes per person per year. Also with levels of resource use for fossil fuels and many minerals that are far higher and likely to be unsustainable if extended to a larger section of the world's urban population.

1.3 MEASURING THE ENVIRONMENTAL PERFORMANCE OF URBAN AREAS

The original definition of sustainable development was a combination of two goals: meeting the needs of the present without compromising the ability of future generations to meet their needs [22]. For urban centres, this requires that they meet their residents' needs (being healthy, enjoyable, resilient places to live and work for all their inhabitants) while ensuring that the draw of their populations' consumption and enterprises' production on local, regional and global resources and sinks is not disproportionate to their finite capacities. Anthropogenic climate change adds to the urgency of the second goal and requires no disproportionate contribution to GHG emissions. It also requires urban adaptation to achieve the resilience needed for the new hazards and increased risk levels that climate change will bring.

How the term 'disproportionate' gets interpreted has long been debated, as have the metrics to be used for measuring sustainability [23]. Initially, the concerns for sustainability focused on the depletion of non-renewable resources, especially oil; then there was a realization that key resources such as fertile soils, freshwater, forests and fisheries that are renewable need to be included because their supply is finite—and can be reduced (through deforestation, soil degradation, pollution or over-fishing). To this were added concerns of global systems degraded or disrupted, including the depletion of the stratospheric ozone layer, global warming and the loss of biodiversity. But there are no easy ways to measure whether a city, smaller urban centre or rural area has a 'disproportionate' draw on local, regional and global resources and sinks. In part, this is because much of its draw on resources and sinks may be embedded in the goods that are sold or used there but manufactured elsewhere. For instance, wealthy cities can retain parks, forests and wilderness sites within and around them because most goods used by their citizens that are water-, energy-, pollution-, waste- and land-intensive are drawn from beyond their surrounds—and much of it from abroad [12].

To manage any city's draw on resources and sinks, there need to be measurements that indicate the scale and nature of this draw [24,25], including draws from beyond the city and its surrounds. Various ways have been developed for mapping the energy and material flows into cities and the wastes out of cities [26,27]. There are also a range of ways of measuring some of the ecological impacts of particular goods or services consumed in urban areas that consider these from the point of production (and its material needs) to the point of consumption or disposal—for instance the natural material input into a product in relation to the service it provides [28] or the measuring of 'virtual water' (for instance to consider the water-using implications of goods imported into a city in their original production) or food miles (the distance that food is transported from where it is produced to where it is consumed). There is also a particular interest in the ecological impacts of food systems; these have successfully met rising demands worldwide, including the rapid increase in the proportion of non-agricultural workers to agricultural workers that accompanies urbanization and including diets that are far more meat-intensive. But there are the environmental and ecological consequences of a far more energy-, chemical- and carbon-intensive food production to consider [29].

Perhaps the best-known and most widely used measure for cities and nations is the ecological footprint [30]. This calculates the productive land area that a city or nation draws on continuously to provide the food and resources that its inhabitants, enterprises and institutions use, to process the wastes and to absorb the carbon dioxide produced. From this can be calculated the average ecological footprint of each person within a city or nation in terms of productive land area. This can be compared with the land area estimated to be the global average that is sustainable. Figures for wealthy nations or cities in wealthy nations have been shown to be several times this average. Ecological footprints can also be calculated for individuals or households—for instance to highlight the much larger ecological footprint of wealthy households when compared with low-income households [24].

But getting a precise and complete measure of resource use and waste generation within a city is limited by data availability. In many nations, much of the data needed are not available or only available for nations. There are also the complications as to what to include—for instance for a city, do the material, energy and waste implications of the consumption of tourists and of commuters who work in the city but live outside get included in the figures for the city? Is it possible to take out of the accounting for a city the material, energy and waste generated in the fabrication of goods or services that are sold outside the city? If this is not subtracted, the cities that are centres for the manufacture of key goods for lowering ecological footprints such as photovoltaic cells, windmills or hydrogen buses will appear to be unsustainable. There are also difficulties with avoiding double counting—for instance with energy consumption that may be accounted for within goods (for instance food and water) that then has to be subtracted from energy statistics [13].

There are data showing the much higher levels of resource use per person in high-income nations when compared with low- and middle-income nations and the higher levels of domestic waste [15]. But it is difficult to judge whether a nation's, city's or individual's draw on planetary resources and sinks is 'disproportionate' when there is uncertainty about the size of the resource or the capacity of the sink. For instance, for many resources, the quantity available depends on the price (as prices rise so a greater stock becomes available as higher prices allow the exploitation of lower quality ores or oil-bearing materials). Higher prices encourage more exploration that can lead to finding new sources that expand the global stock.

For global warming, the metrics for measuring the global impacts of nations or sub-sets of nations including cities are more straightforward—and done through GHG emission inventories. For cities, these have been based on the methods and sources used for national GHG emission inventories, and follow guidelines set by the Intergovernmental Panel on Climate Change (IPCC) [31]. City GHG emission inventories have now been done for over a 100 cities [21] and as noted above, these show the enormous range in the scale of average per capita emissions of GHGs.

An inventory of emission sources can allocate their emissions between cities, other urban centres and rural areas but this is not a simple exercise. For instance, the places with large coal-fired power stations are very high GHG emitters although most of the electricity they generate may be used elsewhere. This is why GHG emission inventories generally assign cities the emissions generated in providing the electricity consumed within their boundaries, even when the electricity is produced elsewhere. This helps explain why some cities have surprisingly low per capita emissions relative to their wealth—for instance cities supplied with electricity from hydropower or nuclear power.

There are other difficulties in assigning GHGs to particular locations. For instance, transport is one of the main sources of GHG emissions in all cities but do the emissions from the fuels used by car-driving or rail or bus using commuters get attributed to the city where they work or the place where they live (which may be a suburb or a rural area outside the city where they work)? Which locations get assigned the carbon emissions from air travel? Total carbon emissions from any city with an international airport are much influenced by whether or not the city is assigned the fuel loaded onto the aircraft. Total carbon emissions for cities such as Sao Paulo, Rio de Janeiro, London or New York are much influenced by whether or not these cities are assigned the fuel loaded onto aircraft in their airports [32,33]. If aviation was included in New York's GHG emission inventory, it would constitute 14 per cent of total CO_2e emissions [32]. Then there are the complications of city airports that are outside city boundaries—so should London be allocated the carbon emissions from planes fuelled just at Heathrow and City airports (that are within its boundaries) or also Stansted, Gatwick and Luton airports that serve Londoners (even though a proportion of the passengers flying from these airports are in transit and do not come to London or do not live in London)?

But at least for global warming, there is an information base on the GHG emissions taking place within each nation's boundaries and thus a guide to their contribution to the global problem. This can guide action now to reduce GHG emissions globally. But here too, there is debate about how responsibility should be allocated between nations. Should it take account of historic contributions to GHGs in the atmosphere (which would further increase the responsibility of high-income nations)? Should the system that sets limits on emissions be based on where the emissions are produced or the homes of the consumers whose consumption caused the emissions (again the latter greatly increases the responsibilities of high-income nations as emissions within their borders are kept down by importing most goods whose fabrication involved large emissions)? But whichever method is used for assigning responsibilities, how urban centres perform as places to live and as centres of production and consumption has great relevance. So the issue is whether urban centres are or could be centres of 'good' development and environmental management and centres with carbon emissions or ecological footprints that are sustainable. This also means centres of lifestyles that keep average per capita GHG emissions below whatever level is calculated as the average level for the world's population that would avoid dangerous climate change—sometimes called the fair share level. These goals have to be considered together. As noted above, most urban centres in low-income nations have very low levels of GHG emissions per person because there is little or no industry and consumption levels are so low (including a high proportion of the population that do not get enough to eat). Many urban centres in high-income nations are among the world's healthiest and safest places but with ecological footprints that are vastly disproportionate, including per capita GHG emissions that are 3–15 times the global fair share level. So for low-income nations, the issue is whether urban centres with small ecological footprints can become healthy, safe, desirable places to live and work without vastly increasing this footprint. This would mean going against the general trend in the past which has been for a strong association between better health and larger ecological footprints and carbon emissions [12]. For high-income nations, the issue is whether urban centres can retain or enhance their quality of life and economic base while radically reducing their carbon dioxide emissions or more generally their ecological footprint. The performance of urban centres in both of these has even greater relevance in a world where the proportion of the world's population living in cities and towns is growing and likely to continue growing at least in the next decade or two [4].

For anthropogenic GHG emissions, it is possible to estimate an average figure for each person's emissions (usually expressed in terms of CO_2e) which if achieved globally should avoid dangerous climate change (the 'fair share' figure). This will be substantially below current figures for global emissions (in 2004, this was around 7.5 tonnes per person [34]), although there are still uncertainties as to the extent to which and the speed with which global emissions must fall. But to bring down global emissions by 2050 sufficiently to avoid dangerous climate change, it needs a global average of around 2 tonnes of CO_2e per person [35]. There are also other issues that need to be resolved in setting this fair share level—for instance in setting it, is consideration given to the maintenance or expansion of carbon sinks (so do people in nations where carbon sinks can be expanded get higher allowances [36])? Also do people who have more carbon-intensive needs—for instance much larger space heating needs during the year—get higher annual allowances [36]?

However, highlighting the fact that it is individuals' consumption patterns and lifestyles that are the real drivers of most GHG emissions means that it is possible to set a fair share level to which those above this level must contract and those below this level are allowed to increase to this level (the concept of contraction and convergence) [37]. This highlights how avoiding dangerous climate change depends on the personal responsibility of consumers. It highlights the large draw of wealthy people's lifestyles on fossil fuel reserves and on the world's forests (as fossil fuel consumption and deforestation are such important contributors to GHG emissions). It also indirectly brings in the use of some mineral resources in that the GHG emission accounts for any individual would include the goods they purchase and thus the fossil fuels used in the mining, refining and use of the resources in manufacturing these goods and transporting, promoting and selling them.

Another reason for focusing on the performance of individuals or households rather than their settlements (cities, small urban centres, rural areas) is the large differentials in GHG emissions per person within such settlements. For instance, the city of Toronto has average per capita emissions far above the fair share level—but there are citizens in Toronto and small pockets in Toronto where per capita emissions are below the fair share level [21]. A study looking at residential and transport related GHG emissions per person (in CO_2e) in Toronto's 832 census tracts found that the top 10 tracts had average annual emissions of 12.3 tonnes per capita compared with 3.6 for the bottom 10 tracts. All the top 10 tracts were outside the central city and private automobile use accounted for 61 per cent of emissions. For the bottom 10 tracts, only 30 per cent of emissions were for private automobile use and all but one of these tracts was in the central city [38].

The differentials are likely to be even larger in cities in many low- and middle-income nations. Per capita emissions among low-income groups will often be between 0.1 and 0.5 tonnes of CO_2e [39] while per capita emissions in many high-income low-density suburbs may be comparable to those in the high-income nations.

1.4 USING THE OPPORTUNITIES/ADVANTAGES THAT URBANIZATION BRINGS FOR HEALTH

Concentrating people, enterprises, motor vehicles and their wastes can make cities very dangerous places—but this same concentration brings many potential advantages for health. That cities have economies of scale, proximity and agglomeration that bring substantial benefits for most businesses is well known; indeed, that is why the world urbanized. But less discussed are the economies of scale and proximity for public goods and services. High densities and large population concentrations usually lower the costs per household and per enterprise for the provision of infrastructure (all-weather roads and paths, good quality piped water, sewers, drains, electricity) and services (including all forms of schools and healthcare, emergency services and access to the rule of law and to government). The concentration of industries reduces the unit cost of making regular checks on plant and equipment safety, as well as on occupational health and safety, pollution control and the management of hazardous wastes [15]. There are also economies of scale or proximity for reducing risks of disasters, and generally a greater capacity among city dwellers to pay for these, or at least to contribute towards the costs. Disasters are much less frequent in well-governed cities with good quality housing, infrastructure and services and when disasters occur in such cities, fatalities are usually much lower [40]. For instance, fatalities from cyclones are far higher in low- and middle-income nations than in high-income nations, even though high-income nations such as Japan and the USA have high exposure to cyclones [41] and even though there are exceptions such as Hurricane Katrina's devastating impact on New Orleans.

But in the absence of needed infrastructure and services and city governments with the needed competence and accountability, cities become very dangerous places to live and work. There are many cities in Africa with life expectancies that are half what they should be—and as low as those in the industrial cities in Europe 160 years ago before key reforms of local governments and of water, sanitation, healthcare, housing access, minimum wages and occupational health. In cities in low- and middle-income nations, it is common for a third to half the population to live in poor quality, overcrowded housing in informal settlements that lack adequate provision for water, sanitation, healthcare, schools, the rule of law and often even the right to vote (as this requires a legal address that houses in informal settlements lack). In such settlements, it is common for infant and child mortality rates to be 10–20 times what they should be [15].

Urbanization is also associated with a growing health burden from chronic non-communicable diseases for adults, related to changes in diet and less exercise and often from large health impacts from air pollution and traffic accidents [42]. In some nations, cities are also associated with high levels of violence [43]. But these too are issues that can be addressed and have been addressed; a well-governed city should also be a city that addresses these health risks.

1.5 RESOURCE USE AND DENSITY

Before suburban trains and high levels of private car ownership, cities were dense largely because of the value of close physical proximity for city businesses and institutions. Low-income groups value central locations for the easier and cheaper access it provides to employment opportunities. Close physical proximity is still valued by most businesses and many people, even though improvements in road, rail and air transport in most cities have allowed a larger physical separation between home and work. The shift of residences from the city to the suburbs and beyond that these improvements brought was in part driven by the possibilities for those who moved to get more space and escape urban pollution and congestion. But the urban sprawl that this often produced brought with it a set of health problems [44] and generally a greater dependence on private automobile use and so higher GHG emissions.

Dense cities have great potential for limiting the use of motor vehicles (and the associated use of fossil fuels, the generation of air pollution and GHGs). This might sound contradictory, as most large cities have problems with congestion and motor-vehicle-generated air pollution but these are problems that can be addressed. As air pollution levels in wealthy cities came down and as commuting times, costs and inconveniences increased for those living in the suburbs or beyond, so the advantages of living in more dense cities became apparent. Dense cities allow many more journeys to be made by walking or bicycling, and they make a greater use of public transport and a high-quality service more feasible. Many prosperous European cities, with among the world's highest quality of life, have one-fifth of the gasoline use per person of the USA's less compact, more car-dependent cities [45]. Most European cities have high-density centres where walking and bicycling are preferred by much of the population, especially where good provision is made for pedestrians and bicyclists (including public transport that can accommodate bicycles). Many European cities also have high-quality public transport that keeps down private automobile ownership and use, and many cities here and in other regions have invested in improving public transport, including greater use of bus-based rapid transit (that is much cheaper and easier to install than light rail or metros). In dense cities where there is good provision for walking, bicycling and public transport, automobile ownership and use can be low, even in prosperous cities and among prosperous households. Singapore has one-fifth of the automobile ownership per person of most cities in other high-income nations, yet also a higher income per person [46]. Among the limited number of cities for which there are data on GHG emissions by sector, the percentage generated by transport varies from 11 to 59 per cent [47]. It is generally the higher density cities that have the lower proportions; the North American cities have among the highest levels of transport energy per person, the highest GHG emissions per person and the lowest densities. However, some of the differences in the contribution of transport may be related to other factors that include the extent to which space heating is needed, the role of industry in the city's economy, the extent to which air conditioning is used and the carbon intensity of electricity production. Some of the differences may also be the result of differences in methods in undertaking emission inventories and their assignment to different sectors.

Meanwhile, in some cities in low- and middle-income nations, the disadvantages of high density have lessened as strategies to improve housing and living conditions in high-density low-quality tenements and informal settlements (often termed 'slum' or 'squatter' upgrading) have transformed the quality of housing and living conditions there, while retaining the advantages of high-density residential areas. For instance, in Thailand, many high-density informal settlements with poor quality housing and a lack of basic infrastructure and services have been transformed without expelling the low-income population and without reducing the density. This has been supported by the central government's Community Organizations Development Institute (CODI) that channels infrastructure subsidies and housing loans direct to savings groups formed by low-income inhabitants in informal settlements. It is these savings groups who plan and manage the improvements to their housing or develop new housing on the same site. They also negotiate with the landowner to sort out legal tenure and work with local governments or utilities to improve infrastructure and services. From 2003 to early 2008, within the Baan Mankong (secure housing) programme, CODI approved 512 projects in over 200 urban centres covering 53 976 households and it plans a considerable expansion in the programme within the next few years. In terms of density, many of the upgraded settlements had 150–300 units per hectare, achieved with two or three storey terraces [48].

For most cities in low- and middle-income nations, one of the most important issues is retaining the high-density low-income residential areas but improving the quality without expelling the low-income population who need to keep close proximity to livelihoods. Generally, the redevelopment of these areas drives out the original inhabitants. But there are a growing number of examples like those of CODI above that have supported the low-income inhabitants to develop their own upgrading. It is also possible to do this at very high densities—as demonstrated in some high-density central city locations in Karachi [49]. In Mumbai, India, there is the long-running debate about how to redevelop Dharavi, a long-established very dense informal township with around 600 000 residents and a high concentration of jobs squeezed into a 2 km^2 site. From a resource-using perspective, Dharavi is very efficient; most residents have low consumption levels and walk to work and many are engaged in reusing or recycling wastes. GHG emissions per person would also be very low. But from a health perspective, it is mostly poor quality and very overcrowded housing and a lack of infrastructure and services. But even with densities this high, Dharavi can be redeveloped incrementally in ways that accommodate residents and livelihoods and that keep the area's economic dynamism and that greatly improve living conditions—and thus keep the resource-using advantages without the health disadvantages [50,51].

For regions that require space heating for part of the year, high-density residential areas with high-quality housing make possible less energy-intensive homes. Space heating constitutes one of the main energy users in most high-income nations and high-density buildings can have an inherent advantage for lower energy use from less external wall area and less indoor space per person. For instance, three to six storey terraces can combine high densities with very high-quality living environments with much less energy use for space heating than detached housing. High density is often seen as one of the problems in cities—but it depends on how it is accommodated. Some of the world's most expensive and desirable housing is in four to six storey terraces in European cities. These have high densities, often between 150 and 300 units per hectare and thus higher than most informal settlements with one-storey buildings, but the indoor space and often outdoor space per person are much higher. Energy use per dwelling can be much lower than in detached housing in suburban or rural areas. High-quality high-density dwellings can also be within city districts with good provision for open space—from small household gardens and neighbourhood squares and playgrounds to larger parks. There are also examples of new high-density residential developments that cut energy and water use, carbon dioxide emissions and the carbon footprints of materials used for their construction—as in the Beddington Zero Energy Development in South London [52]. High-density residential areas and cities generally have lower rates of private automobile ownership and use in relation to household income—and provide more possibilities for car clubs through which members have quick and easy access to car use when they want at a far lower cost than car ownership.

1.6 CITIES AND RESOURCE USE

Cities also have many potential advantages for reducing resource use and waste. It is worth recalling the example of Toronto cited earlier showing the tenfold difference in GHG emissions per person between a dense inner city neighbourhood with good access to public transport and a low density suburb distant from shops [21]. Many of the innovations discussed in *Factor Four*, the study of how to double wealth while halving resource use [28] are facilitated by the potential material, transport and water using efficiencies that more dense cities provide.

In cities, the close proximity of so many water consumers gives greater scope for recycling or directly re-using wastewaters. The concentration of consumer and business wastes in cities cheapens the cost of recovering or re-using material from these waste streams—for instance in separate household and business collections for recyclable or compostable wastes. In most cities in low- and middle-income nations, there is a large and diverse 'waste economy' through which materials are recovered from waste streams for recycling or reuse and which greatly reduces the volume of wastes that need to be disposed of [15]. Although there are many practices here that need improvement, especially in reducing the health risks facing those engaged in these informal waste economies, there is also a need to keep their very large environmental and livelihood-related advantages.

Urbanization has been accompanied by changes in diets, including increased meat consumption, more processed food and more expensive and exotic food items that generally imply more carbon-intensive production and transport to markets. But studies in a range of nations have shown that this has been driven by higher incomes, not urbanization, as can be seen by the consumption patterns of more prosperous rural dwellers that are similar to those of more prosperous urban dwellers [53].

1.7 CITIES' PHYSICAL EXPANSION

Cities are often portrayed as being 'bad' for rural areas but city dwellers' demand for agricultural produce is a large part of the underpinning for farmers' incomes. In addition, most farmers (and their families) depend on markets, goods and services provided by urban enterprises. Most urban centres also concentrate populations in ways that reduce the demand for land relative to population [5]. There is no evidence that agricultural productivity has dropped in nations that are urbanizing and a declining proportion of land used for agriculture around a city may also be accompanied by more intensive production for land that remains in agriculture [54].

Various studies have sought to establish the proportion of land area globally and within each continent or nation that can be considered urban.

One recent study suggests that the urban built-up area represents only 0.5 per cent of global land area and it is only in Europe that it exceeds 1 per cent [55]. Other sources using different methodologies suggest higher figures—for instance the Global Rural–Urban Mapping Project suggested 2.7 per cent of global land area [56] although this is the land areas of urban settlements and so includes open land of various types within urban boundaries. Of course, if most urban expansion has been over high-quality agricultural land and only a small proportion of a nation's land area is high-quality agricultural land, urban areas expanding from (say) one to three per cent of total land area could mean a serious loss of agricultural land. But it is also likely that there is much agricultural production within a proportion of the area defined as urban, particularly under the second study mentioned above. In cities where the role of urban agriculture has been studied, its scale and its contribution to livelihoods and food supply have often been found to be significant [57].

However, many cities in low- and middle-income nations expand without any land-use plan or strategic planning framework to prevent sprawl and unnecessary loss of agricultural land and to protect watersheds and other areas that provide key ecological services. The expansion is determined by where different households, enterprises and public sector activities locate and build, legally or illegally [15]. This also produces a patchwork of high- and low-density land uses that no longer have the advantages noted above for reducing infrastructure costs and resource use.

In most locations, governments could and should restrict the loss of agricultural land to urban expansion. But this can also bring serious social consequences as a government policy that restricts the conversion of land from agricultural to non-agricultural uses around a growing city will push up land and house prices and often reduce still further the proportion of households that can afford a legal housing plot with infrastructure.

1.8 USING THE ADVANTAGES CITIES HAVE FOR REDUCING GREENHOUSE GAS EMISSIONS

Many sources claim that cities are responsible for 75–80% of all GHG emissions, although from the production perspective and drawing on IPCC figures [58], the figure is likely to be 40–45% with agriculture, land-use changes and deforestation contributing to 31 per cent and the rest coming from energy supply, industry, transport, buildings and wastes in rural areas and in urban centres that are not cities [59]. But it is not cities (or small urban centres or rural areas) that are responsible for anthropogenic GHG emissions but particular activities by particular people, enterprises and institutions, a proportion of which are in cities. So one fundamental question is whether the GHG emissions used in producing goods or services are allocated to the enterprises that made, transported, promoted and sold the goods and services or the person or household that consumed these. If emissions are assigned to the final consumer, some of the allocation issues discussed earlier become clearer. For instance, GHG emissions from aviation are allocated to the person who flies. So a flight by a Londoner travelling from New York to South America would be allocated to London. Under this consumption-based accounting, emissions from agriculture, deforestation and industry could be assigned to the people who are the consumers of the industrial goods, wood products and food.

The choice of which system to use in setting limits and targets on GHG emissions has great significance for how responsibilities are assigned between and within nations and cities. Dhakal's [60] study of Asian cities showed that Beijing and Shanghai had higher per capita GHG emissions than Tokyo, when considering the emissions produced within these cities, but Tokyo had much higher per capita emissions if emission inventories for the cities included emissions that went into the goods purchased by city residents. If China's manufacturing cities are assigned all the GHG emissions that were generated producing goods for export and transporting them to their final market, this implies a much larger responsibility for these Chinese cities (and China as a nation) in moderating and eventually reducing such emissions than if these emissions were allocated to the people who consumed these goods (and by implication to the nations or cities where they live) [61]. For London, a shift from production- to consumption-based accounting for GHG emissions increases the average Londoner's responsibility for GHG emissions from 6 to 12 tonnes of CO_2e a year [62].

Both production- and consumption-based accounting are useful. Assigning GHG emissions to the range of enterprises involved in production highlights more resource-intensive centres of goods production while assigning emissions to consumers highlights more affluent places [63]. Under the consumption-based accounting, the contribution of cities to GHG emissions would increase, although it is not cities in general that are the problem but cities (or other settlements) where there are concentrations of high-consumption individuals and households.

If responsibility for GHG emissions is allocated to the consumer, very large differentials become evident. The world's wealthiest high-consumption individuals are likely to be contributing tens of thousands of times more to global warming than many of the poorest individuals (although this is in part because the poorest individual's contribution can be close to zero). For any individual to contribute to global warming, they have to consume goods and services that generate GHG emissions. Perhaps as many as 1.2 billion rural and urban dwellers worldwide have such low consumption levels that they contribute almost nothing to climate change. Their use of fossil fuels is very low (most use wood fuel, charcoal or dung for fuels) and they use no electricity. Most of these 'very low-carbon' people will use transport that produces no carbon dioxide emissions (walking, bicycling) or low emissions (buses, mini-buses and trains, mostly used to more than full capacity [39]).

A recent study examined global carbon dioxide emissions from fossil fuel use based on the individuals whose consumption caused these emissions rather than the nations within whose boundaries the emissions took place [64]. The wealthiest 700 million people's consumption is responsible for around half of all GHG emissions while the poorest 2.4 billion people contribute very little. What the paper emphasizes is the need to focus emission reduction policies on the high emitters (wherever they live). The paper also shows that allowing the growing number of people with very low emissions per person to achieve a good living standard (and its implications for increased emissions) does not add much to needed global emissions reduction targets. It also notes how many of the lowest-cost opportunities for reducing carbon dioxide emissions will be among the many millions who move to cities for the first time and could be housed in well-built energy-efficient accommodation with efficient appliances and well served by public transport.

There is the obvious concern in regard to what urbanization implies for population growth as most of the world's growth in population is taking place in urban centres in low- and middle-income nations and UN projections suggest that this is likely to continue [4]. But the GHG emissions generated by a person added to the world's population varies by a factor of more than 1000 depending on the circumstances into which they are born and their life possibilities and chances. In addition, in low- and middle-income nations, urbanization is associated with falls in population growth rates as fertility rates in urban areas are generally significantly lower than in rural areas [42,65]. In terms of future worries about resource constraints and GHG emissions, it is not the growth in population but the growth in consumption that is the primary concern. Most of the growth in GHG emissions from 1980 to 2005 occurred in nations with low population growth rates [39]. London's current population is smaller than it was in 1941 yet Londoners' total consumption of resources and their implications for GHG emissions are likely to have multiplied many times since then.

1.9 URBAN GOVERNANCE IN A RESOURCE CONSTRAINED WORLD

None of the potential advantages that urban centres have for high living standards (and good health) or for keeping down resource use, waste and GHG emissions happen automatically. They depend on governance structures—local governments and their relations with the population and civil society groups within their boundaries—making and implementing appropriate choices. This includes choices serving those with low incomes so they do not suffer profound health disadvantages. For wealthier cities, this includes the interests of people outside their boundaries and future generations [66]. So it depends on an acceptance by citizens and businesses of everyday practices that do not directly serve them (and may limit their consumption choices) to make local and regional resource use and waste disposal and global systems sustainable for future generations. It also depends on modifying so many aspects of local government—for instance the regulations governing land use and buildings and ensuring that all public buildings and government contracts for goods and services address these multiple goals. It also depends on incentives, regulations and standards set at higher levels of government—although it often falls to local government to implement them. It will be particularly difficult for many local governments in wealthy nations or wealthier districts of low- and middle-income nations to address these goals because it is difficult and expensive to modify the buildings, infrastructure and settlement patterns that developed in the era of cheap oil and no concern for GHG emissions. This is especially the case for low-density sprawl.

Table 1 illustrates some of the different implications of a concern for environmental health and a concern for sustainable use of resources and sinks and the potential conflict between many of these is obvious.

There are many examples of innovation and better practice from city governments in low- and middle-income nations where the need for improved governance is most evident. Many come from local initiatives that arise from more competent and democratic urban governments in nations where decentralization programmes have given more power and resources to such governments [68]. Some address regional and global environmental issues as well as local environment and development issues—for instance Manizales in Colombia with its comprehensive monitoring of the city's environmental performance and its much-copied 'environmental traffic lights' for keeping inhabitants informed of this [69]. Many others come from innovative local civil-society groups—usually a combination of grassroots organizations and local non-governmental organizations—and increasingly from partnerships that these groups form with local governments, which in turn contributes to more accountable and democratic local governments.

It is common to see city problems blamed on rapid city growth. But there are cities that have grown rapidly in the last 50 years that have avoided most of the problems noted above. For instance, Porto Alegre in Brazil has grown from under half a million inhabitants in 1950 to around 3.5 million in its metropolitan area today. It has high-quality living environments and innovative environmental policies. Its inhabitants enjoy an average life expectancy and many indicators of environmental quality that are comparable to cities in Western Europe and North America—and also a city government that during the 1990s was well known for its commitment to supporting citizen participation, greater government accountability and good public health and environmental management. Porto Alegre also integrated a wide-ranging environmental management policy into its participatory budgeting, rooted in a comprehensive regional environmental analysis [70,71].

TABLE 1: Comparing a concern for environmental health and for the sustainable use of resources and waste sinks for cities [67].

	environmental health	sustainable use of resources and waste sinks
key impact	human health within the city	ecosystem health, damage to ecosystem services and climate change much of it outside the city
timing of human impact	immediate	delayed; often indirect
scale and spatial focus	local and city wide	the region around the city and global
worst affected	lower income groups	future populations
aspects emphasized in relation to:		
water	need for increased quantities and better quality to address inadequate access facing much of the population	over-use; need to reduce use, protect water resources and implement water demand-management
air	high human exposure to hazardous pollutants at home and work	acid precipitation beyond city boundaries, GHG emissions
solid waste	inadequate provision for collection and removal of household wastes	excessive generation of waste and poorly managed waste disposal
land	inadequate availability inhibiting supplies of housing that low income groups can afford	loss of natural habitats and agricultural land to urban development and expansion
human wastes	inadequate provision for safely removing faecal matter (and waste water) from homes and living environments	loss of nutrients in sewage and damage to water bodies from release of sewage into waterways
typical proponent	urbanist	environmentalist

Urban governance in a resource constrained world also needs programmes to tackle the backlog in infrastructure and services in the poorer and worst-served areas of urban centres and to support ways in which lower income households can get better quality housing. This is also needed to build resilience to climate change. Many of the most serious risks from climate change in cities arise because of poor quality housing built on sites at risk that lack basic protective infrastructure. Many cities in Latin America, Africa and Asia may have low GHG emissions per person but they house hundreds of millions of people who are at risk from the increased frequency and/or intensity of floods, storms and heat waves and water supply constraints that climate change is bringing or likely to bring [72]. It is generally low-income groups that are most at risk as they live on sites at risk of flooding or landslides, lacking the drains and other needed protective infrastructure. The costs of addressing this very large backlog in basic infrastructure and housing which underpins vulnerability to climate change are at present not included in estimates for the costs of climate change adaptation [73]. Perhaps more worryingly, even if the funding was available, for many cities, the capacity and willingness to address the risks, especially those faced by lower income groups, is not there.

However, there are many co-benefits between improving housing and living conditions and building resilience to climate change. Many cities have had major 'upgrading' programmes to improve provision for water, sanitation, drainage and garbage collection in inner-city tenement districts and in squatter settlements. Initially, these were seen as one-off projects in 'targeted' neighbourhoods; now there is a recognition that city and municipal governments need the capacity and competence to support continuous upgrading programmes throughout the city, working in partnership with their inhabitants. The example of CODI given earlier has particular significance in three aspects: the scale; the extent of community involvement; and the extent to which it seeks to institutionalize community-driven solutions within local governments [48]. Although this programme was never intended as a response to climate change, it has reduced risk levels because of better quality housing with needed infrastructure and services.

In many other nations, national organizations or federations of 'slum' dwellers are working with local governments to improve housing conditions and reduce risks from disasters [40,74]. What is unusual about these federations is that they recognized that making demands on governments that those governments cannot fulfil did not serve them. Many had tried the conventional approaches of protest, strikes, barricades and marches. They came to recognize that they had to change their relations with politicians and civil servants, especially to show that 'slum' dwellers and their settlements were not 'the problem' and how, with local government support, they could generate solutions [75]. Federations in many nations have demonstrated to governments their capacity to design and build housing and infrastructure that is cheaper and better quality than if governments get these built by contractors. These federations have also demonstrated a capacity to undertake the enumerations and mapping of informal settlements needed for planning upgrading. With these demonstrations of their competence and capacity, they offer local governments partnerships—and where local governments work with them, the scale of what can be achieved increases greatly. Partnerships between local governments and these federations can address the critical health issues and contribute to resilience to local climate change impacts. The kind of high-density upgrading or new house development that these federations prioritize are also compatible with high-density and resource use efficiency.

1.10 DESIRABLE URBAN CENTRES WITH LOW ECOLOGICAL FOOTPRINTS

Cities concentrate so much of what contributes to a very high quality of life that need not imply high material consumption levels (and thus high GHG emissions)—theatre, music, museums, libraries, the visual arts, dance, festivals, the enjoyment of historic buildings and districts, diverse choices for eating, easy access to many other services or simply the enjoyment of being in a diverse and vibrant place. Cities have also long been places of social, economic and political innovation at local and national levels. Indeed, in high-income nations, many city politicians have demonstrated a greater commitment to GHG reduction than national politicians. Achieving the needed reduction in GHG emissions globally and more sustainable patterns of resource use depends on understanding this potential of cities to combine a high quality of life with less material resource use and waste and low GHG emissions. New technologies may help but the scale of the needed reduction in GHGs suggests that, as Rees notes, the wealthy may have to accept lower material standards for enhanced geopolitical and ecological security [76]. It will also need the expertise of ecologists applied to urban systems [2] and urban governments recovering control over land-use changes and integrating ecological concerns and climate resilience into this [77].

This paper has also highlighted how it is not cities or urbanization but high-consumption lifestyles that underpin unsustainable or potentially unsustainable levels of resource use, waste and GHG emissions—whether or not those who have such lifestyles live in cities or other urban centres or rural areas. In high-income nations and in many parts of middle-income nations, most of the rural population no longer work in agriculture, forestry or fishing. They also enjoy levels of infrastructure and service provision that used to be associated with urban areas. A high proportion of those of working age commute to urban centres and also travel there for recreation. Advanced communications systems allow much work to be done from people's homes that need not be in cities. It is the resource use and waste generation implications of income levels and consumption choices that need consideration much more than the proportion of people living in cities.

REFERENCES

1. Grimmond S. 2007 Urbanization and global environmental change: local effects of urban warming Geogr. J. 173 83 88 doi:10.1111/j.1475-4959.2007.232_3.x (doi:10.1111/j.1475-4959.2007.232_3.x)
2. Grimm N. B., Faeth S. H., Golubiewski N. E., Redman C. L., Wu J., Bai X., Briggs J. M. 2008 Global change and the ecology of cities Science 319 756 760 doi:10.1126/science.1150195 (doi:10.1126/science.1150195)
3. Bairoch P. (1988) Cities and economic development: from the dawn of history to the present (Mansell, London, UK).
4. United Nations, Department of Economic and Social Affairs, Population Division. (2010) World urbanization prospects: the 2009 revision, . CD-ROM edition, data in digital form POP/DB/WUP/Rev.2009.
5. Satterthwaite D., McGranahan G., Tacoli C. 2010 Urbanization and its implications for food and farming Phil. Trans. R. Soc. B 365 2809 2820 doi:10.1098/rstb.2010.0136 (doi:10.1098/rstb.2010.0136)
6. Sassen S. (2006) Cities in a world economy (Pine Forge Press, Thousand Oaks, CA).
7. Douglas I. (1983) The urban environment (Edward Arnold, London, UK).

8. Douglas I. (1986) in A handbook of engineering geomorphology, Urban geomorphology, eds Fookes P. G., Vaughan P. R. (Surrey University Press (Blackie and Son) Glasgow, UK), pp 270–283.
9. UN-Habitat. (2003) Water and sanitation in the world's cities: local action for global goals (Earthscan Publications, London, UK).
10. Kovats R. S., Akhtar R. 2008 Climate, climate change and human health in Asian cities Environ. Urb. 20 165 176 doi:10.1177/0956247808089154 (doi:10.1177/0956247808089154)
11. Kjellstrom T., Mercado S. 2008 Towards action on social determinants for health equity in urban settings Environ. Urb. 20 551 574 doi:10.1177/0956247808096128 (doi:10.1177/0956247808096128)
12. McGranahan G. (2007) in Scaling urban environmental challenges: from local to global and back, Urban transitions and the spatial displacement of environmental burdens, eds Marcotullio P. J., McGranahan G. (Earthscan Publications, London, UK), pp 18–44.
13. Best Foot Forward Ltd. (2002) City limits: a resource flow and ecological footprint analysis of Greater London (Chartered Institution of Wastes Management Environmental Body, London, UK).
14. Anton D. J. (1993) Thirsty cities: urban environments and water supply in Latin America (IDRC, Ottawa, Canada).
15. Hardoy J. E., Mitlin D., Satterthwaite D. (2001) Environmental problems in an urbanizing world: finding solutions for cities in Africa, Asia and Latin America (Earthscan Publications, London, UK).
16. Connolly P. 1999 Mexico City: our common future Environ. Urb. 11 53 78 doi:10.1177/095624789901100116 (doi:10.1177/095624789901100116)Abstract
17. Romero Lankao P. 2010 Water in Mexico City: what will climate change bring to its history of water-related hazards and vulnerabilities Environ. Urb. 22 157 178 doi:10.1177/0956247809362636 (doi:10.1177/0956247809362636)
18. Satterthwaite D. (2010) in Urbanization and development: multidisciplinary perspectives, Urban myths and the mis-use of data that underpin them, eds Beall J., Guha-Khasnobis B., Kanbur R. (Oxford University Press, Oxford, UK), pp 83–99.
19. Legros G., Havet I., Bruce N., Bonjour S. (2009) The energy access situation in developing countries: a review focusing on the least developed countries and Sub-Saharan Africa (World Health Organization and United Nations Development Programme, New York, NY).
20. CAIT. (2010) Climate Analysis Indicators Tool, CAIT version 6.0 (World Resources Institute, Washington, DC) See http://cait.wri.org/cait.php.
21. Hoornweg D., Sugar L., Lorena Trejos Gomez C. In press Cities and greenhouse gas emissions: challenges and opportunities Environ. Urb. doi:10.1177/0956247810392270 (doi:10.1177/0956247810392270)
22. World Commission on Environment and Development. (1987) Our common future (Oxford University Press, Oxford, UK).
23. McManus P., Haughton G. 2006 Planning with ecological footprints: a sympathetic critique of theory and practice Environ. Urb. 18 113 128 doi:10.1177/0956247806063963 (doi:10.1177/0956247806063963)

24. Wackernagel M., Rees W. (1995) Our ecological footprint: reducing human impact on the earth (New Society Publishers, Gabriola, Canada).
25. Wackernagel M., Kitzes J., Moran D., Goldfinger S., Thomas M. 2006 The ecological footprint of cities and regions: comparing resource availability with resource demand Environ. Urb. 18 103 112 doi:10.1177/0956247806063978 (doi:10.1177/0956247806063978)
26. Weisz H., Steinberger J. K. 2010 Reducing energy and materials flows in cities Curr. Opin. Environ. Sust. 2 185 192 doi:10.1016/j.cosust.2010.05.010 (doi:10.1016/j.cosust.2010.05.010)
27. Barles S. 2010 Society, energy and materials: the contribution of urban metabolism studies to sustainable urban development issues J. Environ. Plann. Manage 53 439 455 doi:10.1080/09640561003703772 (doi:10.1080/09640561003703772)
28. Von Weizsäcker E., Lovins A. B., Hunter Lovins L. (1997) Factor four: doubling wealth, halving resource use (Earthscan Publications, London, UK).
29. Kennedy G., Nantel G., Shetty P. (2004) Globalization of food systems in developing countries: impact on food security and nutrition, Globalization of food systems in developing countries: a synthesis of country case studies (FAO, Rome, Italy) FAO Food and Nutrition Paper 83, pp 1–25.
30. Rees W. E. 1992 Ecological footprints and appropriated carrying capacity Environ. Urb. 4 121 130 doi:10.1177/095624789200400212 (doi:10.1177/09562478920040 0212)
31. IPCC. (2006) Guidelines for national greenhouse gas inventories (Cambridge University Press, Cambridge, UK).
32. Bloomberg M. R. (2007) Inventory of New York greenhouse gas emissions (Mayor's Office of Operations, Office of Long-Term Planning and Sustainability, New York, NY).
33. Secretaria Municipal do Verde e do Meio Ambiente de São Paulo SVMA. (2005) Inventário de Emissões de Efeito Estufa do Municí pio de São Paulo, . Centro de Estudos Integrados sobre Meio Ambiente e Mudanças Climáticas Centro Clima da Coordenação dos Programas de Pós-graduação de Engenharia COPPE da Universidade Federal do Rio de Janeiro UFRJ.
34. IPCC. (2007) Climate change 2007: synthesis report (Intergovernmental Panel on Climate Change, Geneva Switzerland).
35. Stern N. (2009) A blueprint for a safer planet (The Bodley Head, London, UK).
36. Caney S. 2009 Justice and the distribution of greenhouse gas emissions Glob. Ethics 5 125 146 doi:10.1080/17449620903110300 (doi:10.1080/17449620903110300)
37. Meyer A. (2007) in Surviving climate change: the struggle to avert global catastrophe, The case for contraction and convergence, eds Cromwell D., Levene M. (Pluto Press, London, UK), pp 29–56.
38. VandeWeghe J. R., Kennedy C. 2007 A spatial analysis of residential greenhouse gas emissions in the Toronto census metropolitan area J. Ind. Ecol. 11 133 144 doi:10.1162/jie.2007.1220 (doi:10.1162/jie.2007.1220)
39. Satterthwaite D. 2009 The implications of population growth and urbanization for climate change Environ. Urb. 21 545 567 doi:10.1177/0956247809344361 (doi:10.1177/0956247809344361)

40. IFRC. (2010) World disasters report 2010: focus on urban areas (International Federation of Red Cross and Red Crescent Societies, Geneva, Switzerland).
41. United Nations. (2009) Global assessment report on disaster risk reduction: risk and poverty in a changing climate (United Nations, Geneva, Switzerland).
42. Dye C. 2008 Health and urban living Science 319 766 769 doi:10.1126/science.1150198 (doi:10.1126/science.1150198)
43. Moser C. O. N. 2004 Editorial. Urban violence and insecurity: an introductory roadmap Environ. Urb. 16 3 15 doi:10.1177/095624780401600220 (doi:10.1177/09562 4780401600220)
44. Frumkin H., Frank L., Jackson R. (2004) Urban sprawl and public health: design, planning and building for health communities (Island Press, Washington, DC).
45. Newman P. 2006 The environmental impact of cities Environ. Urb. 18 275 295 doi:10.1177/0956247806069599 (doi:10.1177/0956247806069599)
46. Newman P. 1996 Reducing automobile dependence Environ. Urb. 8 67 92 doi:10.1177/095624789600800112 (doi:10.1177/095624789600800112)Abstract
47. Dodman D. 2009 Blaming cities for climate change?: An analysis of urban greenhouse gas emissions inventories Environ. Urb. 21 185 201 doi:10.1177/0956247809103016 (doi:10.1177/0956247809103016)
48. Boonyabancha S. 2005 Baan Mankong: going to scale with 'slum' and squatter upgrading in Thailand Environ. Urb. 17 21 46 doi:10.1177/095624780501700104 (doi :10.1177/095624780501700104)
49. Hasan A., Sadiq A., Ahmed S. (2010) Planning for high density in low-income settlements: four case studies from Karachi (IIED, London, UK) . Human Settlements Working paper.
50. Patel S., Arputham J. 2008 Plans for Dharavi: negotiating a reconciliation between a state-driven market redevelopment and residents' aspirations Environ. Urb. 20 243 254 doi:10.1177/0956247808089161 (doi:10.1177/0956247808089161)
51. Patel S., Arputham J. 2007 An offer of partnership or a promise of conflict in Dharavi, Mumbai Environ. Urb. 19 501 508 doi:10.1177/0956247807082832 (doi:10.1177/0956247807082832)
52. Chance T. 2009 Towards sustainable residential communities; the Beddington Zero Energy Development (BedZED) and beyond Environ. Urb. 21 527 544 doi:10.1177/0956247809339007 (doi:10.1177/0956247809339007)
53. Stage J., McGranahan G. 2010 Is urbanization contributing to higher food prices Environ. Urb. 22 199 215 doi:10.1177/0956247809359644 (doi:10.1177/0956247809359644)
54. Bentinck J. (2000) Unruly urbanization on Delhi's fringe: changing patterns of land use and livelihood (KNAG, Netherlands Geographical StudiesUtrecht, The Netherland).
55. Schneider A., Friedl M. A., Potere D. 2009 A new map of global urban extent from MODIS satellite data Environ. Res. Lett. 4 044003 doi:10.1088/1748-9326/4/4/044003 (doi:10.1088/1748-9326/4/4/044003)
56. Balk D., Pozzi F., Yetman G., Deichmann U., Nelson A. (2004) The distribution of people and the dimension of place: methodologies to improve the global estimation of urban extents. GRUMP, New York, NY: CIESIN. See http://sedac.ciesin.columbia.edu/gpw/documentation.jsp.

57. Redwood M. (2009) Agriculture in urban planning: generating livelihoods and food security (Earthscan Publications, London, UK).
58. IPCC. (2007) Climate change 2007: mitigation, Technical summary, eds Metz B., Davidson O. R., Bosch P. R., Dave R., Meyer L. A. (Cambridge University Press, Cambridge, UK).
59. Satterthwaite D. 2008 Cities' contribution to global warming: notes on the allocation of greenhouse gas emissions Environ. Urb. 20 539 550 doi:10.1177/0956247808096127 (doi:10.1177/0956247808096127)
60. Dhakal S. (2004) Urban energy use and greenhouse gas emissions in Asian cities: policies for a sustainable future (Institute for Global Environmental Strategies, Kitakyushu, Japan).
61. Walker G., King D. (2008) The hot topic: how to tackle global warming and still keep the lights on (Bloomsbury Publishers, London, UK).
62. Bioregional and London Sustainable Development Commission. (2010) Capital consumption: the transition to sustainable consumption and production in London (Greater London Authority, London, UK) .See www.londonsdc.org.uk and www.bioregional.com.
63. McGranahan G. (2005) An overview of urban environmental burdens at three scales: intra-urban, urban-regional and global. Int. Rev. Environ. Strat. 5:335–336.
64. Chakravarty S., Chikkatur A., de Coninck H., Pacala S., Socolow R., Tavonia M. 2009 Sharing global CO2 emission reductions among one billion high emitters Proc. Natl Acad. Sci. USA 106 11884 11888 doi:10.1073/pnas.0905232106 (doi:10.1073/pnas.0905232106)
65. Montgomery M. R., Stren S., Cohen B., Reed H. E. (2003) Cities transformed: demographic change and its implications in the developing world (The National Academy Press, Washington, DC).
66. Haughton G. 1999 Environmental justice and the sustainable city J. Plan. Educ. Res. 18 233 243 doi:10.1177/0739456X9901800305 (doi:10.1177/0739456X9901800305)
67. McGranahan G., Satterthwaite D. (2000) in Sustainable cities in developing countries, Environmental health or ecological sustainability?: Reconciling the brown and green agendas in urban development, ed Pugh C. (Earthscan Publications, London, UK).
68. Campbell T. (2003) The quiet revolution: decentralization and the rise of political participation in Latin American cities (University of Pittsburgh Press, Pittsburgh, PA).
69. Velasquez L. S. 1998 Agenda 21; a form of joint environmental management in Manizales, Colombia Environ. Urb. 10 9 36 doi:10.1177/095624789801000218 (doi:10.1177/095624789801000218)
70. Menegat R. 2002 Participatory democracy and sustainable development: integrated urban environmental management in Porto Alegre, Brazil Environ. Urb. 14 181 206 doi:10.1177/095624780201400215 (doi:10.1177/095624780201400215)Abstract
71. Menegat R. (1998) Atlas Ambiental de Porto Alegre (Universidade Federal do Rio Grande do Sul, Prefeitura Municipal de Porto Alegre and Instituto Nacional de Pesquisas Espaciais, Porto Alegre, Brazil) .(main coordinator).

72. Satterthwaite D., Huq S., Reid H., Pelling M., Romero Lankao P. (2009) in Adapting cities to climate change, Adapting to climate change in urban areas; the possibilities and constraints in low- and middle-income nations, eds Bicknell J., Dodman D., Satterthwaite D. (Earthscan Publications, London, UK), pp 3–47.
73. Parry M., et al. (2009) Assessing the costs of adaptation to climate change: a review of the UNFCC and other recent estimates (IIED and Grantham Institute, London, UK).
74. Rayos Co J. C. (2009) Community-driven disaster intervention: the experience of the Homeless People's Federation Philippines (HED, London, UK) Human Settlements Working Paper, no. 25.
75. Arputham J. 2008 Developing new approaches for people-centred development Environ. Urb. 20 319 337 doi:10.1177/0956247808096115 (doi:10.1177/0956247808096115)
76. Rees W. E. 1995 Achieving sustainability: reform or transformation? J. Plann. Lit. 9 343 361 doi:10.1177/088541229500900402 (doi:10.1177/088541229500900402) Abstract
77. Roberts D. 2008 Thinking globally, acting locally—institutionalizing climate change at the local government level in Durban, South Africa Environ. Urb. 20 521 538 doi:10.1177/095624780809612.

CHAPTER 2

Greenhouse Gas Emissions Accounting of Urban Residential Consumption: A Household Survey Based Approach

TAO LIN , YUNJUN YU, XUEMEI BAI, LING FENG, AND JIN WANG

2.1 INTRODUCTION

More than half of the world's population are living in cities and urbanization is transforming the global environment at unparalleled rates and scales [1], [2]. Cities are estimated to account for about 78% of total global greenhouse gas (GHG) emissions, but are also the loci for innovative solutions to reduce emissions [3]–[8]. Household lifestyle has been recognized as a major driver of energy use and related GHG emissions besides technology efficiency [9]–[14]. Carbon management in cities is increasingly focusing on individuals, households, and communities due to population growth and improved living standards of urban residents [14]–[19]. A better understanding of urban residential consumption patterns in relation to urban system structure and processes, and their linkages to GHG emissions emission profiles, will enable cities to develop tailor-made planning and policy measures towards low carbon cities.

The present accounting methods of GHG emissions can be roughly categorized into production-based and consumption-based accounting approaches [20], [21]. Production-based approaches are always exemplified in national-scale inventories and tracks mainly the direct GHG emissions across all production sectors and the residential sector within the political or geographical boundary [20], [22]. These approches do not include energy embodied in imported goods and services. Strict boundary-limited GHG accounting is unsuitable for cities because they don't include embodied emissions in imported goods and services. Theoretically, consumption-based accounting provides the most rigorous GHG estimation incorporating trans-boundary emissions. Consumption-based approaches link the consumption levels and patterns of urban residents with the associated direct and embodied GHG emissions whether those occur inside or outside the city boundary, through the proxy of local household expenditure. As a result, in cities with significant export-related industrial activities and relatively low resident populations, the consumption-based accounting approach will likely lead to lower GHG emissions estimates compared to production-based accounting approaches. Conversely, for residence and service-oriented cities that typically import all energy and energy-intensive materials and goods, consumption-based accounting approach will more likely yield substantially higher estimation compared to production-based accounting approaches [22]. Production-based accounting approaches often based on top-down statistical data which uses same categories and definitions and is internally consistent to allow comparisons and benchmarking. While consumption-based accounting approaches are always based on an extensive city wide survey and only a limited number of consumption-based accounts for cities are available [23]. Sampling errors in consumption surveys may add some degree of uncertainty [24]. However, it can reflect consumption choices and empower households and governments to redirect a low-carbon lifestyle [20].

The last three decades have seen unprecedented urbanization in China, from 19% in 1980 to 51% in 2011, and this rapid urbanization is expected to continue in the coming decades. Currently, the 35 largest cities contain 18% of the national population, but account for 40% of China's energy use and GHG emissions [25]. The socioeconomic development in Chinese cities and large numbers of new urban migrants has driven significant increases in energy use and related GHG emissions, because urban communities have a greater per capita energy demand than rural settlements [26]. Changing urban lifestyles will play an increasingly important role in shaping China's energy demand and GHG emissions. However, existing research on GHG emissions accounting in China mostly employ production-based accounting using top-down government statistics, and embodied energy use and GHG emissions driven by residential consumption are often omitted or underestimated.

In this paper, we present a survey based GHG emissions accounting methodology for urban residential consumption, and apply it in Xiamen City, a rapidly urbanizing coastal city in southeast China. Based on our results, we explore the current main influencing factors determining residential GHG emissions at the household and community scale, and present typical profiles of low and high GHG emission households and communities. Based on the results, policy implications for developing a low GHG emissions urban consumption pattern are discussed.

2.2 METHODS

Our study consists of four steps: (1) designing a city-wide questionnaire survey; (2) defining the system boundary, establishing consumption categories and GHG emissions accounting methodology for seven consumption categories; (3) conducting the survey; and (4) data processing and analysis of the survey results, including influencing factor analysis and profiling of low, medium and high GHG emission households and communities. Our study obtained ethical approval from the Academic Committee of the Institute of Urban Environment (IUE), Chinese Academy of Sciences.

2.2.1 SURVEY DESIGN

In our study, all the data for GHG emissions accounting of urban residential consumption and influencing factor analysis are derived from an on-site questionnaire survey. The questionnaire consists of two components:

household information and residential consumption. The survey variables of each component are listed in Table 1. GHG emissions accounting of urban residential consumption focuses on seven categories including electricity use, fuel consumption, transportation, solid waste treatment, wastewater treatment, food, and housing (which is treated as a consumable durable good). The quantity consumed in each category was collected directly or converted from the surveyed residential consumption variables, for example, we calculated the actual water consumption by dividing the surveyed water rate of household by current water price. The influencing factors of urban residential GHG emissions in our study were classified into variables at household and community scale. Residential status (permanent population or transient population), marital status, household size, age, household income, housing area, education, building age, and number of houses were considered to be potential influencing factors at the household scale. Average housing area, building age, average household income, building height, and average household size were considered to be potential influencing factors at the community scale.

TABLE 1: Components and survey variables in residential consumption questionnaire.

Components	Survey Variables
Household Information	Residential status; marital status; household size; age; education; household income
Residential Consumption	Number of houses; housing area; building height; building age; water fee; power fee; gas fee; waste production; food consumption; transportation destination; mode of transport; trip frequency; travel time

In view of the heterogenous spatial demographics of households and residential communities, we applied the spatial sampling method, which takes the spatial distribution characteristics of the object into account. The principle of this method is to balance the cost of sampling with the desired sampling precision, depending on study objectives and spatial variation [27], [28]. The spatial distribution characteristics in our study included topography, population density, standard land price, and administrative division.

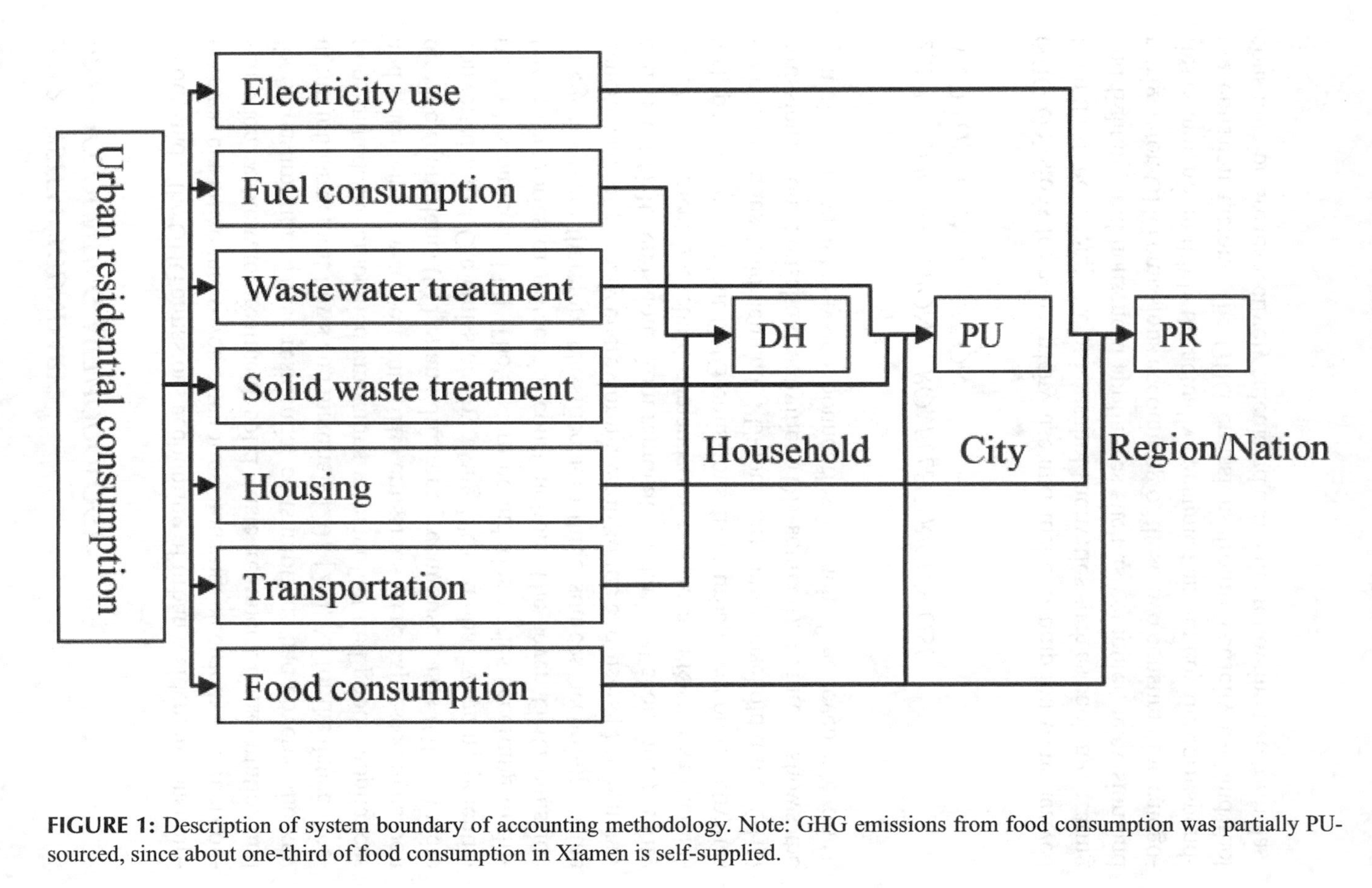

FIGURE 1: Description of system boundary of accounting methodology. Note: GHG emissions from food consumption was partially PU-sourced, since about one-third of food consumption in Xiamen is self-supplied.

2.2.2 SYSTEM BOUNDARY AND ACCOUNTING METHODOLOGY

In our study, the GHG emissions accounting of urban residential consumption was classified into seven categories: housing, electricity use, fuel consumption, wastewater treatment, solid waste treatment, transportation and food consumption. Those residential consumptions had covered the key urban infrastructural flows and materials [20], [29]. As for the data collection limited, the embodied emissions in manufactured goods, appliances and water supply were left out. GHG emissions were expressed in carbon dioxide equivalents (CO_2e) and different greenhouse gases (GHGs) were converted into CO_2e emissions by using IPCC global warming potential (GWP) parameters [30]. The system boundaries varies according to different categories of residential consumption. The seven GHG emission categories were therefore classified into three sources according to the general path of primary energy or materials to the end-users [22]: primary equivalent GHG emissions from regional and national economic activity supplied to meet household demand (referred to as PR-sourced hereafter), primary equivalent GHG emissions from urban economic activities supplied to household demand (PU-sourced), and household direct GHG emissions from household activities (DH-sourced). Figure 1 shows the spatial extension of the system boundaries for each of the seven categories.

2.2.2.1 GHG EMISSIONS FROM ELECTRICITY USE AND FUEL CONSUMPTION

GHG emissions from electricity use and fuel consumption were derived from the direct energy use of household activities such as cooking, heating and lighting, and household appliances such as computer, television and refrigerator. GHG emissions accounting of these two consumption categories commonly multiply the actually consumed amount by the corresponding emission factors. The GHG emissions from electricity use and fuel consumption are respectively calculated using the following two formulas:

$$E_E = E_C \times EF_e = E_c \times (EF_q \times W_q + EF_c \times W_c) \quad (1)$$

Where, E_E is GHG emissions from residential electricity per month; Ec is amount of residential electricity consumption per month; EF_e is the emission factor of electricity. EF_q and EF_c are the marginal emission factor of electrical quantity and marginal emission factor of electrical capacity of the East China Power Grid in 2009, which represent the the emission factors of currently running plants and newly built plants charged by East China Power Grid respectively [31]; W_q and W_c are respective weights of the emission factors for electricity. Here, we assign the same value to the two weights.

$$E_F = F_c \times EF_g = F_c \times (EF_{lpg} \times W_{lpg} + EF_{ng} \times W_{ng}) \quad (2)$$

Where, E_F is GHG emissions from residential gas consumption per month; F_c is amount of residential gas consumption per month; EF_g is the emission factor of gas. EF_{lpg} is emission factor of liquefied petroleum gas; EF_{ng} is emission factor of natural gas; W_{lpg}, W_{ng} are weights of the emission factors for liquefied petroleum gas and natural gas respectively. Here, we assign the values to the two weights according to the gas consumption proportion in Xiamen City (0.63 for liquefied petroleum gas and 0.37 for natural gas). The emission factors are referenced from the 2006 IPCC guidelines for national greenhouse gas inventories [30].

2.2.2.2 GHG EMISSIONS FROM TRANSPORTATION

The GHG emissions from transportation were estimated according to different modes of transport and corresponding consumption of diesel, petrol, gas or electricity. According to the GHG emissions accounting method for mobile sources [30], we calculate the GHG emissions by multiplying GHG emissions intensity per unit time with the travel time of each travel mode, according to the following two formulas:

$$E_T = \sum_j EF_j \times T_j \times f_j \tag{3}$$

Where E_T is total GHG emissions from transportation per month. EF_j is emission factor per unit time of travel mode j; T_j is average travel time of travel mode j; Travel mode j represents walking, cycling, private car, taxi, public bus, bus rapid transit (BRT), shuttle bus, or motorcycle respectively. f_j is frequency of travel mode j. The EF of walking and biking is 0; the motorcycle EF is estimated through electricity consumption per unit time, as most motorcycles in Xiamen City are electric powered. The EF of private car, taxi, public bus, BRT, and shuttle bus are estimated as follows:

$$EF_j = S_j \times E_j/Q_j/V_j \times a \times G \times ef \tag{4}$$

Where S_j is total operation mileage per unit time of mode j; E_j is fuel consumption per unit distance; Q_j is passenger volume per unit time of travel mode j; V_j is travel time per capita of travel mode j; a is fuel density of diesel or gasoline; G is net heat value of diesel or gasoline; ef is emission factor of diesel or gasoline (see Table 2).

2.2.2.3 GHG EMISSIONS FROM FOOD CONSUMPTION

The GHG emissions from food consumption mainly consist of direct emissions from human metabolism and indirect emissions from food processing and supply. As the direct GHG emissions from food consumption by human metabolism will overlap the GHG emissions from wastewater treatment, here only the indirect GHG emissions from food processing and supply are calculated using the following formulas:

$$EF_i = EF_d \times K \tag{5}$$

Where EF_i is indirect GHG emissions from food consumption; EF_d is direct GHG emissions from food consumption; K is proportion of EF_i to EF_d and refers to the proportion of indirect GHG emissions to direct GHG emissions from Chinese residential food consumption in 2006 [32]. EF_d is estimated as follows:

$$EF_D = \sum_i W_i \times R_i \quad (6)$$

$$R_i = C_{pi} \times P_i + C_{fi} \times F_i + C_{ci} \times C_i \quad (7)$$

Where W_i is consumption amount of food i; R_i is carbon content of food i; C_{pi}, C_{fi}, and C_{ci} are contents of protein, fat, and carbohydrate of food i respectively; P_i, F_i, and C_i are the carbon content of protein, fat, and carbohydrate respectively; C_{pi}, C_{fi}, and C_{ci} refers to China food composition [33].

TABLE 2: Parameters for estimating the emission factors of different travel modes in Xiamen City.

Travel Mode	$S_j^{a,d}$	E_j^a	Q_j^a	V_j^a	Fuel Type	Calorific value[b]	EF[c]
	(100 km/a)	(L/100km)	(P/a)	(minute)		(kJ/kg)	(tC/TJ)
Taxi	62,055,780	10.5	22,813	25.46	gasoline	43,124	69.2
Bus	1,763,045	25	41,180	25.46	diesel	42,705	74.0
BRT	26,825	36	2,375	25.46	diesel	42,705	74.0
Shuttle	536,954	23	15.243	25.46	diesel	42,705	74.0

[a]The data of Sj, Ej, Qj, and Vj are derived from Xiamen City's Transportation Committee and Xiamen Transportation Company. bCalorific values are taken from "General calculation principles for total production energy consumption (GB/T-2589-2008)" (in Chinese). cEmission factors were extracted from the Technology and Environmental Database (TED) in Lin's study [6]. dThis equation will always underestimate the total emissions due to transport because the parameter Sj does not record fuel use while a vehicle is stationary.

2.2.2.4 GHG EMISSIONS FROM HOUSEHOLD SOLID WASTE TREATMENT

In 2009, household solid waste disposal and treatment in Xiamen City included landfill (80%) and incineration (20%) The GHG emissions from landfill diposal mainly considered to be emissions of CH_4 and CO_2 from the landfill yard, which can be estimated as follows [34]:

$$E_{CH_4} = \left[MSW \times \eta_1 \times \sum_j (DOC_j \times W_j) \times r \times \frac{16}{12} \times F - R \right] \times (1 - OX) \times GWP_{CH_4} \tag{8}$$

$$E_{CO_2} = SMW \times \eta_1 \times \sum_j (DOC_j \times W_j) \times r \times F \times \frac{44}{12} \tag{9}$$

Where E_{CH4} and E_{CO2} are amount of CH_4 and CO_2 emitted from solid waste disposal respectively; MSW is mass of solid waste deposited in Xiamen City in 2009; η_1 is proportion of solid waste deposited to landfill; DOC_j is fraction of degradable organic carbon to degradable component j; W_j is fraction of degradable component j to total solid waste deposited; 16/12 is molecular weight ratio CH_4/C; r is fraction of degradable organic carbon that can decompose; F is volume fraction of CH_4 in generated landfill gas; R is the recovery rate of CH_4; OX is the oxidation rate of CH_4; GWP_{CH4} is the global warming potential of CH_4.

According to 2006 IPCC guidelines for national greenhouse gas inventories [30], the GHG emissions from landfill incineration is mainly from CO_2 and N_2O and can be estimated as follows:

$$E_{CO_2} = MSW \times \eta_2 \times \sum_j (W_j \times dm_j \times CF_j \times OF_j) \frac{44}{12} \tag{10}$$

$$E_{N_2O} = MSW \times \eta_2 \times EF_{N_2O} \times 10^{-3} \times GWP_{N_2O} \quad (11)$$

Where E_{CO2} and E_{N2O} are amount of CO_2 and N_2O emitted from solid waste incineration; η_2 is proportion of solid waste deposited by incineration; dmj is dry matter content of degradable component j; CF_j is fraction of carbon in degradable component j; OF_j is oxidation factor; 44/12 is the molecular weight ratio CO_2/C; EF_{N2O} is emission factor of N_2O from waste incineration; GWP_{N2O} is global warming potential of N_2O.

2.2.2.5 GHG EMISSIONS FROM HOUSEHOLD WASTEWATER TREATMENT

In our study, all the household water used was assumed to be transformed to wastewater. The calculation of GHG emissions from wastewater treatment mainly considered sewage plant emissions of CH_4 which are produced from anaerobic treatment process and can be calculated as follows:

$$E_{CH4} = W \times P_{COD} \times \eta \times EF_{CH4} \times GWP_{CH4} \quad (12)$$

Where E_{CH4} is production of CH_4 from wastewater treatment; W is mass of wastewater; P_{COD} is content of chemical oxygen demand in wastewater; η is fraction of wastewater through anaerobic treatment; EF_{CH4} is emission factor of CH_4.

2.2.2.6 GHG EMISSIONS FROM HOUSING

In our study, housing is considered to be a durable consumable good with a lifetime of fifty years, as this is the maximum service life of residential housing regulated by the Ministry of Housing and Urban-Rural Development, PR China. Lifecycle GHG emissions result from material mining and processing, construction, house operation and demolition, but material mining and processing and house operation are responsible for most of the emissions. In principle, GHG emissions from housing operation should be the same to those from household electricity use and fuel consumption, so the GHG emissions from housing mainly considered lifecycle GHG emissions from building materials. There are two types of residential buildings (masonry-concrete and steel-concrete) in Xiamen. Liu et al. estimated the energy consumption and environmental emissions of the two types of residential building using life cycle analysis and the Boustead Model [35]. Based on her estimation of GHG emissions per unit area for the two types of residential building (see Table 3), GHG emissions from the two types of housing can be calculated as follows:

$$E_{HC} = \sum_j (EF_j \times GWP_j) \times BA \quad (13)$$

Where E_{HC} is GHG emissions from housing; EF_j is emission amount of greenhouse gas j; j represents CO_2, CO, and N_2O respectively; GWP_j is global warming potential of gas j; BA is the building area.

TABLE 3: GHG emissions per unit area in the lifecycle of building materials.

GHGs	GHG emissions in the lifecycle kg/m^2		GWP_j
	Steel-concrete[a]	Masonry-concrete[a]	
CO	20.1	7.5	2
CO_2	954.2	828.51	1
NO_x	6.2	2.68	310

[a]*The emissions factors of steel-concrete and masonry-concrete refer to Liu's study [33].*

2.2.3 STUDY AREA AND SURVEY IMPLEMENTATION

Xiamen is a typical coastal city located in southeast China (24°25′N-24° 55′N, 117°53′E-118°27′E). It has a land area of 1,565 km^2 and a sea area of 390 km^2 [36]. The rapid urban expansion and economic development of Xiamen was not triggered until the implementation of China's 'reform and opening-up' policy in 1980, when the Xiamen Special Economic Zone was established on the island. Since then, Xiamen has undergone rapid urbanization and its urban population has grown at an remarkable speed. Its regional GDP reached 173.72 billion yuan in 2009, having been just 1.72 billion yuan (comparable GDP value) in 1980. Meanwhile the urbanization ratio increased rapidly from 35% to 80%, and in 2009 the population of Xiamen reached 2.52 million with a population density of 1,602 people per km^2. Average urban disposable income and consumption expenditure reached 26,131 yuan and 17,990 yuan respectively. Residential electricity consumption was 2.75 billion kWh in 2009, up from 0.64 billion kWh in 1999. Residential water use was 9900 million ton, compared to 8300 million ton in 1999. In 2009, Xiamen became one of the first ten pilot cities of the 'COOLCHINA-2009 civil low-carbon action pilot project'. Understanding the characteristics of GHG emissions from urban residential consumption is in urgent needed to reduce residential GHG emissions and develop a low carbon city.

The downtown area is located in Xiamen Island and off island districts are mainly peri-urban areas. According to the spatial sampling, 44 typical communities, including 28 from Xiamen Island (I1-I28) and 16 from Xiamen mainland (O1-O16), were determined as the survey sites (see Figure 2). The onsite questionnaire surveys were conducted in the targeted communities in October 2009 and July 2010. 1,485 questionnaires were completed, of which 714 questionnaires satisfied all the information needed in this study. This represented a sampling of about 0.1% of the total households in the targetted area.

2.2.4 DATA PROCESSING AND STATISTICAL ANALYSIS

Some questionnaire variables are quantitative (e.g. household size, water use per month) while other variables are qualitative (e.g. residential status, marriage, and education). However, all qualitative variables were transformed into ordinal variables to facilitate statistical analysis in SPSS (IBM Corporation). The transformation standards are shown in Table 4. Analysis of variance (ANOVA) which is able to test whether data from several groups have a common mean, was applied to test which potential factors would cause a significant difference ($P<0.05$) in urban residential GHG emissions. Second, a stepwise linear regression analysis was applied to identify the major influencing factors, taking the potential factors as independent variables and urban residential GHG emissions as dependent variables. Finally, taking the main influencing factors identified from regression analysis as the analysis variables, the 714 households and 44 communities of Xiamen City were respectively clustered into three GHG emission categories through K-means cluster analysis. This allowed the characteristics of low, medium and high GHG emission households and communities to be summarized and compared.

TABLE 4: Standards to transform qualitative variables into ordinal variables.

Qualitative variables	Transform Standards
Residential status	Registered resident = 1; Non-registered resident = 2
Marital status	Unmarried = 1; Married = 2; Divorced = 3
Age	<25 =1; 25~30 =2; 31~40 = 3; 41~50 = 4; 51~59 = 5; >59 = 6
Education	Elementary = 1; Junior = 2; Senior = 3; College = 4; Graduate = 5; Others = 6
Household income (yuan/month)	<2,000 = 1; 2,000~5,000 = 2; 5,000~10,000 = 3 10,000~20,000 = 4; >20,000 = 5
Housing area m^2	<40 = 1; 40~69 =2; 70~89 = 3; 90~119 = 4; 120~149 = 5; >149 = 6
Number of houses	None =1; 1 house = 2; 2 houses = 3; >2 houses =4
Building age	Before 1980s = 1; 1980–1990 = 2; 1990–2000 = 3; after 2000 = 4
Building height	1–7 = 1 (low-rise building); >7 = 2 (high-rise building)

2.3 RESULTS

2.3.1 GHG EMISSIONS FROM URBAN RESIDENTIAL CONSUMPTION

In 2009, the average GHG emissions of urban residential consumption per household in Xiamen City were 1042.31 kg CO_2e/month. The emission intensities per household of the seven categories of residential consumption activities could be ranked in decreasing order as: housing (32.98%)>electricity use (26.84%)>food (15.17%)>transportation (9.21%)>solid waste treatment (6.44%)>wastewater treatment (5.20%)>fuel consumption (4.16%). The average per capita GHG emissions from Xiamen urban residential consumption were 323.37 kg CO_2e/month. The order of the emission intensities per capita of the seven categories of residential consumption activities was same as for households: housing (34.11%)>electricity use (26.17%)>food (15.23%)>transportation (8.51%)>solid waste treatment (6.61%)>wastewater treatment (5.17%)>fuel consumption (4.20%) (see figure 3).

According to the system boundary classification, the majority of the GHG emissions from urban residential consumption in Xiamen City were derived from national or regional energy and material supply (PR-sourced), including building materials, electricity, and most food, which accounted for 70.43% of total GHG emissions. Urban economic activities supporting residential consumption (PU-sourced), including waste water treatment, solid waste treatment and a small fraction of food supply, accounted for 16.86% of total GHG emissions. The direct household GHG emissions (DH-sourced) only accounted for 12.71% of the total GHG emissions.

At the household scale, the per household and per capita average GHG emissions from urban residential consumption of Xiamen island (downtown) communities were 991.78 kg CO_2e/month and 321.21 kg CO_2e/month respectively. The per household and per capita average GHG emissions from urban residential consumption per household and per capita of Xiamen mainland communities (peri-urban areas) were 1098.32 kg CO_2e/month and 335.54 kg CO_2e/month respectively (see figure 4). The per household GHG emissions of the downtown communities were not significantly different from of the peri-urban communities (P = 0.243). However, the per capita GHG emissions of the downtown communities were significantly lower than of the peri-urban communities (P = 0.031). The major difference between the downtown and peri-urban communities were in household electricity use and transportation. In addition, differences in average household size meant that the communities with the highest and lowest GHG emission per household were not the same as the communities with the highest and lowest GHG emissions per capita. For example, the community I18 had the highest per household GHG emissions in the downtown but its per captia GHG emissions were lower than I22 because the latter have a smaller average houshold size.

TABLE 5: One-way ANOVA analysis of potential influencing factors.

Survey Variables	Total GHG Emissions		Consumption Categories	
	Per household	Per capita	Per household	Per capita
Residential status[a]	Yes	Yes	4,5,6	2,6
Marital status[a]	Yes	No	4,5,7	4,5,7
Household size[a]	/	Yes	/	1,2,3,4,5,6
Household income[a]	Yes	Yes	1,4,5,6,7	1,5,6,7
Housing area[a]	Yes	Yes	1,2,4,5,6,7	1,2,4,5,6,7
Education[a]	Yes	Yes	1,3,5,6,7	1,3,5,6,7
Age[a]	Yes	Yes	1,5,6,7	1,5,6,7
Building age[a]	Yes	Yes	1,2,4,5,6	1,2,4,5,6
Number of houses[a]	Yes	Yes	1,4,5,6,7	1,4,5,6,7
Average housing area[b]	Yes	Yes	1,4,6,7	1,4,6,7
Building age[b]	Yes	Yes	1,4,6	1,4,6
Building height[b]	Yes	Yes	1,4,5,6	1,4,5,6
Average household income[b]	Yes	Yes	4,5,6	4,5,6
Average household size[b]	/	No	/	4

[a]represents the variables at household scale and [b] represents va riables at community scale. Yes means the survey va riable caused a significant difference in total GHG emissions and No means not significant. Numbers 1-7 respectively represent GHG emissions from the following seven residential consumption categories: Electricity use, Fuel consumption, Solid waste treatment, Wastewater treatment, Food consumption, Housing, and Transportation.

2.3.2 INFLUENCING FACTORS OF URBAN RESIDENTIAL GHG EMISSIONS

Analysis of variance (ANOVA) was applied to test each survey variable to see whether it caused a significant difference ($P<0.05$) in total GHG emissions per household and per capita. If it did, then ANOVA was further used to test which consumption category showed a significant difference corresponding to the survey variable. The results are shown in Table 5. At the household scale, residential status, marital status, household income, housing area, education, age, building age, and number of houses can affect per household GHG emissions. Residential status, household size, household income, housing area, education, age, building age and number of houses can affect GHG emissions per capita. At the community scale, average housing area, building age, building Height and average household income can affect GHG emissions per household and per capita.

The results of regression analysis are shown in Table 6. At the household scale, housing area, household income, building age, household size, marital status, and age present in the regression formula of GHG emissions per household, indicating they are the influencing factors of GHG emissions per household in the statistical sense. Housing area is the main influencing factor with the largest standard regression coefficient of 0.475. Household size, housing area, building age, household income, and residential status present in the regression formula of GHG emissions per capita. Household size and housing area are the main influencing factors, with standard regression coefficients (the relative importance of the independent variables to the dependent variable) of −0.479 and 0.456 respectively. At the community scale, average housing area and building story present in both the regression formulas of GHG emissions per household and per capita. Their standard regression coefficients are respectively 0.519 and 0.497 per household and 0.455 and 0.656 per capita.

2.3.3 CHARACTERISTICS OF URBAN RESIDENTIAL GHG EMISSIONS

At the household scale, taking housing area, household size, building age, household income, and GHG emissions per household and per capita as the analysis variables, the 714 surveyed households are categorized into three groups (low, medium and high GHG emission households) using K-means cluster analysis (see Table 7). The final cluster centers are computed as the mean for each variable within each final cluster and reflected the typical characteristics of the three household categories. A high GHG emission household is always characterized as consisting of 4 persons with more than 150 m^2 of housing area, living in a building constructed after 2000, and with a monthly household income of 10,000–15,000 yuan. A low GHG emission household is characterized as 3–4 persons with about 80–90 m^2 of housing area, living in a building constructed in the 1990s, and with a monthly household income of 6000 yuan. High GHG emissions households emit 4.86 times more than low GHG emissions households. Comparing low and high GHG emissions households, the increase in GHG emissions from electricity use per household is the most significant, followed by increases from housing, transportation and wastewater treatment. Increases are also observed in the other three categories of residential consumption, but the growth rates are very small (see figure 5a).

At community scale, taking average housing area, building height, and GHG emissions per household and per capita as the analysis variables, the 44 surveyed communities are categorized into low, medium, and high GHG emission communities (see Table 7). The final cluster centers show that high GHG emission communities are usually characterized by an average housing area of about 120 m^2 and buildings usually with eight floors or more. Communities with a lower level of GHG emissions are characterized by an average housing area of about 70–80 m^2 and buildings with seven floors or fewer. The difference between low and high GHG emissions communities is less than at the household level, but high GHG emissions communities emitt 2.09 times as much as low GHG emissions communities. From low to high GHG emissions communities, the increase in emissions from housing is the most significant, followed by electricity use and transportation. An increase is also observed in the other four residential consumption categories, but the growth rates are very small (see figure 5b).

2.4 DISCUSSION

2.4.1 CHARACTERIZING GHG EMISSIONS FROM URBAN RESIDENTIAL CONSUMPTION

The lifestyles of city residents are influenced by physical, social, economic factors, as well as the cultural background which affect GHG emissions in various ways. A bottom-up social survey can directly connect lifestyle factors to the GHG emissions from residential consumption and provide potential breakthrough points for carbon reduction policymaking. For example, housing area was the main influencing factor of residential GHG emissions at the household scale in Xiamen City. If other factors remained constant, larger housing area would result in larger GHG emissions, so policies to reduce housing area per urban household would be an effective measure to control residential GHG emissions for Xiamen City. Currently low-storey buildings are being rapidly replaced by high-rise buildings in Chinese cities to increase compactness [37] and this is also believed to have the co-benefit of reducing GHG emissions [38]. However, our results show that high-storey residential buildings and spacious housing both tend to increase GHG emissions from urban residential consumption. It is hard to develop a low-carbon city simply by increasing the compactness of residential buildings. Effective carbon reduction policies must therefore consider other ways to reduce emissions from residential consumption, as will be discussed below.

Another advantage of the survey based approach is that it offers the possibility to further break down the underlying factors. Household size is widely recognized as a major factor influencing residential GHG emissions[10], [17], [39]–[41], and larger households tend to be more efficient in terms of per capita energy use [10], [40], [42]. Our study did find that residential GHG emissions per capita tended to decrease with increasing household size, but only to an optimum household size of four persons. A four-member family could be comprised of, for example, a middle-aged couple with two children, a middle-aged couple with one child and an elderly parent, or an elderly couple living with a child and his/her spouse. Family composition may be a key underlying factor in determining residential GHG emissions and merit further study.

TABLE 6: Stepwise linear regression of the potential influence factors.

Independent variables	Unstandardized Coefficients	Standardized Coefficients	Independent Variables	Unstandardized Coefficients	Standardized Coefficients
Household scale	per household		Household scale	per capita	
Constant	–457.746	/	Constant	205.982	/
Housing area	201.671	0.475	Household size	–81.058	–0.479
Household income	97.823	0.178	Housing area	67.961	0.456
Household size	68.934	0.143	Building age	29.499	0.127
Building age	76.693	0.116	Household income	25.329	0.131
Marital status	130.792	0.101	Residential status	24.666	0.061
Age	–31.109	–0.072			
R^2	0.650		R^2	0.669	
F	84.470		F	126.068	
P	<0.001		P	<0.001	
Community scale	per household		Community scale	per capita	
Constant	122.132	/	Constant	28.502	/
Average housing area	226.844	0.519	Building height	107.818	0.565
Building height	294.515	0.497	Housing area	64.074	0.455
R^2	43.855		R^2	0.692	
F	<0.001		F	45.954	
P			P	<0.001	

Residential consumption will play an increasingly important role in future to shape China's energy demand and GHG emissions. It is necessary to understand the tendencies of Chinese urban lifestyles to achieve low-carbon city development. In our study, the influential factors of residential GHG emissions presented similar trends from low to high GHG emissions households and communities (see Table 7). This GHG emissions gradient existing among present households and communities can provide valuable information on the likely future changes in Chinese urban residential consumption. Currently, most urban households and communities in Xiamen are low or medium carbon emitters (see Table 7). Future urbanization and socioeconomic development is likely to result in increasing income levels, housing renovation, an increase in housing area and the replacing of low-storey buildings with high-rise apartment blocks. As a result, the proportion of low GHG emissions households and communities will gradually reduce while high-carbon households and communities is likely to increase rapidly. At the same time, the composition of residential GHG emissions will change, and GHG emissions from housing and transportation may grow significantly.

2.4.2 POLICY MAKING TOWARD A LOW-CARBON URBAN CONSUMPTION PATTERN

Jones and Kammen argued that realizing GHG emissions reduction required behavior change at the household level through personalized feedback [14]. This makes theoretical sense, because the most effective measure to reduce GHG emissions from household consumption is to cut unnecessary material or energy use directly. However, our results suggest that from a lifecycle perspective, the largest carbon reduction potentials are beyond the control of individual consumers. For example the majority of urban residential GHG emissions in Xiamen City are mainly derived from urban (17%) and regional and national (70%) economic activity. As a result, policy measures such as extending building lifespan and the recycling of building wastes could contribute more significantly to GHG emissions reduction than simply targeting individual consumer choices alone. The percentage of clean primary energy in the total energy use is only 3% in China [43]. Adjusting the composition of primary energy to produce electricity may have a greater potential for carbon reduction than simply reducing household electricity use.

Policymaking for a low-carbon city must therefore adopt a holistic approach in terms of policy scope, priority and timing of implementation. Taking Xiamen City for example, the policy scope should cover the entire path of primary energy or materials to end-users, including household behavior and urban, regional and national activity. Specific policies should include: promoting energy saving appliances and greater use of public transportation at the household scale, promoting low-carbon techniques of pollution control, such as clean coal technology, catalytic combustion technoloy, increasing the proportion of food that is locally sourced at the city scale, adjusting the primary energy mix for electricity production and developing green building materials and technologies at the regional or national scale.

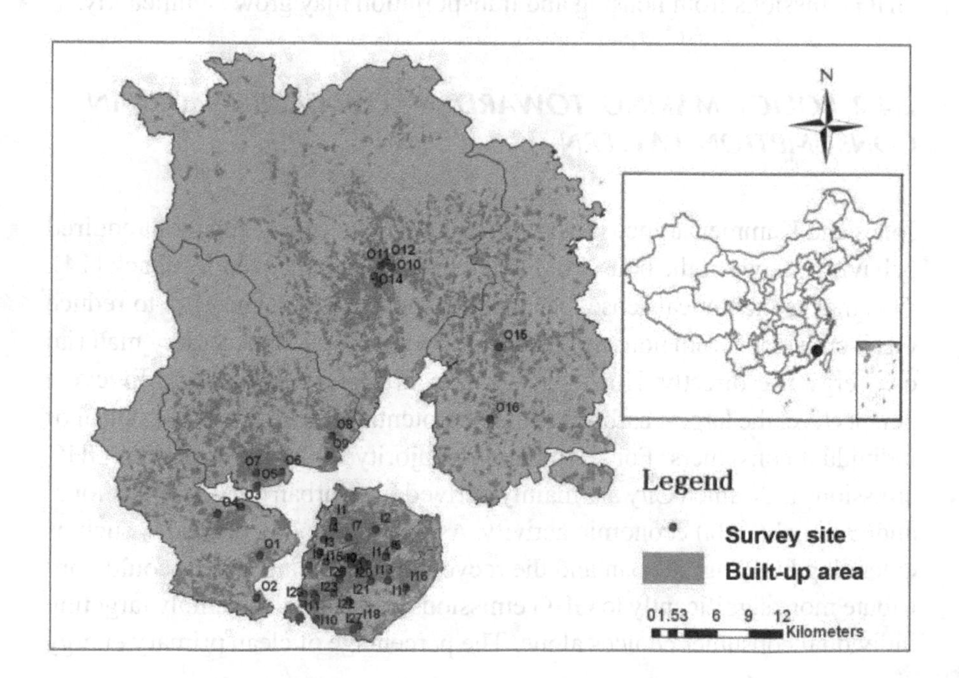

FIGURE 2: Location of Xiamen City and survey site selection.

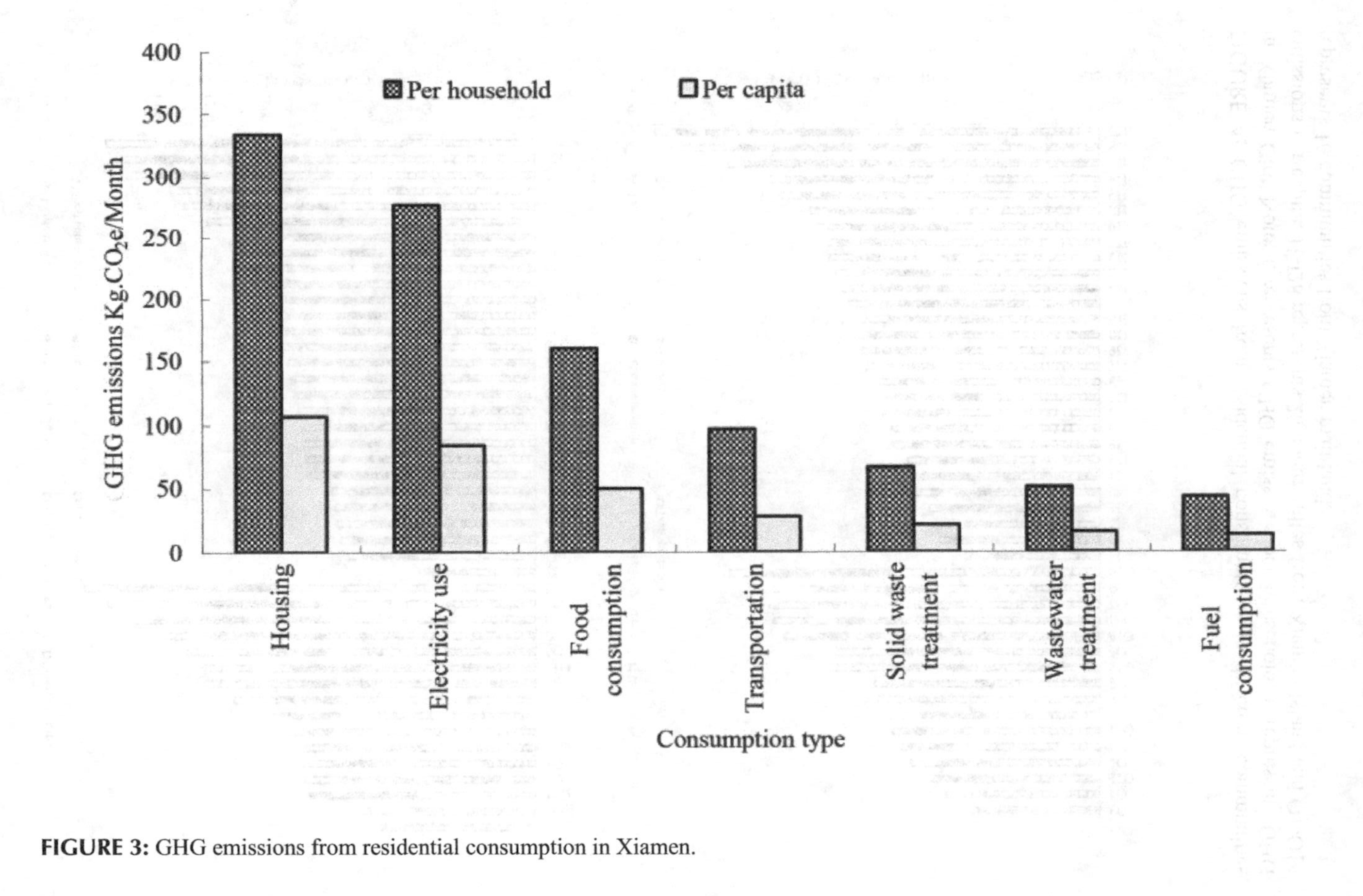

FIGURE 3: GHG emissions from residential consumption in Xiamen.

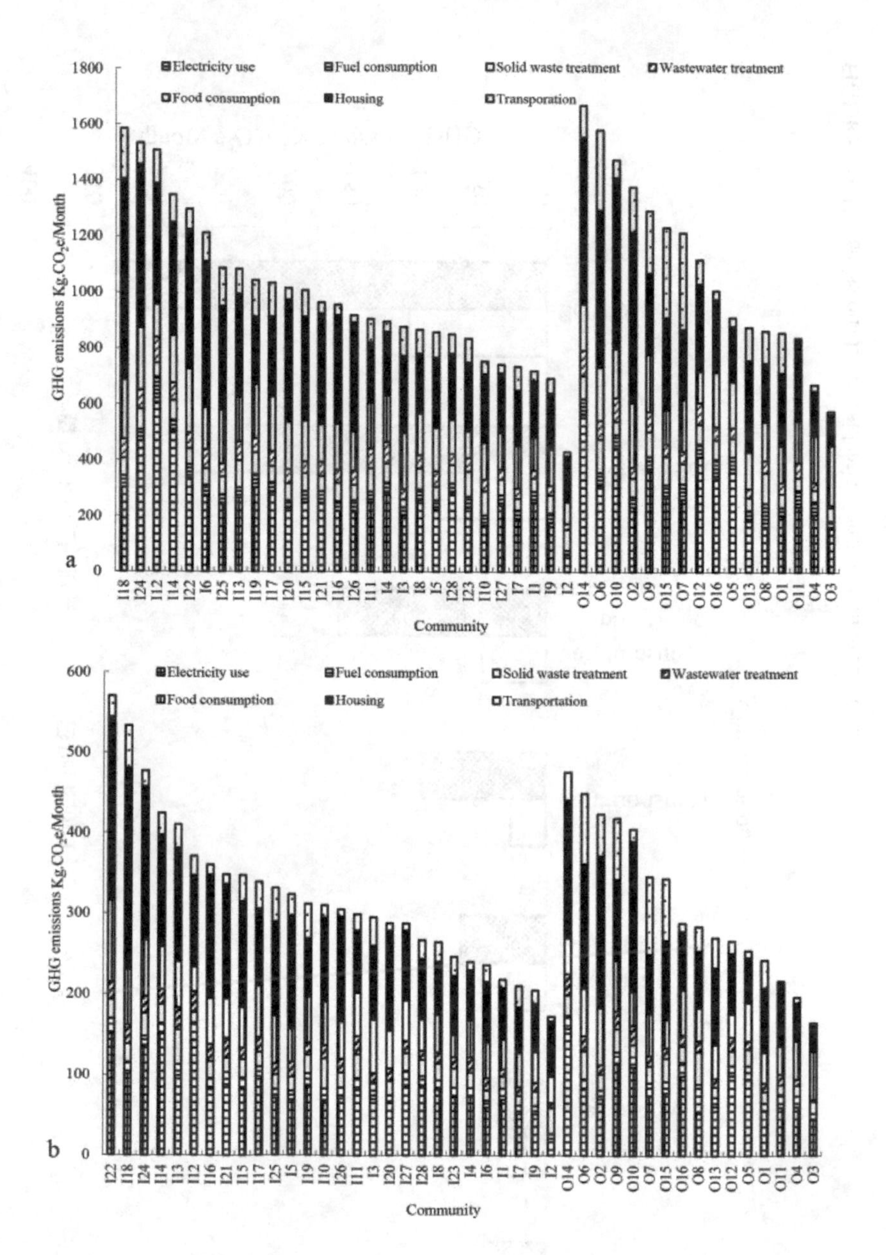

FIGURE 4: GHG emissions from residential consumptions in different communities in Xiamen City. Note: a represents GHG emissions per household; b represents GHG emissions per capita. I1-I28 represents 28 communites from Xiamen Island and O1-O16 represents 16 communities from Xiamen mainland.

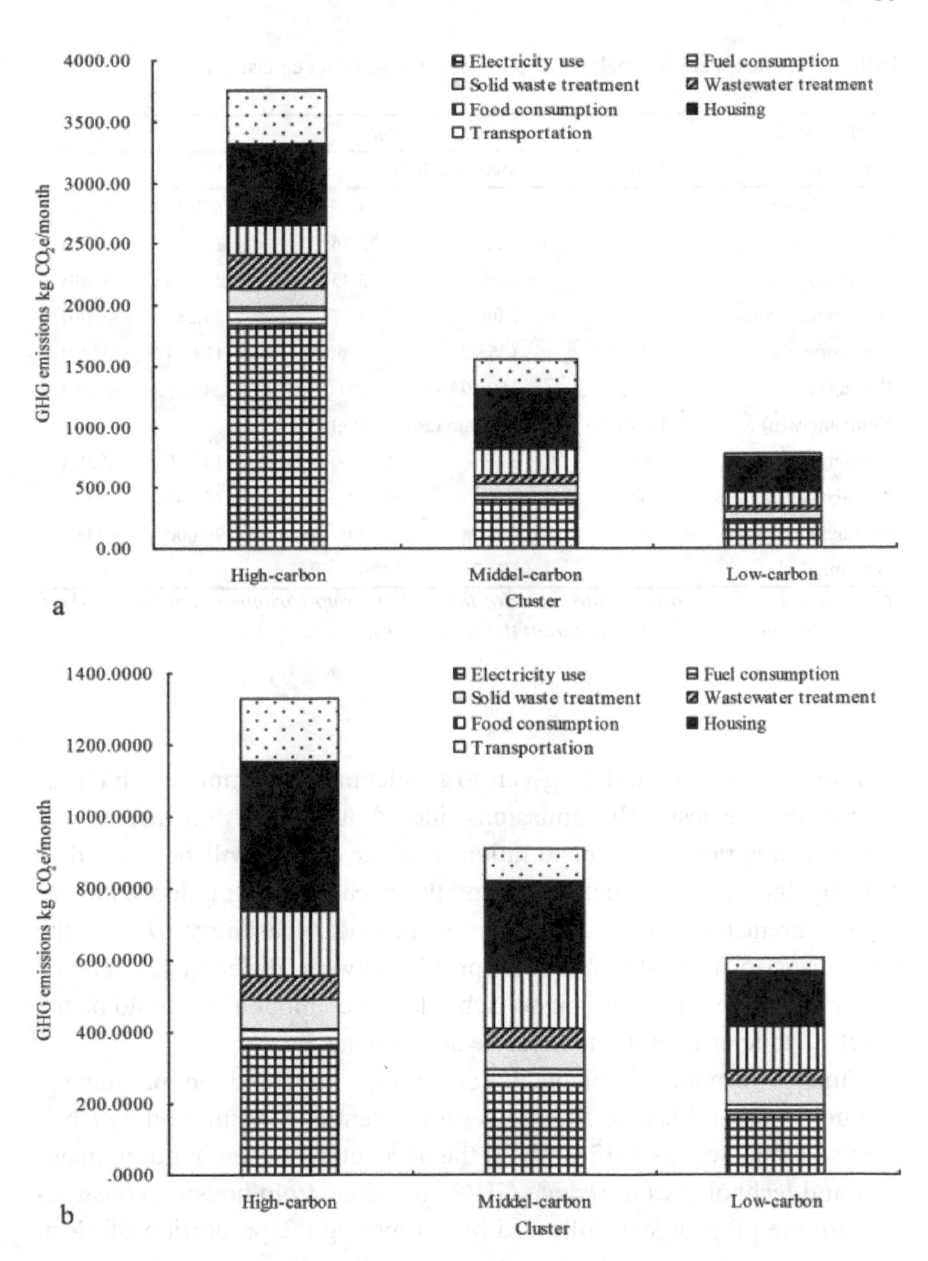

FIGURE 5: GHG emissions from residential consumptions in the high, medium and low carbon household (a) and community (b) of Xiamen City. Note: a represents households; b represents communities.

TABLE 7: K-Means cluster analysis of urban residential GHG emissions.

Analysis Variables	Final Cluster Centers			ANOVA	
Household (n)	Low (497)	Medium (206)	High (11)	F*	P
Household size	3.4	3.77	4	8.714	<0.001
Housing area	2.79	4.23	5.36	140.285	<0.001
Building age	2.72	3.24	3.45	32.565	<0.001
Household income	2.2	3.04	3.27	63.282	<0.001
Per household	770.60	1553.25	3750.46	1133.478	<0.001
Per capita	251.79	460.49	991.28	244.855	<0.001
Community (n)	Low (10)	Medium (24)	High (10)		
Building height	2.44	3.18	4.00	13.810	<0.001
Average housing area	1.00	1.33	1.90	12.340	<0.001
Per household	701.04	986.03	1466.79	99.600	<0.001
Per capita	223.27	302.78	454.69	59.370	<0.001

**F = variance of the group means/mean of the within group variances. The bigger the F value is, the more significantly different the sample groups are.*

Policy priority should be given to residential consumption which results in the greatest GHG emissions, including housing, electricity use, food consumption and transportation. Further studies will be needed to quantify the carbon reduction potentials in each consumption category given current technology and to assess practical feasibility. Due to the large disparity in GHG emissions profile between different households and communities, high-carbon households and communities should be the target of policies to promote lifestyle adjustments.

Timing of policy implementation should be based on predictable changes in urban lifestyle and focus on residential consumption which is expected to increase significantly in the near future. Green building materials and technologies to reduce GHG emissions from housing construction are the most urgent, followed by promoting the proportion of clean energy in electricity production, increasing the efficiency of household electricity use, and encouraging the use of public transportation.

2.5 CONCLUSIONS

As cities become the primary habitat of human beings, GHG emissions from urban residential consumption and the role of urban lifestyle has become increasingly significant. We present a survey-based GHG emissions accounting methodology for urban residential consumption and apply it in Xiamen City, China. According to our results, reducing the GHG emissions from urban residential consumption is often beyond the control of individual consumers. Housing, electricity use and food consumption whose GHG emissions are from regional and national economic activities (PR-sourced) and wastewater treatment and solid waste treatment whose GHG emissions are from urban economic activities (PU sourced) accounted for about 70% and 17% of total residential GHG emissions in Xiamen City respectively. The entire energy or materials pathway to the end-users should be included in the policymaking scope. A large disparity in carbon profile between different households, with the high carbon households emitting about five times as much GHG as low carbon households. High carbon communities emit about twice as much GHG as low carbon communities. Residential consumptions which resulted in the majority of GHG emissions and which would likely increase significantly in the near future including housing, electricity use, and transportation, should be the key points for policymaking of low-carbon urban residential consumption in China. The survey-based GHG emissions accounting method of household consumption developed in this study can be readily applied to other cities. It provides a useful tool to understand and profile residential groups, and makes it possible to design tailored and targeted policies for GHG emissions reduction.

REFERENCES

1. Seto KC, Fragkias M, Güneralp B, Reilly MK (2011) A meta-analysis of global urban land expansion. PLoS ONE 6: e23777. doi: 10.1371/journal.pone.0023777
2. Grimm NB, Faeth SH, Golubiewski NE, Redman CL, Wu JG, et al. (2008) Global change and the ecology of cities. Science 319: 756–760. doi: 10.1126/science.1150195

3. Pataki DE, Alig RJ, Fung AS, Golubiewski NE, Kennedy CA, et al. (2006) Urban ecosystems and the North American carbon cycle. Global Change Biology 12: 2092–2102. doi: 10.1111/j.1365-2486.2006.01242.x
4. Bai XM (2007) Integrating global environmental concerns into urban management - The scale and readiness arguments. Journal of Industrial Ecology 11: 15–29. doi: 10.1162/jie.2007.1202
5. Kennedy C, Steinberger J, Gasson B, Hansen Y, Hillman T, et al. (2010) Methodology for inventorying greenhouse gas emissions from global cities. Energy Policy 38: 4828–4837. doi: 10.1016/j.enpol.2009.08.050
6. Lin J, Cao B, Cui S, Wang W, Bai XM (2010) Evaluating the effectiveness of urban energy conservation and GHG mitigation measures: The case of Xiamen City, China. Energy Policy 38: 5123–5132. doi: 10.1016/j.enpol.2010.04.042
7. Dhakal S (2010) GHG emissions from urbanization and opportunities for urban carbon mitigation. Current Opinion in Environmental Sustainability 2: 277–283. doi: 10.1016/j.cosust.2010.05.007
8. Kaye JP, Groffman PM, Grimm NB, Baker LA, Pouyat RV (2006) A distinct urban biogeochemistry? Trends in Ecology and Evolution 21: 192–199. doi: 10.1016/j.tree.2005.12.006
9. Lenzen M, Cummins RA (2011) Lifestyles and well-being versus the environment. Journal of Industrial Ecology 15: 650–652. doi: 10.1111/j.1530-9290.2011.00397.x
10. Bai XM, Dhakal S, Steinberger J, Weisz H (2012) Drivers of urban energy use and main policy leverages. In: Grubler A, Fisk DJ editors. Energizing sustainable cities. EarthScan.
11. Weisz H, Steinberger JK (2010) Reducing energy and material flows in cities. Current Opinion in Environmental Sustainability 2: 185–192. doi: 10.1016/j.cosust.2010.05.010
12. Schipper L, Bartlett S, Hawk D, Vine E (1989) Linking life-styles and energy use: a matter of time? Annual Review of Energy 14: 273–320. doi: 10.1146/annurev.eg.14.110189.001421
13. Wei Y, Liu L, Fan Y, Wu G (2007) The impact of lifestyle on energy use and CO2 emission: An empirical analysis of China's residents. Energy Policy 35: 247–257. doi: 10.1016/j.enpol.2005.11.020
14. Jones CM, Kammen DM (2011) Quantifying carbon footprint reduction opportunities for US households and communities. Environtal Science and Technology 45: 4088–4095. doi: 10.1021/es102221h
15. HM Government (2006) The UK climate change programme 2006. London, UK: The Stationery Office.
16. Dietz T, Gardner GT, Gilligan J, Stern PC, Vandenbergh MP (2009) Household actions can provide a behavioral wedge to rapidly reduce US GHG emissions. Proceedings of the National Academy of Sciences 106: 18452–18456. doi: 10.1073/pnas.0908738106
17. Druckman A, Jackson T (2009) The carbon footprint of UK households 1990–2004: A socio-economically disaggregated, quasi-multi-regional input-output model. Ecological Economics 68: 2066–2077. doi: 10.1016/j.ecolecon.2009.01.013

18. Wang Y, Shi M (2009) CO2 emission induced by urban household consumption in China. Chinese Journal of Population, Resources and Environment 7: 11–19. doi: 10.1080/10042857.2009.10684933
19. Feng L, Lin T, Zhao Q (2011) Analysis of the dynamic characteristics of urban household energy use and GHG emissions in China. China Population, Resources And Environment 21: 93–100.
20. Ramaswami A, Chavez A, Ewing-Thiel J, Reeve KE (2011) Two approaches to greenhouse gas emissions foot-printing at the city scale. Environtal Science and Technology 45: 4205–4206. doi: 10.1021/es201166n
21. Kanemoto K, Lenzen M, Peters GP, Moran DD, Geschke A (2012) Frameworks for comparing emissions associated with production, consumption, and international trade. Environtal Science and Technology 46: 172–179. doi: 10.1021/es202239t
22. Baynes T, Lenzen M, Steinberger JK, Bai X (2011) Comparison of household consumption and regional production approaches to assess urban energy use and implications for policy. Energy Policy 39: 7298–7309. doi: 10.1016/j.enpol.2011.08.053
23. Grubler A, Bai XM, Buettner T, Dhakal S, Fisk DJ, et al.. (2012) Urban energy systems. In: Global energy assessment. Cambridge University Press.
24. Baynes T, Wiedmann T (2012) General approaches for assessing urban environmental sustainability. Current Opinion in Environmental Sustainability 4(4): 458–464. doi: 10.1016/j.cosust.2012.09.003
25. Dhakal S (2009) Urban energy use and GHG emissions from cities in China and policy implications. Energy Policy 37: 4208–4219. doi: 10.1016/j.enpol.2009.05.020
26. Feng Z, Zou L, Wei Y (2011) The impact of household consumption on energy use and CO2 emissions in China. Energy 36: 656–670. doi: 10.1016/j.energy.2010.09.049
27. Gao L, Li X, Wang C, Qiu Q, Cui S, et al. (2010) Survey site selection based on the spatial sampling theory - a case study in Xiamen Island. Journal of Geoinformation Science 2: 364–385. doi: 10.3724/sp.j.1047.2010.00358
28. Wang J, Liu J, Zhuan D, Li L, Ge Y (2002) Spatial sampling design for monitoring the area of cultivated land. Intenational Journal of Remote Sensing 23: 263–284. doi: 10.1080/01431160010025998
29. Ramaswami A, Hillman T, Janson B, Reiner M, Thomas G (2008) A Demand-centered, hybrid life-cycle methodology for city-scale greenhouse gas inventories. Environtal Science and Technology 42(17): 6455–6461. doi: 10.1021/es702992q
30. Eggleston HS (2006) 2006 IPCC guidelines for national greenhouse gas inventories. Forestry 5: 1–12.
31. National Development And Reform Commission (2009) China grid baseline emission factors 2009. Available: http://qhs.ndrc.gov.cn/qjfzjz/t20090703_289357.htm.
32. Zhi J, Gao J (2009) Analysis of carbon emission caused by food consumption in urban and rural inhabitants in China. Progress in geography 3: 429–434. doi: 10.1109/icbbe.2009.5162345
33. Yang Y, Wang G, Pan X (2009) China food composition. Beijing: Peking University Medical Press.
34. Ngnikam E, Tanawa E, Rousseaux P, Riedacker A, Gourdon R (2002) Evaluation of the potentialities to reduce greenhouse gases (GHG) emissions resulting from various treatments of municipal solid wastes (MSW) in moist tropical climates:

Application to Yaounde. Waste Management and Research 20: 501–513. doi: 10.1177/0734242x0202000604
35. Liu J, Wang R, Yang J (2003) Environmental impact of two types of residential building. Urban Environment and Urban Ecology 2: 34–35.
36. Xiamen Statistics Bureau (2009) Yearbook of Xiamen Special Economic Zone 2009. Beijing : China Statistics Press.
37. Zhao J, Song Y, Tang L, Shi L, Shao G (2011) China's cities eeed to grow in a more compact way. Environtal Science and Technology 45: 8607–8608. doi: 10.1021/es203138c
38. You F, Hu D, Zhang H, Guo Z, Zhao Y, et al. (2011) GHG emissions in the life cycle of urban building system in China–A case study of residential buildings. Ecological Complexity 8: 201–212. doi: 10.1016/j.ecocom.2011.02.003
39. Bin S, Dowlatabadi H (2005) Consumer lifestyle approach to US energy use and the related CO2 emissions. Energy Policy 33: 197–208. doi: 10.1016/s0301-4215(03)00210-6
40. Druckman A, Jackson T (2008) Household energy consumption in the UK: A highly geographically and socio-economically disaggregated model. Energy Policy 36: 3177–3192. doi: 10.1016/j.enpol.2008.03.021
41. Martinsson J, Lundqvist LJ, Sundström A (2011) Energy saving in Swedish households. The (relative) importance of environmental attitudes. Energy Policy 39: 5182–5191. doi: 10.1016/j.enpol.2011.05.046
42. Lenzen M, Wier M, Cohen C, Hayami H, Pachauri S, et al. (2006) A comparative multivariate analysis of household energy requirements in Australia, Brazil, Denmark, India and Japan. Energy 31: 181–207. doi: 10.1016/j.energy.2005.01.009
43. Rühl C (2008) BP Statistical Review of World Energy. Available: http://eugbc.net/files/13_47_749294_BPStatisticalReviewofWorldEnergy-Brussels,September2008.pdf.

CHAPTER 3

Exploring the Attitudes-Action Gap in Household Resource Consumption: Does "Environmental Lifestyle" Segmentation Align with Consumer Behavior?

PETER NEWTON AND DENNY MEYER

3.1 INTRODUCTION

Trajectories of consumption across a wide spectrum of urban services and resources in high income societies continue to trend upwards [1,2]. At the same time, attempts to create "wedges" linked to the introduction of social and economic policies, new processes and products, etc. [3,4], capable of underpinning a significant reduction in consumption, underpin much contemporary research. Many prospective sustainability "wedges" are found in the literature and include:

- Dematerialisation: a reduction in the resource inputs required to produce consumer products, such as building materials, domestic appliances and automobiles via recycling, re-use and closed loop manufacturing;
- Substitution of renewable for non-renewable resources, e.g., use of solar energy rather than fossil fuels, of public transport or active transport rather than private car, of low energy lighting for higher energy products;
- Efficiency in use of materials, spaces, time, etc., capable of being achieved via the (re-)design of a product or system, so as to be reflected in its operating performance, e.g., energy-efficient housing design; and
- Conservation of resources capable of being seen simply as a lowered absolute consumption of a resource (such as energy, water, petroleum, land, housing space and kilometres travelled) achieved by any of the pathways listed above, as well as by a change in habits and behaviours by those involved as consumers.

They cover the spectrum of research from sustainable production to sustainable consumption—from the supply side to the demand side. A conceptual framework capable of being used to "map" consumption research is presented in Figure 1. It suggests that demand-side studies—the focus of this paper—will centre on better understanding the role of social and demographic, as well as behavioural attributes associated with individuals and households and the contextual settings related to their dwelling, urban location or social milieu. Demand-side research also needs to probe why studies of urban resource use [5,6,7,8] are finding a gap between attitudes and actions; individuals are not behaving in a manner that is congruent with their stated attitudes and intentions. Here, a major question mark continues to surround the issue of whether individuals or households in high income societies are prepared to make voluntary changes to their consumption practices in order that the 21st century has some prospect for a more sustainable and equitable world. This necessitates a drilling down into the structural and behavioural attributes of individuals to explore those factors, both intrinsic and extrinsic, linked to consumption [9]. It can be argued that behaviour change on the part of individuals and the households with which they are associated holds the prospect for a much faster rate of sustainability transformation than supply-side technological innovation of key infrastructures and services (e.g., energy, water, waste or in the redesign of built environments). Radical change is possible in both these arenas, but the timescale required is considerable, hence the increasing interest being shown by governments in behaviour change policies and programs [10,11].

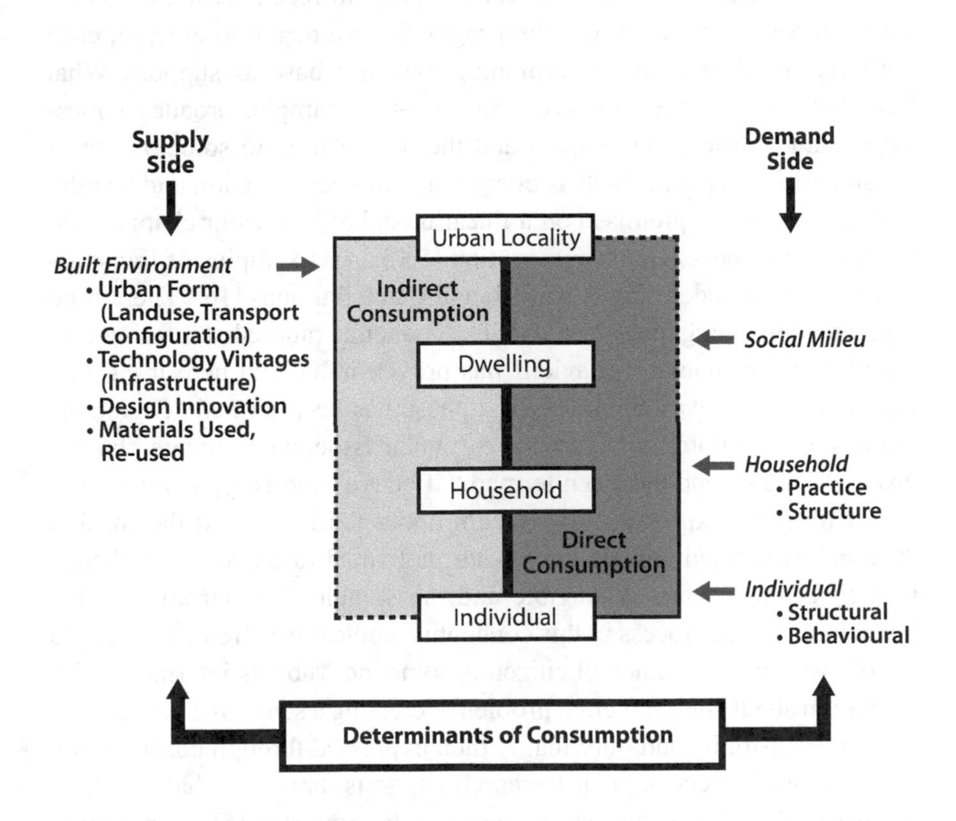

FIGURE 1: A conceptual framework for consumption research.

There is both an applied and a theoretical dimension to the study of household resource consumption and behaviour change. It has been argued that no voluntary changes in consumption practices can be achieved if policy makers, business and consumers alike lack an understanding of how and why individuals and households consume and what factors tend to be associated with behaviour change [12]. In recent years, a significant number of behaviour change programs have begun to emerge, each with varying degrees of underpinning evidence base as support. What have been termed transmissive projects—for example, broadcast messages via television, newspapers and the like relating to some aspect of consumption reduction, such as energy or water conservation and healthy eating—are mostly premised on a linear model of behaviour change: info rmation→awareness→concern→action [5,13,14]. Examples of transmissive projects include Earth Hour [15] and Black Balloons [16]. There is no segmentation or targeting involved. The economic rationalist information-based model of human behaviour has proven influential here in attracting government investment The first phase has been termed "cognitive" (awareness of information about a particular issue, e.g., climate change, drought); the second has been termed "affective" and is typically associated with some expression of concern about the issue; and the third is "action", where individuals have been motivated to make some change in their behaviour that is tangible and measurable. The narrative of the behaviour change process in this conceptual context is therefore portrayed as follows: an environmental citizen is someone who has internalised information about environmental problems, creating a sense of concern, personal responsibility and duty that is then expressed through consumption and community actions [17]. Research suggests, however, that the three facets of behaviour change are not necessarily sequential [5]. Transformative initiatives focus on face-to-face engagement with specific communities in an attempt to achieve the necessary traction for behaviour change. Examples of transformative programs include: Sustainable Consumption Roundtable's "triangle of change" (government, business and population) that involves engaging, exemplifying, enabling and encouraging [18]; the "kitchen table" processes [19]; the social marketing approach [20]; the "seven doors" model [21]; Green Street [22] and other socially (rather than cognitively)-based behaviour change programs [23,24].

Lifestyle segmentation (LSS) has emerged from the marketing and communications fields of research as a means by which the behaviour change programs of governments might be better targeted, whether transmissive or transformative in nature. LSS represents a marriage of two concepts into a single system [25]. Lifestyle refers to a distinctive pattern of a person's social life that straddles notions of individual identity, on the one hand, and community/sociality, on the other, "embodying notions of choice and self-actualisation alongside opportunities for collectivity and attachment" [26]. It is a broad based concept that incorporates everyday facets of individual lives, including their attitudes, opinions, values, feelings, intentions, habits and social contexts. Segmentation refers to a division of the population or market into relatively homogeneous clusters capable of being readily differentiated or labelled—via either quantitative or qualitative routes. The premise underpinning LSS is that it identifies groups of people who might be more likely to behave in similar ways as consumers—of particular products, services, information and messages. The attractiveness of LSS and its link with sustainability is the contention that consumer behaviour is likely to vary across different lifestyle groups comprising individuals with specific combinations of environmental attitudes and practices [27]. The research tasks then become those of establishing a valid and differentiated set of lifestyle groups among a population derived from either quantitative (lifestyle segmentation) or qualitative (e.g., consumer culture or social practice) research methods (or some hybrid) and then examining the extent to which lifestyle "type" can explain differences in some aspect of consumer behaviour. There is an extensive body of publication associated with the former, but there is a dearth of studies that proceed to examine the latter.

As intimated above, two contrasting approaches to LSS have emerged in research literature: the longer established and more widely implemented "traditional" approach, which is characterized by quantitative analysis and data reduction (and which is adopted for this paper—see section on Methods); and the cultural theory or social practices approach, which is qualitative in nature, drawing on a different set of disciplines for its execution. Quantitative LSS obtains responses to self-rating questionnaire items encompassing attitudes to a wide spectrum of statements that are deemed to cover the breadth of domains of interest to contemporary living. They are then subjected to some form of cluster analysis to identify groups that have similar response patterns, after which descriptive labels are assigned capable of evoking the distinctiveness of each group. The strength of this approach is its ability to synthesise a large volume of complex information into meaningful insights that are valued by clients in both private and public sectors. LSS studies of this type have been part of market and consumer research since the 1960s and 1970s for products, such as cars, alcohol, tourism, clothing and furniture, and there are many proprietary lifestyle typologies that have been developed. These include GfK Roper's broad, multi-product lifestyle segments that comprise: dreamers, homebodies, settled, adventurers, rational-realists, open-minded organics and demanding [28]. Other segmentations [27,29] have been developed for specific applications, such as:

- energy saving behaviour: economically modest, open-minded value pluralists, hedonists, conservative environment conscious, alternative environment-conscious, disinterested materialists and potentially sustainable;
- general household consumption: organised eco-families, childless career-oriented, young disinterested, everyday creative, disinterested in consumption, rural traditional, disadvantaged stressed, inconspicuous families, active seniors and status-oriented privileged;
- leisure mobility: fun-oriented, modern exclusive, stressed family-oriented, disadvantaged and traditional domestic; for urban travel: malcontented motorists, complacent car addicts, diehard drivers, aspiring environmentalists, car-less crusaders and reluctant riders.

The weaknesses of this class of LSS studies have generally been the lack of validation or replication and lack of proof as to their capacity to discriminate or predict actual behaviour.

The second, qualitative approach to LSS research to emerge more recently has been termed the social practice or cultural theory approach [30,31,32]. It focuses on the common contexts of everyday life, such as homes and workplaces, and the associated research seeks to understand how observed consumption behaviours reflect daily routines and practices, i.e., habits. It views the traditional LSS as failing to understand the "complex socio-cultural processes and situations that underlie human behaviour" [33]. By undertaking in-depth interviews often combined with observational and participative (ethnographic) research, it provides the basis for fleshing out lifestyle segments as part of a process for better understanding and characterizing contemporary society and the value this holds for developing more successful strategies for behaviour change. Its prime weaknesses have been identified as restriction to a small number of consumption settings and activities, such as water and energy use in the home [34], and a focus on overly complex, in-depth, but small-sample, qualitative studies, uncovering experiences that are often seen as unique to individuals and their spatial and temporal settings and are difficult to generalise or ascribe to a larger population [35].

More recently, LSS research has been directed towards the identification of an evidence base for behaviour change policies and programs associated with the environment and climate change [36,37,38]. In a UK study [39], seven lifestyle clusters were identified: positive greens, waste watchers, concerned consumers, sideline supporters, cautious participants, stalled starters and honestly disengaged. In a rare replication of such studies (albeit with less than half the indicator variables) in New Zealand [40], significant variation was found in the proportions of the population aligned with each of the seven lifestyle segments. In a similar Australian study [41], three segments were derived from self-reported environmental behaviours: committeds, privates and reluctants, "naturally forming groups of people who exhibit similar behavioural patterns regarding the environment". A United States study [42], identifies "six Americas", where each population segment holds different attitudes to the issue of climate change: the alarmed (18%), the concerned (33%), the cautious (19%), the disengaged (12%), the doubtful (11%) and the dismissive (7%). In later surveys [43], the alarmed and concerned proportions had declined somewhat and shifted to "cautious". What all studies reveal is the existence of common environmental lifestyle "types" ranging from the green/engaged to the sceptic/disengaged. None of these studies, however, assessed alignment of segments with target actions or behaviours.

Many lifestyle typologies, such as those listed above, are made up of objective, as well as subjective, attributes. This would appear to be valid for descriptive analyses of a population or sample. To the extent that socio-demographic factors have been found to be particularly significant determinants in their own right of consumer behaviour [44], combining socio-demographic with attitude, opinion and intention attributes in a segmentation for use in explanatory or predictive modelling of consumption risks redundancy. As Gust [27] argues:

> *Socio-demographic variables, such as age, income, family size and lifecycle stage were found to limit any attitude or lifestyle variables. For example, although young mothers are quite environmentally conscious consumers, this consciousness may not go as far as replacing their car with public transport, since the car might be very important in the daily family organisation. Here, the fact that there is a small child in the household will have a greater effect of the likeliness of sustainable consumer behaviour than the environmental orientations of the mother.*

Other weaknesses of the lifestyle approach identified by this author [27] are that most typologies are conceived for specific applications and, as such, lack the prospect for comparative assessment or replication and that attitudes and behaviours are highly dynamic and can change at any point in time, especially as needs and contexts change (e.g., environmental versus economic priorities, before compared to during the global financial crisis of 2008 and 2009).

The consensus is, however, that lifestyle orientation and segmentation can provide additional insight into consumer behaviour that may not be attributable to socio-demographic factors alone. The LSS approach has also been embraced as a tool with a capacity to tailor and better communicate sustainable behaviour diffusion strategies, as well as to market sustainable products and services to specific groups more likely to be receptive to the differentiated message [36].

Against this background, the aim of this paper is four-fold:

1. To establish an environmental lifestyle segmentation based on the cluster analysis of data from a representative sample of Melbourne household's responses to a range of questions that probed environmental attitudes, opinions and intentions;
2. To explore the extent to which this segmentation also reflects major socio-demographic and geographic differences among the population (segmentations in their own right);

3. To establish to what degree the different environmental lifestyle groups displayed contrasting patterns of actual (consumption) behaviour. The methodology has been designed explicitly to test the strength of this nexus, which tends to be absent in most segmentation studies;
4. Based on the level of congruence established between the environmental lifestyles and actual behaviour, reflect on the relevance of segmentation approaches for behaviour change programs.

3.2 METHODOLOGY

Given that the research objective of this paper was to explore the nature of any link between "environmental lifestyle" and consumption, it was deemed necessary to establish a lifestyle typology that derived from an interplay of the behavioural attitudes, opinions, concerns and intentions of a population—not their structural (socio-demographic) attributes, nor their actual patterns of consumption (see Figure 1). In this, it aligns with the traditional LSS approach, sometimes characterized as AIO (activities, interests, opinions), since it was part of a much larger study of the determinants of urban consumption (see Acknowledgements; [44]). The objective of the overarching study was to quantify the respective contributions of individual (behavioural, structural) versus contextual (dwelling, urban location) to an explanation of energy and water use, housing and domestic appliance consumption and carbon emitted on city travel. In this way, the survey provided data enabling an examination of the extent to which there is or is not a gap between the subjective indicators of sustainable living (the lifestyle groupings) and actual consumption behaviour (see Figure 2). This is a "model" for studies that seek to explore the disparity between awareness, concern and attitudes among the population for some significant issue, such as climate change and consumption and the level of action displayed by that population [45]. Literature suggests an explanation for the identified gap can be attributed to a set of barriers, constraints or situations that include the psychological, social and institutional, as well as the informational.

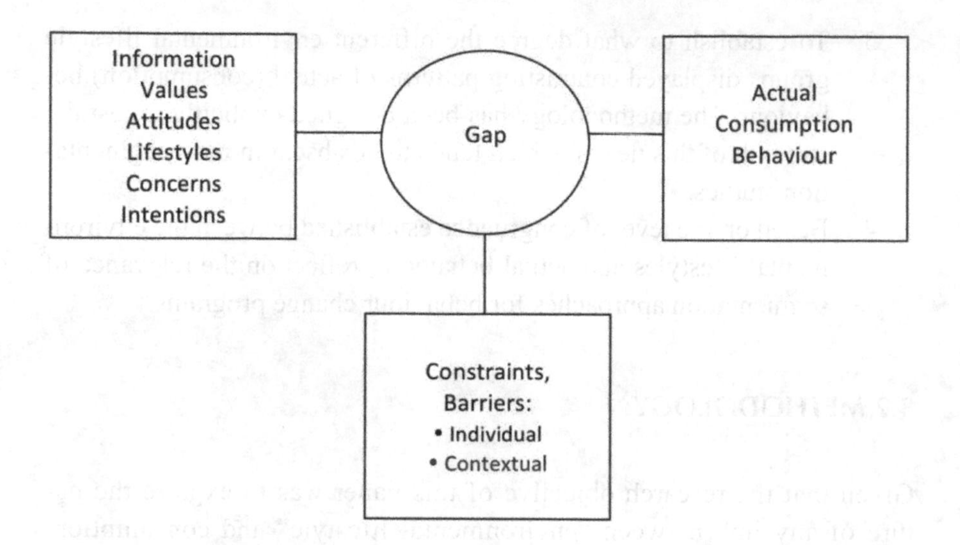

FIGURE 2: The gap between subjective indicators of intent and actual consumption behaviours.

3.2.1 SURVEY

A postal survey undertaken in June 2009 in seven residential precincts across Melbourne, Australia, resulted in data for adult individuals representing 1,250 households at a response rate of 16%. The questionnaire was designed to collect information on the structural and attitudinal attributes of individuals, their household, dwelling and location characteristics, as well as actual household consumption data for electricity, gas and water (based on the most recent utility bills), housing space, urban travel and domestic appliances. The full questionnaire is found in [46]. The level of representativeness of the sample to the metropolitan population was considered to be high and successful in avoiding non-response bias. The respective percentage of sample versus metropolitan area populations was as follows:

- Living in detached or medium density housing: 80.9; 83.3
- Living in high rise apartment: 19.1; 16.7

- Gender (female/male): 60/40; 51/49
- Born in Australia: 65.6; 64.2
- Owner/purchaser of dwelling: 66.5; 67.7
- Family household: 68.8; 68.1
- Person living alone: 25.5; 22.6
- Aged less than 25: 4.2; 23.0
- Aged 25-64; 81.5; 64.2
- Aged 65 and over: 14.3; 12.8

The age of respondents reflects the fact that an adult member of the household was required to complete the survey. Electricity and gas bills were combined in this analysis, because both these sources of energy are used for operating the spectrum of built-in and plug-in domestic appliances in a dwelling. The seven precincts provide representatives of the new outer greenfield suburbs, long-established middle suburbs and redeveloped and gentrified inner city neighbourhoods, with contrasting levels of residential density and public transport access.

3.2.2 STATISTICAL ANALYSIS

After providing a descriptive analysis for environmental attitudes and behaviours, a two-step clustering in SPSS was used to produce lifestyle clusters based on all these measures. This approach [47,48] allows an automatic choice for the number of clusters based on changes in the Bayes Information Criterion. The resulting three clusters were then appropriately named and compared in terms of their demographics and per capita consumption.

3.3 SUBJECTIVE INDICATORS OF ENVIRONMENTAL CONCERN AND ENVIRONMENTAL INTENTIONS AND THE IDENTIFICATION OF ENVIRONMENTAL LIFESTYLE CLUSTERS

3.3.1 CONCERN FOR THE ENVIRONMENT

"Concern for the environment" has long been used as a leading indicator of the level of environmental engagement by an individual or community

[49]. Questions that have examined and reported on levels of environmental concern by households have been used in Australian Bureau of Statistics (ABS) surveys since 1992. As reported by Newton [50], the level of environmental concern at a national level (there has been little variability by state) declined from 75% in 1992 to a low of 57% in 2004. In the latest ABS survey [51], it had risen to 81.3%, a similar level to that of the Sustainability Victoria survey (see Table 1). The *Living in Melbourne* survey was undertaken some 18 months after the ABS and Sustainability Victoria surveys (mid-2009), following the Black Saturday bushfires and during continuing issues of drought. The "concerned" percentage was accordingly very high (92.3%), suggesting a significant level of environmental sensitivity among the surveyed population. It should be noted that the exact wording of the "concern" question in the ABS surveys prior to 2007 was as follows: "Is … [respondent] concerned about any environmental problems?" [52]. For the 2007 survey, the question was modified as follows: "Are you concerned about any environmental problems in Australia?" The *Living in Melbourne* survey adopted the original framing of the question, "Is [respondent] concerned about any environmental problems?".

TABLE 1: Levels of environmental concern. Question: are you concerned about any environmental problem?

	Living in Melbourne survey (2009) (%)	Sustainability Victoria (2008) (%)	Sustainability Victoria (2009) (%)	ABS (2008) Victoria (%)	ABS (2008) Australia
No	7.7	16	14	15.8	18.3
Yes	92.3	84	86	84.2	81.7
A great deal	30.7	38	42		
A fair amount	48.3	31	30		
A little	13.3	15	14		

Sources: [46,51,54,55]

While some authors [53] have indicated that variants of "environmental concern" measures (questions) will evoke different patterns of response, it is unlikely that the sudden surge in concern picked up in the most recent set of surveys is an artefact of the wording of the survey question. At the time of these surveys, the nation's drought of some 15 years had not yet broken and there was a raft of issues surrounding the vulnerability of settlements to bushfires and water restrictions. Even the global financial crisis did not seem to alter attitudes of environmental concern [54,55], although that may have been due to the fact that Australia was one of only a handful of advanced Western economies to avoid slipping into recession. Or it may be reflecting the fact that there is increasing recognition among the population that we are living in a finite world with resource constraints and climate constraints that are becoming increasingly binding [56].

TABLE 2: How important is each of the following to you and your family?

	Not important at all	Neutral			Very important
Recycling household waste	2.4%	1.2%	10.9%	28.1%	57.4%
Saving water	0.6%	0.6%	5.6%	28.0%	65.2%
Saving energy	0.5%	0.6%	8.8%	32.1%	58.1%
Driving more slowly than the speed limit	12.7%	9.3%	35.6%	20.3%	22.1%
Using public transport	12.0%	5.9%	29.9%	25.1%	27.0%
Walking or cycling	3.2%	4.7%	21.4%	28.9%	41.8%
Having a large home with space	23.0%	11.5%	26.9%	21.5%	17.1%

Note: n ranged from 1,231 to 1,249

3.3.2 INTENTIONS AND PREDISPOSITIONS

The proposition from some behaviour change theorists is that high levels of "concern for environment"—as we have identified here—will be translated into a predisposition towards more sustainable forms of consumption. Evidence from Table 2 suggests that representatives of households are strongly disposed towards saving water and energy (both rated "important" or "very important" by more than 90%). And even here, a significant proportion (74.5%; see Table 3) say they can envisage ways to further reduce their energy and water consumption. However, making adjustments to urban travel behaviour is projected to be more challenging.

TABLE 3: Could you please indicate your level of agreement with the following statements?

	Strongly agree/ agree	Neither agree nor disagree	Strongly disagree/ disagree
I have done as much as I can to reduce my household's energy and water consumption	57.4%	18.6%	24.1%
I know some specific things I could do to reduce my household's energy and water consumption	74.5%	15.9%	9.6%
Reducing my household's energy and water consumption would help protect the environment	84.4%	11.5%	4.2%
The main reason for reducing my household's energy and water consumption is to save money	34.7%	29.3%	35.9%
Reducing my household's energy and water consumption is not worth the trouble	6.1%	9.0%	84.9%

Note: n ranged from 1,221 to 1,238

Probing the travel domain more deeply with the question: "Over the next 12 months, how likely are you to make the following changes in your personal travel to reduce your petrol consumption and carbon emissions?", the responses in Table 4 suggest that some transition has already begun in areas, such as switching to a smaller car and increased use of public transport, but there are still significant proportions (49.7 and 46.0%, respectively) where there was no likelihood of change, nor was car pooling seen as an option. Walking more was the most significant "maybe/ likely" response (52.7%).

The areas most likely to be embraced by households as a basis for delivering a clean environment (see Table 5) are recycling more, using substitutes for plastic bags when shopping, buying more local products and buying "green"-labelled products—what could be classed as the "low-hanging fruit", where there is little or no inconvenience or cost associated with any shift in practice. There is less traction associated with modifying car usage and least of all with changes linked to a hip pocket impact, such as paying more tax or paying higher charges for electricity and water (61% "no" to both).

TABLE 4: Over the next 12 months, how likely are you to make the following changes in your personal travel to reduce your petrol consumption and carbon emissions?

	Not likely	Maybe	Likely	Already have
Switch to a smaller (3 or 4 cylinder) vehicle	49.7%	9.5%	6.1%	34.7%
Use car less often	35.4%	22.9%	19.3%	22.3%
Begin to use public transport	42.0%	13.9%	8.5%	35.6%
Use public transport more often than at present	46.0%	23.0%	17.7%	13.2%
Move house so I need to travel less	77.8%	5.6%	6.0%	10.6%
Car pool to work	86.8%	8.5%	2.1%	2.6%
Walk more	21.7%	25.7%	27.0%	25.6%
Drive more slowly than the speed limit	37.4%	25.3%	18.0%	19.3%
Use communication technology to avoid travel	27.8%	15.4%	14.1%	42.8%
Fly less often	49.3%	16.2%	13.0%	21.4%

Note: n ranged from 1,100 to 1,181.

I will if you will was the title of a widely referenced report by the Sustainable Consumption Roundtable [11] in the UK. This revolved around responses to three key positions that people could hold as to their predisposition to change:

- "I am making changes to my behaviour now, irrespective of what other people do"—could be classed as proactive leaders;
- "I will make changes when the community around me begins to change"—could be classed as followers;
- "I will make changes when government introduces some regulation that requires everyone to meet a particular target"—could be classed as laggards.

Based on the profiles of response to these statements (see Table 6), 83% of the population could be placed in the leader group, 5% followers and 12% laggards, which is likely to be overstating or misrepresenting the label of "leader" in this instance. Perhaps this group just represents the "independently minded".

TABLE 5: What would you be prepared to change in order for there to be a clean environment?

	Yes	No	Maybe
Pay more tax	19.7%	61.0%	19.4%
Travel less by motor vehicle	48.9%	24.4%	26.7%
Choose to buy "green" labelled products	70.0%	8.1%	21.8%
Donate an hour or two each month to do voluntary work for the environment	36.8%	29.8%	33.4%
Choose to buy products produced locally	82.9%	4.0%	13.1%
Recycle more	90.9%	2.8%	6.2%
Pay higher charges for electricity and water	22.4%	60.8%	16.8%
Give up using plastic bags	72.2%	10.7%	17.1%

Note: n ranged from 1,219 to 1,237

TABLE 6: I will if you will. "Question: which of the following positions best reflects your view about changing behaviour?"

Option	N	%
I am making changes to my behaviour now, irrespective of what other people do	999	82.6
I will make changes when the community around me begins to change	67	5.5
I will make changes when government introduces some regulation that requires everyone to meet a particular target	144	11.9

In response to statements, such as "I have done as much as I can to reduce my household's energy and water consumption", 57% of the sample believe that they have "maxed out" (see Table 3)—evidence, perhaps, of some "demand hardening" beginning to emerge among the population (see [57] for further discussion of this concept). This was contradicted to a degree in responses by three-quarters acknowledging that they "know some specific things I could do".

A self-assessment of the degree to which the respondent had been active over the past 12 months in responding to environmental issues indicated that 90% saw their behaviour as "very active" or "reasonably active" (see Table 7). A cross-tabulation of "concern for environment" with a self-assessment of the level of activity in responding to environmental issues revealed clear evidence of some link (see Table 8).

TABLE 7: Over the past 12 months, would you say you have been…?

Very active in responding to environmental issues	Reasonably active in responding to environmental issues	Not at all active in responding to environmental issues
13.1%	76.3%	10.6%

Note: n = 1232

TABLE 8: Expressions of concern versus indications of action.

		Over the past 12 months, would you say you have been…?					
		Very/reasonably active in responding to environmental issues		Not at all active in responding to environmental issues		Total	
Are you concerned about any environmental problems?	Yes	1047	92.4%	86	7.6%	1133	100.0%
	No	46	50.5%	45	49.5%	91	100.0%
	Total	1093	89.3%	131	10.7%	1224	100.0%

3.3.3 BARRIERS TO ACTION

Across the battery of potential barriers that have emerged from previous studies, 14 were collectively put to the test in the *Living in Melbourne* survey. Five clusters of barriers were evident (see Table 9) in explaining the gap between intentions and actions. For approximately one-quarter of households, there was no sense of ownership of the problem, driven either by their tenure status (and the issue of split incentives; that is, the landlord could benefit from any investment by the tenant) or by feeling that it is not their responsibility, that any initiative on their part would not make a difference to the wider environment and that there is no regulation requiring action on their part.

TABLE 9: Barriers to actions: reasons advanced for not undertaking domestic energy and water conservation measures.

Barrier	Percentage indicating that the reason either applied strongly or applied somewhat
Ownership of problem	
Not my responsibility	22.5
I rent—it's up to my landlord	28.5
It won't help Melbourne's environment1	9.7
No regulation requiring me	27.9
Information shortage/access	
Lack of information	55.4
Can't work out what's best	47.9
Organisational challenges	
Too difficult to organise	54.6
Can't work out what's best	47.9
Difficult to get right tradespeople	39.3
Time constraints (level of priority)	
Planning to, but haven't got to it yet	54.4
Lack of time	51.1
Financial	
Lack of money	68.2
Expense not worth benefits	52.3
I rent—it's up to my landlord	28.5

For those motivated to change, four barriers crop up that tend to be endorsed by at least half of the surveyed population as being significant: problems of lack of relevant information or how/where to find it; organisational challenges (i.e., how to get it done: identifying best option from alternatives and successfully contracting to get the work done); time constraints, reflecting perhaps that environmental issues are not priority #1 in Australian households; and financial constraints—determining whether the benefits reward the financial outlay and, if so, whether funds are available at the time.

3.3.4 LIFESTYLE TYPOLOGIES

A cluster analysis was performed using responses to five attitudinal and behavioural questions and 39 individual variables emanating from those questions. In particular, the following questions were asked:

- Perceived barriers to undertaking energy- and water-saving measures—respondents were asked how strongly each of the barriers applied to them, with responses obtained on a scale, with 1 = strongly applies, 2 = applies a little and 3 = does not apply (see Table 9 for the list of barriers);
- Attitudes regarding the environment—respondents were asked to rate their agreement with each of 10 attitudes, with 1 = strongly agree and 5 = strongly disagree (see Table 10 for list of attitudes).
- Attitudes to consumption reduction—respondents were asked to rate their agreement with each of five statements, with 1 = strongly agree and 5 = strongly disagree (See Table 3);
- Changes that people are prepared to make in order for there to be a clean environment—respondents were asked to rate their preparedness in eight respects, with 1 = yes, 2 = no and 3 = maybe (see Table 5);
- Positions that best reflect views about changing behaviour—(See Table 6)

  1 = "I will make changes when the community around me begins to change",
  2 = "I will make changes when government introduces some regulation that requires everyone to meet a particular target", and
  3 = "I am making changes to my behaviour now, irrespective of what other people do".

Only 441 observations were considered in the two-step cluster analysis of environmental attitudes and behaviours, because of the large number of missing values. Three clusters emerged, with the 13 most important clustering variables identified using the SPSS Variable Importance Criterion, based on p-values (see Table 11). Only these 13 weakly correlated variables were used to classify all respondents on the basis of their responses. Fewer classifiers resulted in unacceptably high misclassification rates, and more classifiers resulted in an increased number of missing values without a worthwhile reduction in the misclassification rate. A multinomial logistic regression was used to obtain the cluster classification, and the clusters were named in accordance with the response frequencies for the 13 most important classifiers (see Table 12). This model correctly classified 92.3% of the 441 complete observations, while failing to classify only 10% of the data for the original 1,250 observations as the result of missing values.

TABLE 10: Attitudes regarding the environment.

Attitudes	Percentage Agree
We are approaching the limit of the number of people the earth can support	53.7
The earth has plenty of natural resources, if we just earn how to develop them	54.7
When humans interfere with nature, it often produces disastrous consequences	74.9
The balance of nature is very delicate and easily upset	77.7
The "environmental crisis" is greatly exaggerated	21.5
If things continue on their present course, we will soon experience a major environmental catastrophe	58.8
Change is rarely for the good, and I prefer things as they are	10.2
The highest priority should be given to protecting the environment, even if it hurts the economy	52.3
There is nothing we can do about climate change—it is already too late	8.3
When shopping, I rarely think about how much use I'm going to get out of the things I buy	15.8

Source: The first six statements were drawn from the NEP (New Ecological Paradigm) scale [58].

Table 12 shows that the majority (40.3%) of people who responded to this survey were assigned to a "material" greens cluster, with 33.5% assigned to a "committed" greens cluster and 26.3% assigned to an "enviro-sceptics" cluster.

TABLE 11: Analysis for top 13 discriminators based on 1,122 cluster predictions.

	Percentages agree				Percentages disagree			
Lifestyle cluster	1	2	3	All	1	2	3	All
Prepared to pay more tax?	50	1	13	20	13	96	66	60
Prepared to pay higher charges for utilities?	56	4	11	23	18	90	70	60
The environmental crisis is exaggerated	2	20	44	20	89	51	29	58
Choose to buy green-labelled products	86	84	32	71	0	2	25	8
Environment has highest priority even, if it hurts the economy	80	43	32	52	6	20	35	19
The expense is not worth the benefits	24	56	80	52	76	45	20	49
Give up using plastic bags	89	83	35	73	3	6	26	10
I have more important things to do	14	22	55	28	86	78	45	72
Donate an hour or two each month to do voluntary work for the environment	61	30	16	37	10	27	57	29
There is no regulation requiring me to	15	20	54	27	85	80	46	73
The balance of nature is very delicate and easily upset	86	80	61	77	4	4	19	8
Reducing my household's energy and water consumption is not worth the trouble	2	4	13	6	97	90	66	86
It's not my responsibility	8	18	45	22	92	82	55	78

TABLE 12: Cluster descriptions.

Lifestyle cluster	1	2	3	All
Number of complete observations	149	177	115	441
Percentage correct classifications with 13 classifiers	93.3%	92.7%	90.4%	92.3%
Number of original observations classified	376	452	294	1122
Percentage of original classified observations	33.5%	40.3%	26.3%	100%
Cluster name	"Committed" greens	"Material" greens	"Enviro-sceptics"	

3.3.4.1 CLUSTER 1: "COMMITTED" GREENS

This is the only group prepared to pay more tax if it would benefit the environment (50%), as well as higher utility charges (56%), that is, indicating a willingness to personally outlay hard cash if environmental benefits will accrue. There is a high percentage agreeing that the environment should be the highest priority, even if it hurts the economy (80%). This group also strongly disagrees (76%) that the expense is not worth the benefits, affirming the need for the environment to take higher priority over the economy.

This group is highly consistent with its "green choice" stated behaviours related to the purchase of green-labelled products, declining use of plastic bags and volunteering time for green projects. In terms of environmental beliefs, it strongly disagrees with statements such as "The environmental crisis is exaggerated", "I have more important things to do", "There is no regulation requiring me to", "Reducing my household's energy and water consumption is not worth the trouble" and "It's not my responsibility".

On the basis of the response profile of this group of respondents, the cluster is labelled as "committed" green: strongly pro-environment in beliefs, in behavioural preferences and indicating a preparedness to sacrifice economically for an environmental benefit.

3.3.4.2 CLUSTER 2: "MATERIAL" GREENS

Among this group, there is a moderate level of support for the view that the environment should have a high priority vis-à-vis the economy and a sense that the balance of nature is delicate and easily upset, but 56% agree that the expense is probably not worth the benefits and—as a bottom line position—they are not willing to pay! This group is vehemently opposed to paying more taxes or higher utility charges (96% and 90%, respectively) from their household budget.

The group will be pro-purchase of green-labelled products and will avoid use of plastic bags, but is unlikely to donate the hours to voluntary environmental work that characterises the "committed" green cluster of households—although more generous than the "enviro-sceptics". In terms of environmental beliefs, it tends to fall between the two. The group tends to be one that views the environment as important, but not worth paying for, especially by themselves as individuals, either in terms of dollars or time—basically, only when it does not "cost" them. This cluster has been termed the "material" greens.

3.3.4.3 CLUSTER 3: "ENVIRO-SCEPTICS"

This group indicates a low level of preparedness to make higher personal payments for the environment, and there is a high level of agreement that the expense would not be worth the benefits. It also has the lowest level of agreement with propositions that the environment should be the highest priority. This translates into attitudes and practices against what could be termed "green choices", having the lowest proportions choosing to buy green-labelled products, giving up plastic bags and donating time for voluntary environmental projects.

In terms of environmental beliefs, a relatively high percentage believe (agree) that the environmental crisis is exaggerated (44%), that they have more important things to focus on (55%), that there is no regulation requiring them to (54%) and that it's not their responsibility (45%). This cluster has been labelled "enviro-sceptics".

3.4 LIFESTYLE AND CONSUMPTION

How does actual consumption vary for these three clusters? A multivariate general linear model analysis performed using the five per capita consumption measures defined previously suggested some small significant differences between the clusters ($F(10,1430) = 2.99$, $p = 0.001$, partial eta-squared = 0.021), but only in the case of the carbon intensity of travel and per capita appliance consumption (see Table 13). Appliance consumption was highest for the "enviro-sceptics" and lowest for the "material" greens, while carbon intensity of travel was highest for the "material" greens and lowest for the "committed" greens. However, a univariate general linear model analysis for the total of these consumption measures (after standardisation) showed no significant difference between the clusters (see Table 13). Lifestyles characterised by pro-environmental values, attitudes and intentions did not reflect actual low consumption behaviour.

TABLE 13: Lifestyle cluster comparisons in regard to per capita consumption.

	ANOVA test of significance			Lifestyle cluster means		
Per capita consumption measure	F-value	p-value	Effect size (eta-squared)	"Committed" greens	"Material" greens	"Enviro-sceptics"
Water ($)	F(2,843) = 0.87	0.419	0.002	58.7	55.0	55.1
Energy ($)	F(2,814) = 2.36	0.095	0.006	123.4	123.9	137.5
Appliances (number)	F(2,1119) = 13.60	<0.001	0.024	4.82	4.78	4.95
Carbon intensity of travel score	F(2,1099) = 6.27	0.002	0.011	17.0	19.4	17.8
Housing space (m2)	F(2,1103) = 0.74	0.479	0.001	97.2	91.6	101.3
Total consumption score	F(2,719) = 0.23	0.799	0.001	0.244	0.129	0.312

Next, the clusters were compared in terms of urban location. Table 14 suggests that there is some geographical segregation (Chi-Sq = 35.7, df = 4, $p < 0.001$). Half of the "committed" greens live in the Inner City suburbs (where, in recent years, the Greens Party has become politically dominant), while the "material" greens tend to live in greenfields and outer suburban areas—the group that political commentators have termed as "aspirational" in terms of their ambition to "get ahead" in life. The "enviro-sceptics" cluster is dispersed across the city.

Are there significant socio-demographic differences for these three clusters of people? Indeed, there are. Table 14 suggests significant differences for age, gender, level of education, household income, family structure and suburb location. However, there was no significant difference for country of birth (Australia versus overseas).

TABLE 14: Demographic profiles for each lifestyle cluster.

Lifestyle Cluster	Percentage of cluster members				Degrees Freedom	Chi-Square
	Committed greens	Material greens	Enviro-sceptics	All		
University graduates	62	35	43	46	2	57.5***
Not university graduates	38	65	57	54		
Females	64	66	44	60	2	39.3***
Males	36	34	56	40		
Under 45	51	54	41	49	8	25.7**
45 and over	49	46	59	51		
Live alone	27	22	24	24	18	35.1**
Couple with no children	36	29	38	34		
Family with children	30	42	32	35		
Household income under $60,000 p.a.	28	39	39	34	10	19.0*
Household income $60,000–$120,000 p.a	44	40	40	41		
Household income over $120,000 p.a	29	22	22	25		
Inner city suburbs	50	34	34	42	4	35.7***
Middle city suburbs	26	24	24	25		
Outer city suburbs	23	42	42	33		

The "committed" greens cluster contained more university graduates and households with higher incomes. They know what behaviours are likely to be required in a climate- and resource-constrained future and have a financial capacity to embark on that transition. The "material" greens households, by comparison, had the lowest proportion of university graduates, were the youngest category of households and also tended to be on lower household incomes. The "enviro-sceptics" cluster contained more men and those aged 45 and over than either of the other clusters. Although the "enviro-sceptics" and "material" greens clusters tend to have similar incomes, the latter cluster is more likely to consist of households with children, which could have had some influence on their pro-environment attitude.

Overall, it seems that the "committed" greens cluster can afford to have "green" attitudes and "green" behaviours. Low consumption in the case of carbon intensity of personal travel suggests that this cluster is making an effort to reduce consumption in some areas—by virtue of where they live in the city—but in other areas appears to be doing no better than the other clusters. The "material" greens cluster also has sympathy with the environment, but these households feel that their limited budget does not allow them to pay higher rates or taxes in order to help the environment. As indicated in Table 13, they have the lowest per capita consumption of appliances (reflecting the "averaging out" effect of families with children identified earlier), but in other areas, their consumption is no lower than the other clusters; in fact, their carbon intensity of travel is particularly high—reflecting the car dependency of households in the outer suburbs. Finally, the older, male-dominated "enviro-sceptics" cluster tends not to care about the environment as much as the other clusters, but, surprisingly, the total consumption score does not appear to be significantly higher for these households, although their mean total consumption score is highest overall.

3.5 CONCLUSIONS

Changing behaviour, which involves winding back currently unsustainable levels of resource consumption, has the prospect of making an impact much more rapidly than most, if not all, supply-side responses, although both clearly need to be operating in tandem if the 21st century is to deliver a sustainable platform of urban living. It is the relative speed with which individuals and households can potentially decide to stop or significantly alter a range of domestic practices—related to their choice of housing, mode and frequency of travel, energy and water use habits, appliance purchases (number and environmental performance)—that makes "social marketing" so attractive to governments. An ability to get households to change aspects of their behaviour voluntarily via engagement processes that seek to provide the social, environmental and economic narrative for change and the processes ("steps") by which this can be achieved would obviate governments having to pursue less politically attractive and costly paths (such as taxation, regulation and incentives) to secure more sustainable patterns of consumption. This paper has found that voluntary change will be no easy task. Actions speak louder than words.

The research has demonstrated that there is a gap between intentions and actions. Segmentation of the *Living in Melbourne* sample of households into lifestyle groups based on their responses to a battery of questions probing environmental sustainability-related values, attitudes and intentions produced three archetypal clusters: "committed" greens, "material" greens and "enviro-sceptics". Each cluster was also found to be distinctive on a range of socio-demographic variables that further validated the typology. Commonalities with similar typologies elicited for the UK, USA and NSW (Australia) are strong. On the basis of the *Living in Melbourne* survey, only one-third of households (the "committed" greens) would currently appear to be prepared to voluntarily change their consumption behaviour and bear the direct economic consequences. However, when it came to examining the extent to which actual consumption behaviour varied across the three clusters, there was little or no differentiation evident.

The reasons why there is currently a gap between intentions, on the one hand, and action, on the other, are to be found in the range of current barriers to sustainable consumption that need to be overcome. These include: problems of lack of relevant information or how/where to find it; organisational challenges, i.e., how to get it done; time constraints; and financial constraints—determining whether the benefits reward the financial outlay. At a pragmatic level, there remains a lack of information on what can be done and how best to get it done. For individuals and households, comfort, convenience and cost factors seem to underpin many of the habits and practices that currently promote consumption of urban resources in Australia. The issue of sustainable consumption would appear to be more deeply rooted within 21st century society, however. High income societies are now over-reliant on consumption as an engine for growth—and developing economies are being encouraged to follow suit. Social norms relating to sustainable consumption are yet to materialise in high income societies, such as Australia, that would constitute an important influence on the voluntary behaviour of individuals and households. A culture of unsustainable consumption is reflected in a dominant set of behaviours at present. And it appears to be embracing all segments of the population, including those who espouse green attitudes, opinions and intentions. It is here that social practices research can add value to our understanding by probing more deeply into these lifestyle segments to explore whether consumption related habits and activities are also held in common. Whether imminent system failure will be required to trigger a "tipping point" in societal values associated with environment and consumption remains an open question (witness the increasing number of well-credentialed reports emerging of a future "4° world"). This is a major reason why supply-side urban sustainability initiatives need to proceed apace, why governments need to remain actively involved in regulation, pricing and incentive programs and why research that spans the cognitive-social spectrum of consumption must continue to search for triggers for effective behaviour change. Understanding urban consumption is clearly a more complex challenge than a market segmentation approach alone can address.

REFERENCES

5. OECD, Environmental outlook to 2030; Organisation for Economic Co-operation and Development: Paris, France, 2008.
6. Newton, P. Liveable and sustainable? Sociotechnical challenges for 21st century cities. J. Urban Technol. 2011, 19, 81–102.
7. Robinson, J.; Tinker, J. Reconciling ecological, economic and social imperatives. In Proceedings of the Cornerstone of Development: Integrating Environmental, Social and Economic Policies; Schnurr, J., Holtz, S., Eds.; Lewis: Boca Raton, FL, USA, 1998; pp. 9–43.
8. Pacala, S.; Socolow, R. Stabilization wedges: solving the climate problem for the next 50 years with current technologies. Science 2004, 305, 968–972.
9. Lorenzoni, I.; Nicholson-Cole, S.; Whitmarsh, L. Barriers perceived to engaging with climate change among the UK public and their policy implications. Global Environ. Chang. 2007, 17, 445–459.
10. Gilg, A.; Barr, S. Behavioural attitudes towards water saving: evidence from a study of environmental actions. Ecol. Econ. 2006, 57, 400–414.
11. Soderholm, P. Environmental Policy in Household Behaviour; Earthscan: London, UK, 2010.
12. Fielding, K.S.; Louis, W.R.; Warren, C.; Thompson, A. Understanding household attitudes and behaviours towards waste, water and energy conservation. In Urban Consumption; Newton, P., Ed.; CSIRO Publishing: Melbourne, Australia, 2011; pp. 199–214.
13. Curran, S.R.; de Sherbinin, A. Completing the picture: the challenges of bringing "consumption" into the population equation environment. Popul. Environ. 2004, 26, 107–131.

14. UNEP, Visions for Change: Recommendations for Effective Policies on Sustainable Lifestyles; United Nations Environment Programme: Paris, France, 2011.
15. UNEP, Taskforce on Sustainable Lifestyles; United Nations Environment Programme: Paris, France, 2011.
16. Heiskanen, E.; Pantzar, M. Toward sustainable consumption: two new perspectives. J. Consum. Policy 1997, 20, 409–442.
17. Voronoff, D. Community Sustainability: A Review of What Works and How It Is Practiced in Victoria; Environment Victoria: Melbourne, Australia, 2005.
18. Allen Consulting Group, Mandatory Disclosure of Communal Office Builders Energy Efficiency; Report to Commonwealth Department of Environment, Water, Heritage and the Arts: Sydney, Australia, 2008.
19. Earth Hour. Available online: http://www.earthhour.org/ (accessed on 7 March 2013).
20. Black Baloons. Available online: http://www.savepower.nsw.gov.au/about/about-save-power/whats-a-black-balloon-.aspx/ (accessed on 7 March 2013).
21. Hobson, K. Competing discourses of sustainable consumption: does the "rationalisation of lifestyles" make sense? Environ. Politics 2002, 11, 95–120.
22. Sustainable Consumption Roundtable, I Will if You Will: Towards Sustainable Consumption; Sustainable Development Commission: London, UK, 2006.
23. Sarkissian, W.; Hofer, N.; Shore, Y.; Vajda, S.; Wilkinson, C. Kitchen Table Sustainability: Practical Recipes for Community Engagement with Sustainability; Earthscan: London, UK, 2008.
24. McKenzie-Mohr, D. Promoting sustainable behaviour: an introduction to community-based social marketing. J. Soc. Issues 2000, 55, 543–554.
25. Robinson, L. Introduction to the five doors model. Enabling Change, 2007. Available online: http://www.enablingchange.com.au/ (accessed on 18 December 2012).
26. Green Street. Available online: http://www.greenstreet.net.au/ (accessed on 18 December 2012).
27. Earls, M. HERD: How to Change Mass Behaviour by Harnessing our True Nature; John Wiley and Sons: Chichester, UK, 2007.
28. Seyfang, G.; Haxeltine, A. Growing grassroots innovations: exploring the role of community based initiatives in governing sustainable energy transitions. Environ. Plann. C 2012, 30, 381–400.
29. Plummer, J. The concept and application of lifestyle segmentation. J. Marketing 1974, 38, 33–39.
30. Evans, D.; Jackson, T. Towards a Sociology of Sustainable Lifestyles; RESOLVE Working Paper 03–07; Research Group on Lifestyles, Values and Environment, University of Surrey: Guildford, UK, 2007.
31. Gust, I. Strategies to promote sustainable consumer behaviour: the use of the lifestyle approach. MSc Thesis, Lund University, Sweden, 2004.
32. Peichl, T. Between adventurers and realists. Gfk insite 2007, 4, 28–31.
33. Anable, J. Complacent car addicts" or "aspiring environmentalists"? Identifying travel behaviour segments using attitude theory. Transport Policy 2005, 12, 65–78.
34. Maller, C.J. Practices involving energy and water consumption in migrant households. In Urban consumption; Newton, P., Ed.; CSIRO Publishing: Melbourne, Australia, 2011; pp. 237–250.

35. Shove, E. Comfort, Cleanliness and Convenience: The Social Organization of Normality; Berg: Oxford, UK, 2003.
36. Ahuviar, A.; Carroll, B.; Yang, Y. Consumer culture theory and lifestyle segmentation. Innovative Markets 2006, 2, 33–43.
37. Thompson, C.; Troaster, M. Consumer value systems in the age of postmodern fragmentation. J. Consum. Res. 2002, 28, 550–571.
38. Barr, S.; Prillwitz, J. Sustainable travel: mobility, lifestyle and practice. In Urban Consumption; Newton, P., Ed.; CSIRO Publishing: Melbourne, Australia, 2011; pp. 159–171.
39. Connolly, J.; Prothero, A. Sustainable consumption. Consumption Markets Culture 2003, 6, 275–291.
40. Collier, A.; Cotterill, A.; Everett, T.; Muckle, R.; Pike, T.; Vanstone, A. Understanding and Influencing Behaviours: A Review of Social Research, Economics and Policy Making in Defra; Department for Environment, Food and Rural Affairs: London, UK, 2010. Available online: http://www.defra.gov.uk/evidence/series/documents/understand-influencebehaviour-discuss.pdf (accessed on 18 December 2012).
41. Behavioural Insights Team, Behavioural Insights Team annual update 2010–11; Cabinet Office: London, UK, 2011.
42. Cabinet Office, Government Response to the Science and Technology Select Committee report on Behaviour Change; Cabinet Office: London, UK, 2011.
43. Russell, G. A Framework for Sustainable Lifestyles; Department for Environment, Food and Rural Affairs: London, UK, 2011. Available online: http://sd.defra.gov.uk/2011/10/framework-for-sustainable-lifestyles/ (accessed on 18 December 2012).
44. Ministry for the Environment, Household Sustainability Survey 2008; Part 4 Population Segmentation, 2012, Wellington (www.mfe.govt.nz). In Environmental stewardship for a prosperous New Zealand.
45. Department of Environment, Climate Change and Water, Who cares about the environment in 2009?; NSW Government: Sydney, Australia, 2010.
46. Maibach, E.; Roser-Renouf, C.; Leiserowitz, A. Global warming's Six Americas; George Mason University: Fairfax, VA, USA, 2009.
47. Maibach, E.; Roser-Renouf, C.; Leiserowitz, A. Global warming's Six Americas; George Mason University: Fairfax, VA, USA, 2012.
48. Newton, P.; Meyer, D. The determinants of urban resource consumption. Environment Behaviour 2010.
49. Kollmuss, A.; Agyeman, J. Mind the gap: why do people act environmentally and what are the barriers to pro-environmental behaviour? Environ. Education Res. 2002, 8, 239–260.
50. Newton, P.; Burke, T.; Meyer, D.; Wulff, M. Consuming the Urban Environment: Report on Results from Extension of 2009 ARC Discovery "Living in Melbourne" Survey to Residents of Selected Precincts in the City of Melbourne; Institute for Social Research, Swinburne University of Technology: Melbourne, Australia, 2009.
51. Zhang, T.; Ramakrishnan, R.; Livny, M. BIRCH: an efficient data clustering method for very large databases. In Proceedings of the ACM SIGMOD Conference on Management of Data; Jagadish, H.V., Mumick, I.S., Eds.; ACM: Montreal, Canada, 1996; pp. 103–114.

52. Chiu, T.; Fang, D.; Chen, J.; Wang, Y.; Jeris, C. A robust and scalable clustering algorithm for mixed type attributes in large database environment. In Proceedings of the seventh ACM SIGKDD international conference on knowledge discovery and data mining; Provost, F., Srikant, R., Eds.; ACM: San Francisco, CA, USA, 2001; pp. 263–268.
53. Dunlap, R.E.; Jones, R.E. Environmental concern: conceptual and measurement issues. In Handbook of Environmental Sociology; Dunlap, R.E., Michelson, W., Eds.; Greenwood: Westport, CT, USA, 2002; pp. 482–524.
54. Newton, P. Horizon 3 planning: meshing liveability with sustainability. Environ. Plann. B 2007, 34, 571–575.
55. ABS, Australia's environment: issues and trends, 2007; Cat. No. 4613.0; Australian Bureau of Statistics: Canberra, Australia, 2008.
56. Morrison, B. Wording of "environmental concern" question, Australian Bureau of Statistics, Canberra, Personal e-Mail Communication. 5 September 2007.
57. Van Liere, K.D.; Dunlap, R.E. Environmental concern: does it make a difference how it's measured? Environ. Behav. 1981, 13, 651–676.
58. Sustainability Victoria, Green light report: Victorians and the Environment in 2008; Sustainability Victoria: Melbourne, Australia, 2008.
59. Sustainability Victoria, Green light report: Victorians and the Environment in 2009; Sustainability Victoria: Melbourne, Australia, 2009.
60. Krugman, P. The US coughs and the global economy presses on. The Age, 28 December 2010.
61. Inman, M. The water-efficient city: technological and institutional drivers. In Urban Consumption; Newton, P., Ed.; CSIRO Publishing: Melbourne, Australia, 2008; pp. 495–505.
62. Dunlap, R.; van Liere, K.; Mertig, A.; Jones, R. Measuring endorsement of the new ecological paradigm: a revised NEP scale. J. Soc. Issues 2000, 56, 425–442.

CHAPTER 4

Rethinking Waste as a Resource: Insights from a Low-Income Community in Accra, Ghana

MARTIN OTENG-ABABIO

4.1 BACKGROUND

> *"To waste, to destroy our natural resources, to skin and exhaust the land instead of using it so as to increase its usefulness, will result in undermining in the days of our children the very prosperity which we ought by right to hand down to them amplified and developed."*
>
> Theodore Roosevelt, Seventh Annual Message, December 3, 1907

The admonition from Roosevelt is an indication of society's long cherished dream of living in an environment of resourcefulness rather than wastefulness. Yet, as the world hurtles towards its urban future with the world's urban population increasing by two new people every second, and with 95 per cent of such increases taking place in cities of developing countries, nowhere is the impact more obvious than in society's "detritus," or solid waste (Hoorneweg and Bhada-Tata 2012). Studies show that currently, cities cover only 2% of the world's surface but generate 70% of the world's waste (Zaaman and Lehman 2011). In Ghana, Accra produces conservatively 2,200 tonnes of solid waste daily, and is expected to reach 4,419 tonnes by 2030 (Oteng-Ababio 2010). The relationship between waste and cities is particularly threatening the future, in situation where practically everything is based on a throwaway mentality, which pays little regard to sustainability.

In countries where efficient solid waste management (SWM) policies have been instituted, including effective waste collection, segregation, transportation, storing, treatment and disposal, 'waste' is economically re-circulated (Medina 2005; Sternberg 2013). In such situations, SWM is not only seen as a necessary process to promote health and environmental safety, but also as presenting an opportunity to mine the largely untapped resources embedded in waste. The dilemma in Ghana is how to develop systems that can best utilize these inherent resources, even though the paper believes that the inability of city authorities to develop appropriate SWM policies cannot be wholly attributed to technical or economic reasons (see Grant and Oteng-Ababio 2012). It also has much to do with how waste is conceptualized and operationalized in the first place. Ojeda-Benitez et al. (2000) re-emphasizes the point when she opines that to appreciate the value of 'waste', it must be differentiated from refuse. To her, refuse refers to a situation where different kinds of waste are disposed of or mixed together in the same container, causing unpleasant odour and pollution, making it impossible for re-use. However, when kept separately with a minimum of order and care, such materials become waste to the original owner but may have some value for other persons.

In recent times, it appears that people's socio-cultural practices or throwaway mentality, where activities are designed to take 'resources,' make products and turn them into waste ('take-make-waste' attitude) are impacting negatively on the ecosystems, particularly with increasing population, improved technology and increased consumption. At the same time, waste is also becoming so diverse in its origin and forms, and so pervasive in its impacts, through terrestrial, aquatic, and atmospheric ecosystems that it has the potential to adversely affect both the inhabited and uninhabited parts of the world. The penchant of state officials also to unilaterally accept foreign SWM prescriptions during their international conferences, regardless of local technical or socio-economic circumstances and expecting overnight transformations, has worsened matters (see Ali 2010). The inherent uncertainty about 'local therapies' is part of an age-long debate but such tendency stifles local innovative resourcefulness. It also brings UN Sustainability City Program into question as per the present 'take-make-waste' attitude, neither industry nor the consumer has any incentive to use resources frugally. This creates chronic environmental concerns and pushes cities to grow in a disorderly manner, causing an accumulation of urban problems.

It is beyond the scope of this paper to venture into the debate on "the limits to growth" vis-à-vis resource consumption or the negative environmental impacts that will occur from wastes generated by an increasingly consumerist urban population. That said, the fear about these effects is warranted, particularly since nearly 95% of environmental damage occurs before a product is discarded as solid waste (Hoorneweg and Bhada-Tata 2012). This paper intends to shift the current emphasis from the "end-of-the-pipe approach", where waste is collected and disposed of (without any incentive to recycle), towards encouraging consumption within the waste stream. It is believed that most waste materials can cascade through a series of uses that can enhance job creation, decrease the need for 'natural resources' and eventually, the overall waste flows.

In what follows, the paper discusses the notion of waste and its value chain, highlighting on its implications for urban livelihoods and sustainable environmental management. Particular reference is given to the activities of a local non-governmental organization (NGO), Great Thinkers' Club, which has pioneered a youth enterprise from recycling of solid waste generated in James Town, a low-income community in Accra. The paper draws on the value chain thinking, demonstrating how the project has been designed synergistically to produce employment for the youth, through a range of valuable by-products of waste while, at the same time ensuring efficient SWM. Finally, the paper proposes a paradigm shift, which is intended to help all project stakeholders to appreciate waste as a resource.

TABLE 1: List of definitions of waste

Organization	Definition	Source
Eionet	Waste includes all items that people no longer have any use for, which they either intend to get rid of or have already discarded. Wastes are such items which people are required to discard, for example by lay because of their hazardous properties.	http://scp.eionet.europa.eu/themes/waste
Full cycle	Waste, or rubbish, trash, junk, garbage, depending on the type of material is an unwanted or undesired material or substance. It consists of unwanted materials left over from a manufacturing process or from community and household activities.	http://www.fullcycle.co.za/index.php/what-is-waste-and-why-is-it-a-problem.html
Basel convention	Wastes are substances or objects which are disposed or are intended to be disposed or are required to be disposed of by the provisions of national laws.	http://www.basel.int/Portals/4/Basel%20Convention/docs/text/BaselConventionText-e.pdf
UN Statistics Division	Wastes are materials that are not prime products (that is products produced for the market) for which the generator has no further use in terms of his/her own purposes of production, transformation or consumption, and of which he/she wants to dispose	http://unstats.un.org/unsd/environment/wastetreatment.htm
EU	Waste is any substance or object which the holder discards or is required to discard	http://ec.europa.eu/environment/waste/pdf/WASTE%20BROCHURE .
OECD	Wastes are materials other than radioactive materials intended for disposal	http://www.oecd.org/env/waste/
UNEP	Wastes are substances or objects, which are disposed of or are intended to be disposed of or are required to be disposed of by the provisions of national law	http://www.unep.org/greeneconomy/Portals/88/documents/ger/GER_8_Waste.pdf
European EPA	Waste is any substance, which constitutes scrap materials or any effluent or other unwanted surplus substance arising from the application of a process or any substance or article, which requires to be disposed of as being broken, worn-out, contaminated or otherwise spoiled.	http://www.nwcpo.ie/forms/EWC_code_book.pdf
Institute of safety Professionals (Nigeria)	Waste is when something is no longer useful to the owner or it is used of fails to fulfill its purpose.	http://www.eco-web.com/edi/090901.html

4.2 THE NOTION OF WASTE AND THE VALUE CHAIN

4.2.1 WASTE IS WASTE?

Waste, otherwise termed rubbish, trash, junk or garbage depending on the type of material, or the region in question, is a complex, subjective and perhaps controversial issue. Though waste is generally seen as an unwanted or undesired material and a bothersome problem and sometimes, a health threat if improperly treated, it can in some cases, generate good business for some, when properly handled. The complexity of waste manifests in the difficulty and multiplicity of its definition and description, which is also contingent on who is looking at the subject matter: ordinary citizens, technicians, businessmen, politicians, activists (Progracz 2002). Since different approaches have been used in defining waste, it is notably a challenge to gather comparable data for example, between waste in rich and poor countries. Table 1 presents some definitions of waste.

These definitions do not resolve the uncertainties surrounding what waste is. What runs through them, however, is the fact that waste is something that its holder has disposed of or discarded. This creates the problem of distinguishing between 'dispose of' and discard, as both principally mean getting rid of. Cheyne and Purdue (1995), however, suggest that disposal is a deliberate and thoughtful act to put something in a suitable place (either for sales, transfer of ownership, etc), while discard implies outright rejection, with or without interest in its final destination, since that thing is seen as useless or undesirable.

Other studies (CEFIC 1995) also argue that it is not the nature of the material that determines whether it is a 'waste', but the actions or intentions of the holder. Therefore, only those materials for which the holder has no further use and which he discards or intends to discard are waste. Lox (1994) also see waste as "either an output with (a negative market) no economic value from an industrial system or any substance or object that has been used for its intended purpose (or served its intended function) by the consumer and will not be reused". This definition also appears deficient, since the last part of it creates an impression that a product is designed for a single purpose and therefore, as soon as the purpose is fulfilled, it turns to waste.

Gourlay (1992) exemplifies this deficiency using a dollop of mustard left on a plate, which he argues is neither useless nor lost its properties and queries if such a material only became waste just because the owner failed to use it. In his contribution to unravel the puzzle, Gourlay proposes a working definition for waste as "what we do not want or fail to use". However, his definition also appears more human-related and fails to properly account for the concept of production wastes, especially the issue of by-products that are not necessarily created from carelessness but unavoidably emanating from the production process itself.

Contributing to the debate, Stanbury (2005) attribute the complexity of the term waste to its genesis, which they traced to the Latin word uastus and which evinces several meanings; to ravage, to leave desolate or cultivate. To them, waste connotes inefficiency and mirrors situations where managers of entities fail to minimize cost or maximize output because they are not using the best technology available. Elwood and Patashik (1993) add another dimension when they blithely asserted that waste, like beauty, is in the eye of the beholder. It can thus be argued that since there is no such thing as waste in nature, then waste is a human concept. Within the ecosystem, recycling is rife, hence production and decomposition are well balanced and ensure stability and sustainability in natural systems.

Human engagement with the ecosystem, however, places much emphasis on its economic value, resulting in situations where production and consumption become the dominant activities. Such a system tends to be highly destructive, consuming massive natural capital and energy, which return into the environment as waste, requiring even more natural capital to be consumed in order to feed the system (Pongracz 2002). The presence of waste is therefore an indication of overconsumption and inefficient use of resources, since the capacity of the natural environment to absorb and process these materials is finite and the improper waste disposal results in the loss of valuable resources. From all indications, the main problem confronting most developing countries is the sheer volume of waste being produced and how 'waste' is conceptualized.

4.2.2 RE-DEFINING SOLID WASTE MANAGEMENT

Significantly, the goal of any legislation on waste is to protect the environment and public health. Consequently, the conventional SWM hierarchy adopted by most local authorities and their donor agencies includes a 5-staged process, involving waste reduction, re-use, recycle, incinerate, and safe disposal of the residual at landfills (Forbes et al. 2001). From that perspective, this paper re-affirms that waste and its management involves multi-faceted activities and processes, yet opines that an efficient SWM should include the supervision of such operations and maintenance of disposal sites, which the current hierarchy appears silent. The present models implicitly assume that waste already exists "in space" and needs to be managed and that, SWM is simply a reaction to the presence of something that needs to be eliminated.

Seeing waste with that lens may create a barrier to an efficient and sustainable SWM system and may negatively affect the overall decision related to the transport, reuse, and sale of materials. Conversely, this paper believes that preventing the accumulation of waste from the beginning maximizes resource efficiency and sustainable management. An optimal approach to SWM would be an integrative system-based approach that provides control over processes that generate waste, waste handling, and waste utilisation (IPCC 2007). Thus, the key to any effective SWM involves waste minimization that helps the reduction of the amount of things to be disposed of, and this includes:

- Preventing and/or reducing the generation of waste at source;
- Improving the quality of waste generated, such as reducing the hazard, and,
- Encouraging re-use, recycling and recovery.

It is important to stress that minimizing the amount of things intended for disposal is only one of the options of waste minimization, which then reinforces the perception that 'the wasted thing' is already there to be disposed of. This contravenes the principal meaning of waste minimization, which is to avoid waste generation in the first place (Pongracz 2002). By inference, waste can be seen as a value concept, culturally construed, very subjective and may implicitly have a remote economic value.

Most developed countries have adopted the philosophy of waste minimization as the menu for their SWM systems, but scarcely has this informed policies in developing countries, where pressures of urbanization, land space for waste disposal and wanton destruction of natural resources are creating a bigger challenge (Baud et al. 2001). In Ghana, the individual local authorities (assemblies) are charged with the responsibility to manage waste in their respective jurisdiction and must do so in the most economically, socially and environmentally optimal manner possible (MLGRD 1999). Prior studies (Post and Obirih-Opareh 2002; Oteng-Ababio et al. 2013) have shown that waste management remains the single largest item on the assemblies' budget yet the most pernicious local pollutant (uncollected waste remains the leading contributor to local flooding and pollution). Equally important challenge confronting most authorities relates to their inability to appreciate waste as a resource as to date, and, as a consequence, waste separation remains alien or not properly address due to lack of finances. Even the National Environment and Sanitation Policy of 1996, and revised in 2010, remains loudly silent on waste minimization, creating a waste industry that is vague, dysfunctional and disjointed.

Earlier studies (see Hetherington 2004; Gregson et al. 2007; Adama 2012; Thieme 2013) highlight the fact that urban poverty and waste picking are inextricably linked with the urban poor engaging in scavenging as a survivalist strategy while at the same time, filling the vacuum created by the public sector-led SWM operations. There are also enough evidence showing that historically, communities have been resourceful to use waste as food for animals, fertilizer and as materials for second life products (Wilson et al. 2006). These studies point to a vibrant waste enterprise, devoid of policy or legislative, often tedious, unhygienic, unrecognized and potentially hazardous, whose contributions are rarely documented or quantified. These activities rank lowest in the waste recycling rung and participants are mostly the urban poor, especially women and children (Ahmed and Ali 2006). This paper is hopeful that more lessons can be gleaned from such informal operations in our quest for economically and environmentally cost effective SWM. Figure 1 establishes tractable pathways of making waste a resource through waste segregation at source, based on the value chain concept.

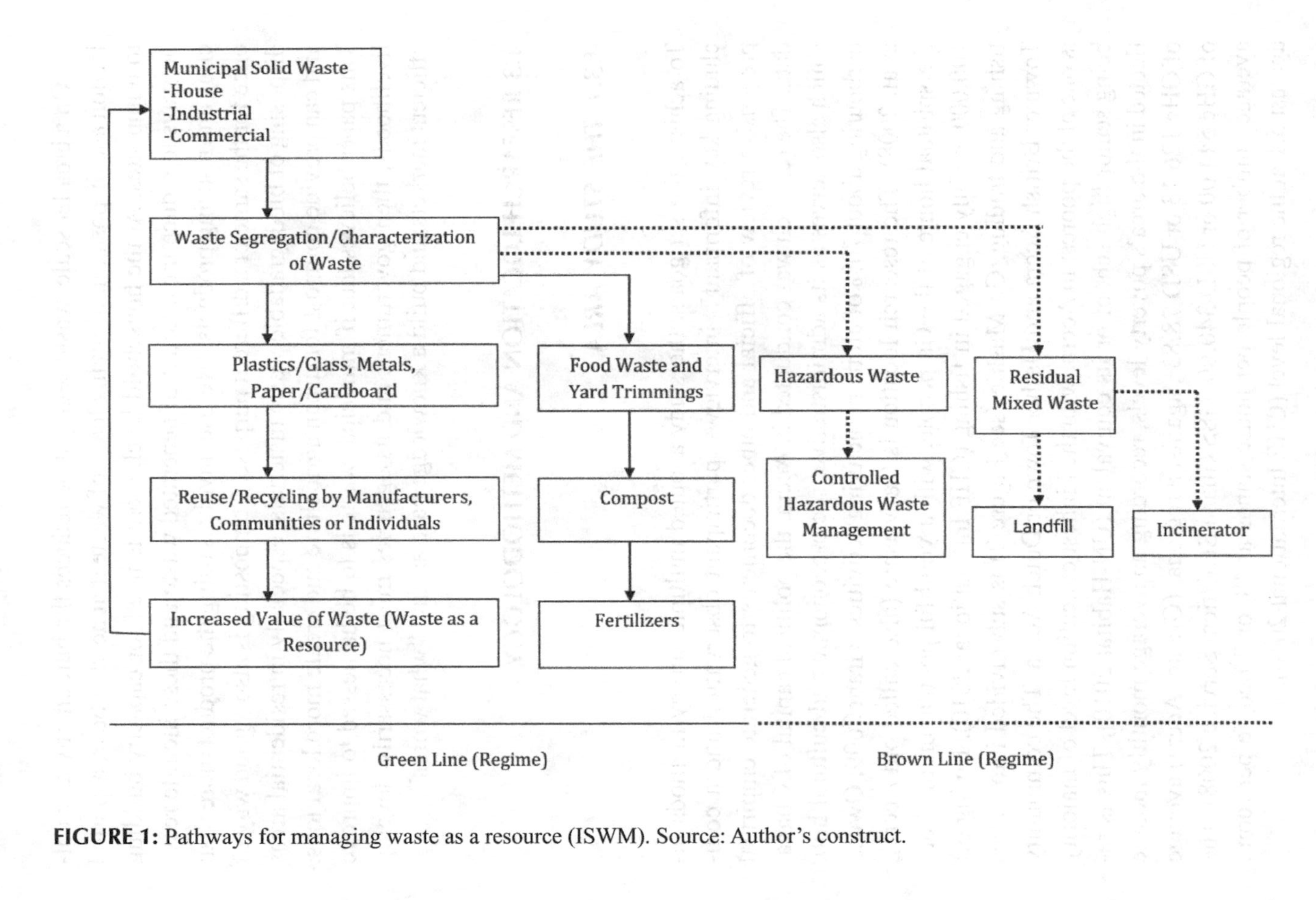

FIGURE 1: Pathways for managing waste as a resource (ISWM). Source: Author's construct.

On a broader scale, waste segregation reduces the burden of waste collection and disposal, minimizing loads of waste that need to be transported to dump sites. At the household level, there is need for capacity building on ways to reduce the amount of generated waste and these include recycling and re-using products. 'Re-use' involves using the product more than once for the same or a different purpose. Composting is also a good way of disposing of biodegradable waste materials, especially in residential areas and can provide compost for urban agriculture and create more local jobs. This paper believes that if municipal waste is to be harnessed to improve livelihoods, then government and its agencies must necessarily develop efficient markets and build a knowledge base around "what works".

4.3 RESEARCH LOCATION AND METHODOLOGY

4.3.1 THE STUDY AREA

To achieve the set goals, the study adopted multiple survey methods including key informant's interviews, participant observation and a comprehensive review of official and other documents to generate empirical data. The research was conducted in Accra, the political capital of Ghana, which also serves as its administrative, socio-economic and cultural hub, harbouring about 70% of all manufacturing activities (Grant 2009; Owusu et al. 2008). The research location is Ga Mashie (also called old Accra), the spiritual home of the Ga people, with a youthful population of about 100,000, mainly engaged in fishing (CHF International 2010). Living off fishing and trading, Ga Mashie (see Figure 2) is sub-divided into James Town or British Accra and Ussher Town or Dutch Accra. The community is one of the poorest in Accra, with the highest education level of majority being senior high school or its equivalent (UN-Habitat 2010). This is reflected in the area's poverty levels, recording an average monthly income of GH¢ 126.13 or USD 78.83[a], against a regional (Greater Accra) average of GH¢ 544.00 or USD 340.00 (GSS Ghana Statistical Service 2008). The average number of people per house stands at 48, or 7 people per room, against 5.1 at the regional level (CHF International 2010).

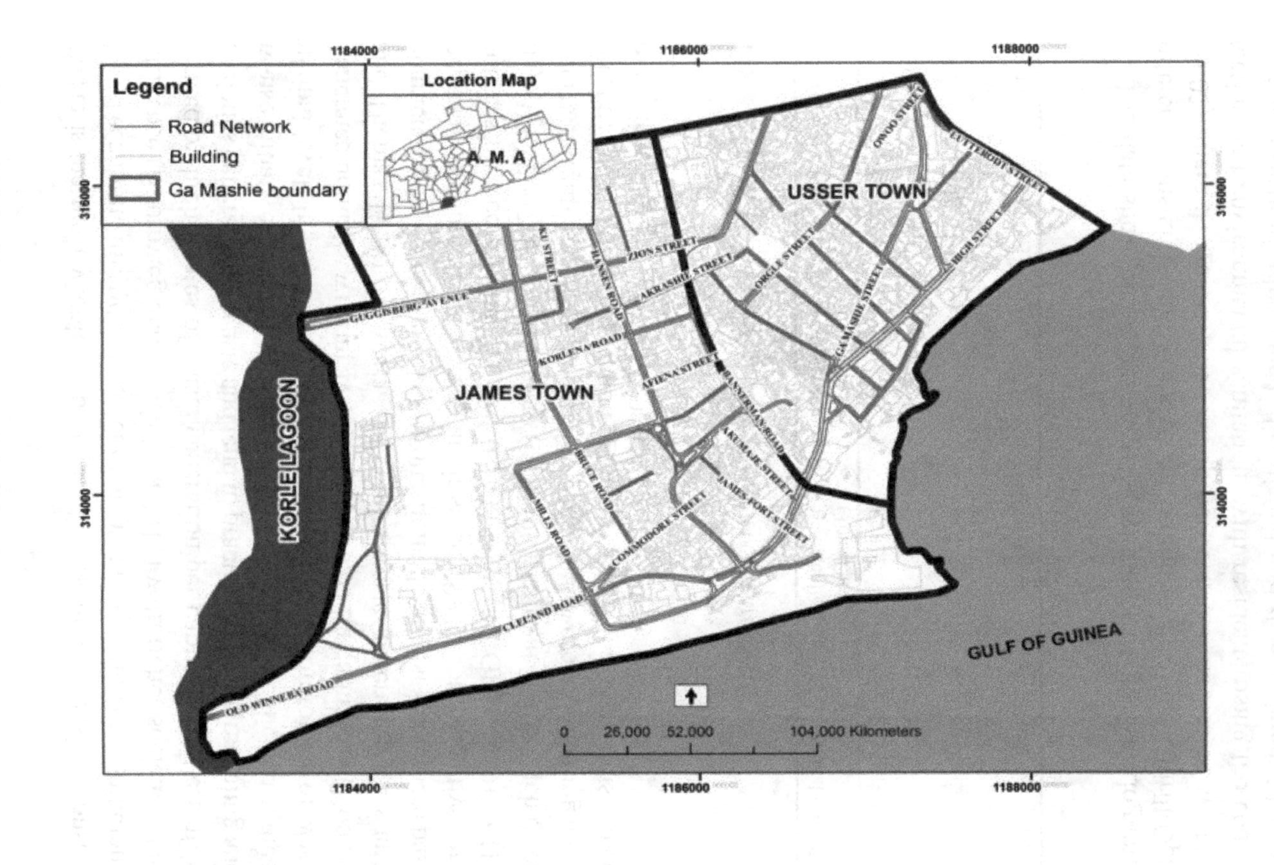

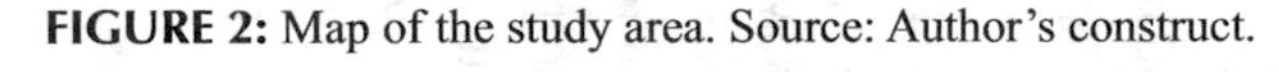
FIGURE 2: Map of the study area. Source: Author's construct.

4.3.2 WASTE GENERATION IN ACCRA

Waste is the most visible by-product of a resource-intensive, consumer-based economic lifestyle. Its generation rate is a function of population and affluence (Hoornweg and Bhoda-Tata 2012) with richer societies characterized by higher rates of waste generation per capita, while less affluent societies generate less waste and tend to practice informal recycling and/or re-use initiatives that reduce the waste per capita to be collected at the municipal level. In Sub-Saharan Africa, about 62 million tonnes of waste is generated annually with an average rate of 0.65 kg/capita/day (Hoornweg and Thomas 1999). In Accra, the correct waste generation rates and its characteristics are non-existent or questionable at best since there has not been any comprehensive waste audit since 1993. Consequently, various sources give conflicting figures: 0.51 kg/day, (MLGRD 1999), 0.41 kg/day (WRI 2007) and 0.81 kg/day (CHF International 2010). Such discrepancies in figures serve as potential impediment to proper planning and this must be taken into consideration in interpreting. Table 2 presents projected municipal waste scenario by 2030.

TABLE 2: Estimated waste generation of Accra (2000-2030)

Years	Population	Waste generation (Tons/Day)	Waste collection (Tons/Day)	Residual (Tons/Day)
2000	1,658,939	2,127	1,702	425
2005	1,960,797	3,369	2,695	674
2010	2,317,583	2,654	2,123	531
2020	3,237,730	3,390	2,712	678
2030	4,523,203	4,419	3,535	884

The data show increasing waste generation so is the backlog and by inference, unsightly environment. According to Post and Obirih-Opareh (2002), there is a direct relationship between the method of waste collection and environmental sanitation, with residents using the container system clearly disadvantaged, dissatisfied and disappointed. Meanwhile studies have shown that for health reasons, the organic nature of waste in tropical regions like Ghana makes it imperative that it is actually collected daily. This makes the challenges and costs of SWM even more daunting. In Accra, it is generally the city center and the wealthier neighbourhoods that receive service when it is available (see Oteng-Ababio 2010). In poorer areas, uncollected wastes accumulate at roadsides, are burned by residents, or are disposed of in illegal dumps, which blight neighborhoods and harm public health (Huober 2010).

In neighbourhood like James Town, roadside accumulation "has reached levels resembling those that spawned epidemics in European cities 500 years ago" (see EGSSAA 2009: 2; Melara, et al 2013), with its impacts not routinely monitored or virtually ignored. On few occasions where attempts have been made to improve dumpsites management, these often tend to be in reaction to threats of their imminent closure by nearby communities. Unless more effective urban SWM programs and public water supply systems are put in place, outbreaks of cholera, typhoid and plague may become increasingly common.

4.3.3 WASTE CHARACTERIZATION

Waste composition of any community is influenced by its geographic location, level of economic development, cultural norms, energy sources, and climate, among others. As countries urbanize and population becomes wealthier, consumption of inorganic materials (plastics, paper, aluminum) increases, while the organic fraction decreases. The study examined a number of baseline surveys (see Table 3) to compute the waste composition, since no official data exists. In all instances, proper permission was sought, including appropriate ethical clearance (No ECH 061/13-14) approved by Institute for Statistical, Social and Economic Research (ISSER), University of Ghana, Legon. The findings yielded an average organic component of 67.68%, but the high rate of illegal disposal in drainages, river channels and open spaces raises lots of concern (Oteng-Ababio 2010).

Be that as it may, the findings show that with efficient institutional arrangements, political will and appropriate infrastructure investment and human resource, most of the household waste can receive value addition. Unfortunately, until the advent of the project of the Great Thinkers' Club in James Town, separation and treatment of organic waste was very rare. While the recovery and reuse of materials was generally for personal use, there were many 'professional' waste pickers—popularly called 'Kaya bola' in the local parlance, whose activities were seriously threatened by disease organisms, sharp objects and other hazards in the waste. They generally lacked protective equipment—a situation that is a clear manifestation of the extent of poverty in the neighbourhood.

TABLE 3: Waste characteristics/ Waste stream analysis

Type of Waste	Zone A1 (%)	Zone B1 (%)	Zone C1 (%)	Nima2 (%)	James-town2 (%)	Newtown3 (%)
Organic	74	71	78	56	60	67
Plastic	10	8	6	17	16	20
Paper	5	5	4	2	1	4
Metal	4	4	4	5	5	2
Glass	2	1	0	2	2	2
Textiles	1	1	1	9	6	5
Inert	4	7	5			
Others		3	2	3	1	

Source: 1-Dagadu, (2007); 2-Kumashie I, (2011); and 3-Earth Institute, (2010). Zone A: High-income, low-density waste zone (Cantonments, Airport Residential Areas). Zone B: Middle-income, medium-density waste zone (Adabraka; Kaneshie). Zone C: Low-income, high-density waste zone (James Town, Nima).

4.4 CASE DESCRIPTION

4.4.1 OVERVIEW OF GREAT THINKERS' CLUB ACTIVITIES

Studies (Medina 2005; Sternberg 2013) have noted that the adverse impacts of SWM are best addressed by establishing integrated programs where all types of waste and all facets of the waste management process are considered together. Though desirable, limited resources normally prevent these programs from being implemented, and only a piecemeal solution tend to be possible. The "Great Thinkers' Club" changed this dynamics in the study area by initiating and developing an integrated SWM system and building of the technical, financial, and administrative capacity of the community to manage and sustain it. Suffice to add that whether pursued holistically or in a piecemeal faction, waste managers must ensure that the program is appropriately tailored to local conditions and that the practical environmental, social, economic, and political needs and realities are balanced.

As the waste audit indicated, a large part of household waste in Accra is potentially recyclable as organic waste accounts for about 60%, while paper and other inorganic recyclables (glass, tin) are fairly represented. Yet, this opportunity remains poorly unexplored. Indeed, within the city, a compost plant built in 1976 at Teshie (another suburb of Accra), with a processing capacity of 200 tonnes a day was decommissioned in 2002 due to operational difficulties (see Oteng-Ababio et al. 2013). In particular, lack of source separation of waste greatly affected its operations. Thus, the NGO's activities demonstrate a bold attempt to both tap into the untapped potential embedded in waste and help advance that scholarship in the area of environmental sustainability.

Broadly, the NGO is a youth community-based organisation established in 1999 in Ga Mashie, a suburb with poor sanitation and rising youth unemployment (UN-Habitat 2011). Its mission includes increasing opportunities to enrich and address youth community needs. With a membership of 100, the club's activities span from maintaining efficient environmental sanitation to a recently instituted youth employment through SWM. The funds for the current project, dubbed Youth Engagement in Service delivery (YES) was secured through Co-operative Housing Foundation (CHF) International under the Bill and Melinda Gates Foundation. The project involves a 4-value chain process of municipal waste (biodegradable organic fractions; metals; plastics; electronic waste or e-waste).

The project was initiated with a dual dividend strategy: to engage the beleaguered youth in gainful employment through the provision of environmental services. As part of the project, a house-to-house waste collection, which hitherto was the preserve of only rich neighbourhoods in the city, was introduced at Ga-Mashie in place of the poorly managed container collection system (Kumashie 2011). Using local resources, teams from the local NGO were selected, trained and equipped to provide house-to-house waste collection services to households who willingly agreed to register with the group. Each youth was equipped with a tricycle, nose masks, safety boots, safety overall, and hand gloves (see Figure 3) provided with sponsorship from CHF international (see CHF International 2010). With the exception of the tricycle, all other equipment and training were provided at no cost to participants. The tricycle however was given to participants on a hire-purchase scheme under which each waste collector was to pay cumulatively GH¢600, (US$310), which beneficiaries unanimously agreed to pay in nine months (i.e. GH¢ 63 [US$ 32.57] a month).

The project, which also had a plastic recycling enterprise (Trashy Bags) and a community compost plant, incorporates two rational SWM principles: it encourages source separation of waste and embodies the polluter-pay-principles. At the time of the research, a total of 770 households or 33% of total households in the study area were covered under the project. Each registered household is given two waste bins with which to separate organic and inorganic waste. Each household is also encouraged to separate plastics, waste electrical and electronic equipment (WEEE) and metals. The bins are given for free as a way of motivating participants to undertake source separation of waste (minimise waste), which is seen as the bedrock of any sustainable SWM system. The waste collectors have the responsibility to collect the segregated organic waste daily for onward transfer to the community compost plant.

It is important to add that the quality of compost produced by the YES project was scientifically certified by Ghana Standards Board and has been successfully piloted by some vegetable farmers in Accra (CHF International 2010). Accordingly, the waste behaviour (composition) and the subsequent certification of end-product give a glimmer of hope; the possibility of converting the huge organic waste into compost and thereby, ensuring safer, liveable environment. Its actualization however needs commitment and participation of all stakeholders—government, local authorities, service providers and beneficiaries. Currently, a private company, Zoomlion Ghana Limited, has initiated a waste recycling project; Accra Compost and Recycling Plant Limited. However, the lack of source separation has compelled the company to be very selective in where to pick waste for recycling, (i.e. only high-income areas), where there is little likelihood of waste contamination with fecal matter, which could be as high as 10% (see Kumashie I 2011).

4.4.2 THE COLLECTION AND ANALYSIS

The fieldwork involves qualitative data collection that took place between August and November 2012. It consisted of personal observation, focus group discussions (FGDs), and semi-structured interviews with 25 key informants, including the executives of the local NGO, service beneficiaries and participants, 5 community opinion leaders including the assembly man and the leaders of 2 prominent women groups. Some public officials from the Municipal Assembly, Ministries of Local Government and Rural Development, and Gender, Children and Social Welfare as well as a representative of project sponsors were also interviewed. Specifically, the key informants were interrogated on the structure, prospects and challenges of the project, how the transformation of perceived waste into some value and the marketability of the products occurs, and present a more nuanced account of the value chain than the over-simplified end-of-pipe approach. Answers were also sought for the role of the NGO as well as the relationship among the local authorities, the NGO and the local community.

To assess the modus operandi of the NGO, participant observation was employed to learn at first-hand, how the club engages with the community and how it strategizes to outwit the authority's most "preferred" players—formal waste collectors. Two separate FGDs were organized between the beneficiaries and the NGO; and secondly, between the NGO and the representatives of the local authority with the sole aim of gathering in-depth understanding of the structure, processes and human behaviour and the reason governing such behaviour. The FGDs were held in the community, had nine participants each and explored issues bothering on efficiency, power relations and affordability of the services.

Participants discussed the project's challenges, proffered solutions and the possibility of scaling up the project. All submissions were recorded, later transcribed and thematically analyzed. To ensure unrestricted access to the NGO's project sites, their commitment was secured through the well-respected Assemblyman, re-enforced by the intervention of a local sub-chief who happens to be an "environmentalist" and has great passion for a good environment. The study benefitted immensely from earlier works, which engage with the challenges of managing waste in low-income communities (Huober 2010; Matter et al. 2013).

4.5 DISCUSSION AND EVALUATION

4.5.1 TAPPING THE UNTAPPED POTENTIALS IN WASTE: REFLECTIONS FROM JAMES TOWN

The YES Project waste flow is depicted in Figure 4 and falls in tandem with the government's 'privatization' agenda which, in principle, involves gradually disassociating state-owned enterprises or state-provided services from government control and subsidies, and replacing them with market-driven entities (Grant 2009). In the context of municipal services, the process generally implies reducing local government activity within a given sector by: involving participation from the private sector including NGOS; or reducing government ownership, through divestiture of enterprises to unregulated private ownership, and commercialization of local government agencies.

In the case of the YES project, which is an example of community participation in SWM, the study shows that each participating household is encouraged to separate both recyclables (either organic, plastics, metals) and non-recyclables. Generally, some of the recyclables (plastics, WEEE, and metals) are sold directly at the household level to prospective itinerant waste buyers who roam the community on daily basis, while the mixed waste and organic materials are collected by the trained community waste pickers for a fee. These service providers haul waste from registered household sources to the composting facility. This is effectively carried out by pre-collection services using the tricycles acquired through the YES project.

According to the system set up by the project and agreed upon by all stakeholders, a household pays between 10p (US$0.05) and GHC 1 (US$0.50) upon a visit by a waste picker depending on the volume of waste available at the time. The waste collector collects (isolates) the organic materials for onward delivery to the compost plant. The waste picker sends collected mixed waste to the transfer stations (mainly communal skips at sanitary sites), where a disposal fee of between GHC1 (US$ 0.50) – GHC2 (US$ 1.00) is paid. During the FGDs, beneficiaries most of whom work in the informal economy, lauded the project's affordability and flexibility "pay-as-you-go" billing system against the formal system which operates a monthly billing option. The NGO has the responsibility of providing a training environment to participants (capacity building) and explain repeatedly the basic concepts and principles upon which the project is founded to the community at large.

FIGURE 3: Youth team with their tricycles on the house-to-house waste collection operation. Source: Field work, 2013.

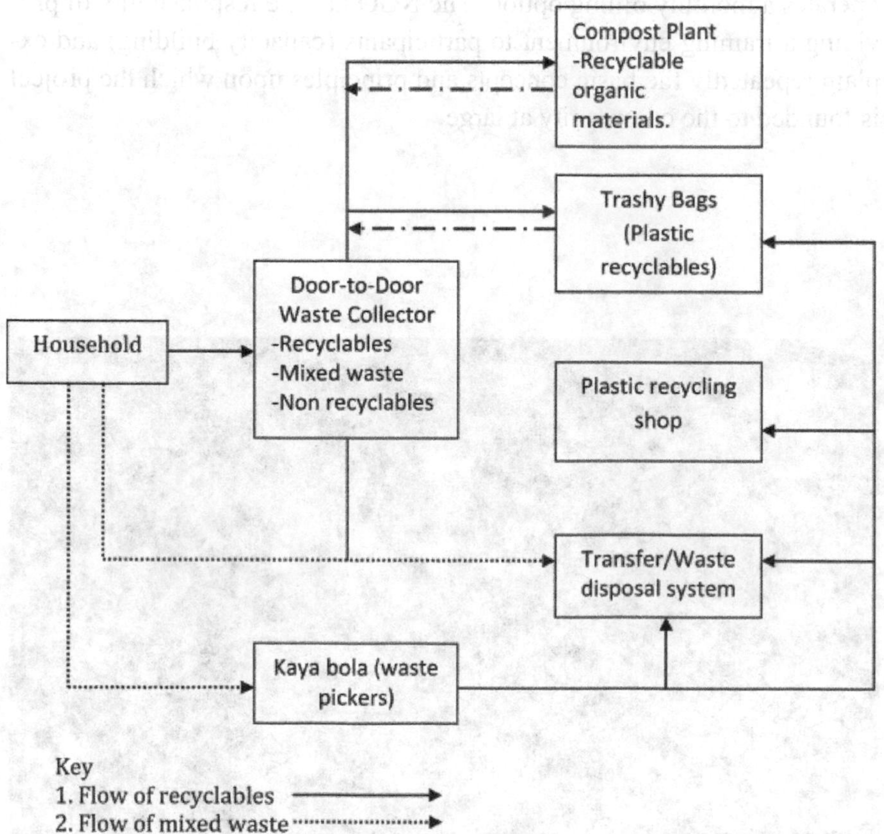

FIGURE 4: Schematic Overview of the YES Waste Management System. Source: Author's construct.

FIGURE 5: Recycling of thin plastics. Source: Field work, 2013.

4.5.2 THE WASTE VALUE CHAIN: PLASTICS RECYCLABLES AND COMPOST FACILITY

Results from the waste audit show that since 1993, there have been a gradual reduction in the proportion of organic materials, but a corresponding increase in plastics, increasing from 1.3% in 1993 to 3.5% in 2010 (Oteng-Ababio 2012). The increasing presence of plastics is attributed to the upsurge in the use of polythene bags for take-away food orders and water sachet bags (see Fobil et al. 2005). It is in recognition of this increase that the NGO established a plastic recycling enterprise (Trashy Bags) and trained some of its members to use thin films collected from the community as feedstock to produce handbags, traveling bags, raincoats, jackets, etc (see Figure 5). The study reveals that 'waste' plastics are bought from members at GH¢ 0. 05 for 500 pieces of clean ice cream sachets (250 ml) and GH¢ 0.30p per kilo for empty water sachets (500 ml).

FIGURE 6: A compost facility showing the processing of waste into manure. Source: Field work, 2013.

Personal observations during the fieldwork confirmed increased presence of plastic waste and how it has been indiscriminately disposed in the drains and other open places. Thus, the move by the project presents positive indicators or economic incentives inherent in thin plastic recycling that the city authority can capitalize on to encourage source separation and minimize the current waste menace. This position was reinforced when it was revealed during the study that 7 new plastic waste purchasing outlets have emerged within the community since the commencement of the project. Indeed, Post and Obirih-Opareh (2002) intimate that most plastic industries operating in the city struggle to get feedstock to meet their installed capacity.

Similarly, the NGO has constructed a community compost plant in the community at the cost of US$22,000 (CHF International 2010). The plant feeds into the employment motivation agenda of the project and creates a synergy between waste collection and disposal. The compost plant operators have also been trained on compost method and facility management as well as occupational safety and health procedures. The facility uses organic waste from the community and the processing of compost normally takes four (4) months (Figure 6).

In terms of profitability analysis, waste picking is an informal activity organized on an individual basis and therefore quantifying the actual labour costs incurred to enable one perform proper cost-benefit analysis is quite daunting, if not impossible. However, during the FGDs, participants in the plastic waste chain claim they earn between GH¢ 0.20 and GH¢ 0.35 per kilo or between GH¢ 8.20 and GH¢ 10.20 daily depending on the level of foreign materials (contaminants), which at the upper end, is 95% higher than the national minimum daily wage of GH¢ 5.24. By inference, each participant collects on average in excess of 20 kilos of plastic waste daily, emphasizing the project's potential not only to generate income but more importantly, contributing to a cleaner and healthier environment.

Concerning the compost facility, the challenge participants identified during the study relates to lack of sustained market for the end product. Prior scholarship had identified potential market to include i) real estate developers—using compost as soil conditioner in landscaping; ii) vegetable farmers—for nursing vegetables, and iii) Waste Management Department (WMD) for disposal operations—as capping material to reduce malodors. Meanwhile, at the time the Teshie plant was decommissioned in 2002, the price of a tonne of compost was just GH¢2 (US$1.43)[b]. After almost a decade, the market price at a private facility (Zoomlion, Ghana Limited) is GH¢30 (US$ 15.5), indicating an increasing opportunity should the question of marketability be resolved (Matter et al. 2013).

On the part of the local government authorities, all respondents from the public service alluded to the fact that the project is a microcosm of what government stands to gain by encouraging community participation in service delivery. They opine that such a process leaves municipal resources available for other equally important urban infrastructure and equipment. The Head of WMD of the assembly asserted, "privatization of such urban services reduces the cost of public services to consumers; relieves the financial and administrative burden on the assembly; increases productivity and efficiency by promoting competition; stimulates the adoption of innovation and new technology; improves the maintenance of equipment; and creates greater responsiveness to cost control measures".

A representative of the local government ministry reveals that community management of service provision as demonstrated by the YES project, is the highest level of community participation, and gives the community authority and control over operation, management and/or maintenance services benefiting its members. He however maintains that such partnership requires institutional support and recognition in order to be successful, which in this case both the Accra Metropolitan Assembly (AMA) and the local government ministry have shown and demonstrated enough commitment.

At the community level, a participant alluded to how the project has ensured a clean environment, as according to him, they could easily identify those who clandestinely litter the environment by the type of waste. An opinion leader also described the project as "important and innovative" emphasizing, "the use of community members to execute community service provides jobs, minimizes crime, poverty, rowdyism and also improves community communication and fosters social acceptability and participation". He called on the local authorities to 'dream' of scaling-up the project into a citywide system than creating new, parallel structures.

4.5.3 SUSTAINABILITY OF YES PROJECT: IS THE YOUTH IN AGRICULTURE PROJECT THE ANSWER?

The Youth in Agriculture Programme (YIAP), which is a public sector commercial entity, initiated by Ghana government in 2006, seeks to motivate the youth into farming. As part of the incentives, the government provides certain services—land preparation, provision of seedlings, fertilizers, market opportunities, etc. In the 2010 national budget statement, the government declared financial commitment of GH¢2 million (US$1.034 m) mainly for procuring seedlings and fertilizers (MOFA 2012). Given the huge debt the assembly incurs (accumulates) in disposing waste and equally huge funds government commits into procuring organic fertilizers, strategically scaling-up the YES project perhaps, provides an opportunity to address the marketability problem facing the YES compost industry. This is imperative not only because of the paucity of locations for dumpsites, but the possibility of building an engineered landfill anytime soon appears economically a remote possibility. Even in cases where donor support has been secured for such construction, poor planning results in community protestations and resistances (Oteng-Ababio 2010; Owusu et al. 2011).

Meanwhile, critical spatial analysis of the YES chain of activities shows that the city can take advantage of its physical and economic synergies and scale it up, particularly in low-income communities within the metropolis where majority remains severely underserved when it comes to waste management services. Currently, the assembly remains heavily indebted to private service providers, standing at GH¢100 million (US$ 51.7 m) by the end of 2011. Additionally, the disposal charge[c] of GH¢12 (US$ 6.2) per tonne is an indication of potential cost savings that can accrue to the city if alternative options for value-addition are identified. The YES project has demonstrated that a sizable portion of what is thrown away as "waste" contains valuable resources such as metals, glass, paper and plastics which can be reprocessed as raw materials. Constrained by mounting national debts in the face of many competing needs and tepid economic growth rates, the assembly will continue to bear more responsibility in managing its public affairs with limited resources. Thus, building greater cooperation will not only help manage the 'waste' sustainability but also provide an avenue for job creation and by extension, poverty reduction.

Assuming that the national economic realities are not going to change in the short-term, then it becomes necessary to adopt consumption and waste management practices that reduce the amount of waste destined for disposal or to be discarded. In other words, envisioning waste as a resource will extend the current dump's life, reduce the pressure on areas needed for new dumpsites while at the same time, creates employment. The role of local authorities in this direction is critical and indeed indispensable. Even interventions at household levels also benefit from involving the local authorities, who are the overall entity that have the responsibility for municipal SWM and represent public interest, to enhance legitimacy, support and acceptance or for up-scaling the initiative.

In summation, the study has demonstrated that establishing an integrated SWM system including waste separation at source, resource recovery and composting of organic waste, at least at the local level such as the YES project, requires representation of waste pickers (Kaya bola) and integration of the community to work with all stakeholders, including representatives of waste pickers. It also shows that in such community-led service delivery, local leaders are often active in management or maintain close contact with the municipality or community management body. Additionally, women and teens can play crucial roles, such as initiators, managers, operators, educators, and watchdogs for the community. In other words, encouraging community-based management of service delivery may help address certain social and management problems in SWM—low participation of household, management and operational problems, financial difficulties, and lack of municipal cooperation and capacity building.

4.6 CONCLUSIONS

The study re-affirms the position that a global depletion of natural resources leads to a global competition for resources and therefore the future competitiveness and vulnerability of cities are dictated by cities' ability to shift to a less resource-intensive production and way of life, with less waste generation and increased recycling (EU 2011). Clearly, there is a compelling need to reduce the volume of waste destined for dumpsites in Accra. The city authorities' apparent addiction to 'landfilling' is damaging for the environment and a manifestation of our failure to see waste as a potential resource. The situation in Accra may not be an isolated case, as most cities countrywide lag behind in terms of policy and practice. The religious enforcement of some 'by-laws' such as the "pay-as-you-dump" policy, which were seemingly enacted without reference to local knowledge, is tantamount to punishing the public. Rather, attempts should be made to give residents incentives to do the right thing and the YES project is a microcosm of such a model.

The project has amply demonstrated that awareness creation amongst all key stakeholder groups, including local government officials, NGOs and the public, is imperative for effective and sustainable SWM services, and that there is a wide range of methods used to raise awareness. It is however, important to choose the communication channels and materials that are appropriate for each target audience and for the specific situation of the municipality. It also building the capacity and adequately equipping participants as waste management, especially landfill operations, is highly technical and normally requires people with specific technical skills and sound scientific background. Bench-marking SWM functions and knowing what expertise is required to perform these functions will assist in drafting appropriate job descriptions and specifications.

By investing and cultivating the interest, enthusiasm and willingness of residents, and adopting acceptable socio-economic initiative-based models that sustainably increase volumes of waste separated at source and recycled, the project has been extremely successful. It demonstrates that putting in place appropriate mechanisms to ensure a reliable and timely waste collection regime and an easy, stress-free, accessible and economical market for recyclables is critical for sustainable SWM system. This is of great importance in Accra, where municipal SWM poses a considerable environmental risk because waste is merely collected and buried with no engineered landfills.

Admittedly, such programmes will not be effective without concurrent community campaigns on the collection and sale of recyclables. Again, the negative preconceptions of city managers and planners about the informal economy in general, play an important inhibiting role in attaining integrated SWM system and this situation needs to be addressed. Seeing informality as a symptom rather than a victim of poor policies and programmes crowds out its inherent potentials, which can be gleaned to ensure sustainable environmental development. Appreciating the waste value chain means that all stakeholders ought to be made part of the governance structure.

Municipal SWM requires planning, foresight and reflexive governance to ensure that sustainable systems are adding value and/or complementary to each other (see Oteng-Ababio 2013). The current situation, where some urban residents are often prevented from participating and benefiting from the development processes and services because of their perceived lack of knowledge, or due to others' lack of knowledge about their situation, conditions and development needs, must be abated. Once knowledge and information is cultivated at the local level, accessed and shared, such tendency can place the community on an equal footing (Veeravatnanond et al. 2012). Such knowledge can be usefully shared between cities, improving prospects for pro-poor investments and development.

Efforts directed towards empirical research must continue, especially regarding waste audits since these will greatly support any future decision by city authorities. It is important to sample other areas to ascertain the possibility of replicating the YES project. In addition, waste separation programme must be accompanied by proper education to create deeper understanding in each community regarding separation, re-using and re-cycling activities so that the exercise is not simply regarded as tedious one imposed by 'bureaucratic, profit-making government authorities'. Rather, the process will be seen as a new approach that encourages residents to regard waste as a potential resource. The findings resonate with earlier studies (UN-Habitat 2011) that have shown that urbanization of poverty would increase unless the governance structure improve urban planning and management, and recognize the right of urban poor to improve their conditions and participate in the distribution of the benefits that the urbanization process has to offer.

ENDNOTES

[a]In 2012, the exchange rate of a cedi to dollar stood at US$ 0.517.

[b]In year 2000, the exchange rate of a cedi to dollar was US$ 0.715.

[c]The private contractors charge a disposal fee of GH¢12 per tonne and with no weighing bridges installed at these sites, city authorities are at the mercy of both haulers and managers of the sites.

REFERENCES

1. Adama O (2012) Urban governance and spatial inequality in service delivery: a case study of solid waste management in Abuja, Nigeria. Waste Manag. Res 30(9):991-998
2. Ahmed SA, Ali SM (2006) People as partners: facilitating people's participation in public-private partnerships for solid waste management. Habitat Int'l 30:781-796
3. Ali A (2010) Wasting time on solid waste in developing countries. Waste Manag. 30:1437-1438

4. Baud I, Grafakos S, Hordijk M, Post J (2001) Quality of life and alliances in solid waste management contributions to urban sustainable development. Cities 18(1):3-12
5. CEFIC (1995) Discussion paper on the definition of waste, European Chemical Industry Council. 22 Feb 1995. http://www.cefic.org/position/Tad/pp_ta029.htm (3. Sept. 2002)
6. Cheyne I, Purdue M (1995) Fitting definition to purpose: The search for a satisfactory definition of waste. Journal of Environmental Law 7(2):149.168
7. CHF International (2010) Accra Poverty Map: A Guide to Urban Poverty Reduction in Accra. CHF International, Accra, Ghana.
8. Dagadu PK (2007) Municipal Solid Waste Source Separation at the Household Level – A Case Study of the Accra Metropolitan Area.
9. An Unpublished Mphil Thesis submitted to the Institute of Environment and Sanitation Studies, University of Ghana
10. Earth Institute (2010) Millennium Cities Initiative Report: Findings of Waste Composition Study for Ayidiki Electoral Area, Accra, Ghana. New York, and University of Ghana, Legon.
11. Elwood JH, Patashik E (1993) In praise of pork. Public Interest 132(Summer):19.33
12. Environmental Guidelines for Small-Scale Activities in Africa (EGSSAA) (2009) Solid waste: generation, handling, treatment and disposal, EGSSAA: Municipal Solid Waste/ March 2009/ EXCERPT. 7 http://www.encapafrica.org
13. EU (2011) Challenges, Visions, Way Forward. European Commission, Directorate General for Regional Policy. Brussels. doi:10.2776/41803
14. Fobil JN, Carboo D, Armah NA (2005) Evaluation of municipal solid waste (MSW) for utilisation in energy production in developing countries. Int'l J of Env. and Tech Manag 5:76-86
15. Forbes DL, McLean RF, Tsyban A, Burkett V, Codignott JO, Mimura N, Ittekkot V (2001) "Coastal Zones and Marine Ecosystems." 264 Geography Reader. In: James J, Canziani MOF, Leary NA, Dokken DJ, White KS (eds) Climate Change 2001: Impacts, Adaptation, and Vulnerability. Contribution of Working Group II to the Third Assessment Report of the Intergovernmental Panel on Climate Change, UK: Cambridge University Press. pp 343-379
16. Gourlay KA (1992) World of Waste: Dilemmas of industrial development. Zed Books Ltd, London.
17. Grant R (2009) Globalizing City: The Urban and Economic Transformation of Accra. Syracuse University Press, Ghana.
18. Grant R, Oteng-Ababio M (2012) Mapping the invisible and real African economy: urban e-waste circuitry. Urban Geog 33(1):1-21
19. Gregson N, Metcalfe A, Crewe L (2007) Moving things along: the conduits and practices of divestment in consumption. Trans of the Inst of Brit Geog 32:187-200
20. GSS (Ghana Statistical Service) (2008) Ghana Living Standards Survey. Report of the Fifth Round (GLSS 5). Accra, Ghana.
21. Hetherington K (2004) Secondhandedness: consumption, disposal, and absent presence. Env and Plan D: Soc and Space 22:157-173

22. Hoornweg D, Bhada-Tata P (2012) What a Waste: A Global Review of Solid Waste Management.
23. Hoornweg D, Thomas L (1999) What A Waste: Solid Waste Management in Asia, East Asia and Pacific Region. Urban and Local Government Working Paper, World Bank, Washington DC, USA.
24. Huober AL (2010) Moving Towards Sustainable Solid Waste Management in Accra: Bridging the Formal-Informal Divide. Dissertation, Amherst College.
25. Intergovernmental Panel on Climate Change (IPCC) (2007) Climate Change 2007: The Physical Science Basis. In: Solomon S, Qin D, Manning M, Chen Z, Marquis M, Averyt KB, Tignor M, Miller HL (eds) Contribution of Working Group I to the Fourth Assessment Report of the Intergovernmental Panel on Climate Change, New York: Cambridge University Press.
26. Kumashie I (2011) Municipal Solid Waste Management in the Accra Metropolitan Assembly: The Case of The Communal Container Collection System. An unpublished Mphil Thesis submitted to the Institute of Environment and Sanitation Studies, University of Ghana
27. Lox F (1994) Waste Management. Life Cycle Analysis of Packaging. Final Report. Vlaamse Instelling voor Technologisch Onderzoek, Belgian Packaging Institute, for the European Commission, DG XI/A/4, Consortium Vrije Universiteit Brussel.
28. Matter A, Dietschi M, Zurbrugg C (2013) Improving the informal recycling sector through segregation of waste in the household – the case of Dhaka, Bangladesh. Habitat Int'l 38:150-156
29. Medina M (2005) Serving the unserved: informal refuse collection in Mexico. Waste Manag 23:390-397
30. Melara JE, Grant R, Oteng-Ababio M, Ayele B (2013) Downgrading - an overlooked reality in African cities: reflections from an indigenous neighborhood of Accra, Ghana. App Geog 36:23-30
31. MLGRD (Ministry of Local Government and Rural Development) (1999) National Environmental Sanitation Policy. Ministry of Local Government and Rural Development, Accra.
32. MoFA (2012) Annual Report on Draft MoFA Strategic Plan for 2013-2015, Ghana. Unpublished data from various directorates' internal reports
33. Ojeda-Benitez S, Armijode-Vega C, Ramirez-Barreto E (2000) The potential for recycling of household waste: a case study from Mexicali, Mexico. Environ and Urb 12(2):163-173
34. Oteng-Ababio M (2010) Private sector involvement in solid waste management in Ghana: the case of the Greater Accra Metropolitan Area (GAMA). Waste Manag and Res 28(4):322-329
35. Oteng-Ababio M (2012) The role of the informal sector in solid waste management in the GAMA, Ghana: challenges and opportunities. Tijdschrift voor Economische en Sociale Geografie 103(4):412-25
36. Oteng-Ababio M (2013) Unscripted (in) justice: unequal exposure to ecological hazards in Metropolitan Accra. J of Enviro and Plan A 45(5):1199-1218
37. Oteng-Ababio M, Melara JE, Gabbay O (2013) Solid waste management in African cities: sorting the facts from the fads in Accra, Ghana. Habitat Int'l 39:96-104

38. Owusu G, Agyei-Mensah S, Lund R (2008) Slums of hope and slums of despair: mobility and livelihoods in Nima, Accra. Norsk Geografisk Tidsskrift. Norwegian J of Geog 62:180-190
39. Owusu G, Oteng-Ababio M, Afutu-Kotey LR (2011) Conflicts and governance of landfills in a developing country city, Accra. Lan and Urb Plan 104(2012):105-113
40. Pongracz E (2002) Re-defining the concepts of waste and waste management: Evolving the theory of waste management. Academic Dissertation to be presented with the assent of the Faculty of Technology, University of Oulu.
41. Post J, Obirih-Opareh N (2002) Quality assessment of public and private modes of solid waste collection in Accra, Ghana. Habitat Int'l 26:95-112
42. Stanbury W (1995) Toward a political economy of government waste: First step, definitions. Public Administration Review 55(5):418-427
43. Sternberg CA (2013) From "cartoneros" to "recolectores urbanos": the changing rhetoric and urban waste management policies in neoliberal Buenos Aires. Geoforum 48:187-195
44. Thieme AT (2013) The 'hustle' among youth entrepreneurs in Mathare's informal waste economy. Journal of Eastern African Studies 7(3):389-412
45. UN-Habitat (2010) Collection of Municipal Solid Waste in Developing Countries. United Nations Centre for Human Settlements.
46. UN-Habitat (2011) Ghana Housing Profile. United Nations Human Settlements Programme. UNON, publishing services section, Nairobi. ISBN 978-92-1-132416-7
47. Veeravatnanond V, Nasa-Arn S, Nithimongkonchai W, Wongpho B, Phookung K (2012) Development of risk assurance criteria to the utilization of natural resources and environment for sustainable development of life quality, economy and society in rural Thai communities. Asian Soc Sci 8(2):189-195 February 2012
48. Wilson DC, Velis C, Cheeseman C (2006) Role of informal sector recycling in waste management in developing countries. Habitat Int'l 30(4):797-808
49. WRI (Water Research Institute) (2007) Assessment of the greater Accra water resources monitoring for sustainable management. Water Research Institute, Accra.
50. Zaaman AU, Lehman S (2011) Challenges and opportunities in transferring a city into a 'zero waste city'. Challenges 2(4):73-93

CHAPTER 5

Challenges to Achieving Sustainable Sanitation in Informal Settlements of Kigali, Rwanda

AIME TSINDA, PAMELA ABBOTT, STEVE PEDLEY,
KATRINA CHARLES, JANE ADOGO, KENAN OKURUT,
AND JONATHAN CHENOWETH

5.1 INTRODUCTION

In Kigali, Rwanda's capital city, like many other cities in developing countries, the most widely used sanitary facilities in the poor neighbourhoods are pit latrines, occasionally supplemented with flushing toilets and septic tanks [1]. Conventional pit latrines provide a cheap way to handle human waste and require little maintenance; however, they provide limited comfort, attract flies and spread diseases such as diarrhoea and dysentery through contamination of the environment [2].

Rapid population growth and urbanization associated with the proliferation of informal settlements are often accompanied by environmental degradation. About 62% of the urban population in sub-Saharan Africa lives in informal settlements [3,4]. The problem of informal settlements remains one of the greatest challenges for city managers. In sub-Saharan Africa, less than half of the population has access to a sanitation facility that meets the WHO/UNICEF Joint Monitoring Program (JMP) definition of improved [5,6].

In Kigali, the population is growing faster than the provision of services. In 1996, the population was 358,200, but by 2012, it had increased to 1,135,428 [7]. Much of the urban growth has taken place in unplanned settlements that now accommodate 62.6% of the population. The 2010 Demographic and Health Survey reported 88.7% of sanitation to be improved; although this number falls to 46.2% if the JMP definition, which excludes shared sanitation, is used [8]. However, this percentage does not point out the disparities in conditions within the formal and informal parts of the urban area.

Pit latrines in the informal settlements are often poorly maintained and rarely emptied; the pits are generally not lined with bricks and can collapse after a period of use [9]. Furthermore, there are few suction trucks available to empty soakage pits and septic tanks, and often sites are not accessible due to the narrow steep roads which lead to the latrines. Even if there is a possibility to empty liquid from pits, the sludge is not always disposed of in a proper manner. However, neglecting pit emptying or employing poor quality emptying services can have serious health and environment consequences. For example, substandard pit emptying services in Freetown, Sierra Leone, have been partly responsible for diarrhoeal disease, cholera outbreaks and high infant mortality, especially in informal settlements and poor unplanned areas [10,11].

This situation might be improved if Kigali was equipped with a sewerage system. However, unlike other cities in East Africa, which have networks of sewer pipes and treatment plants to cover a small percentage of its inhabitants, Kigali has neither a central treatment facility for sewage nor a system of sewers. As a result, the large volumes of wastewater produced in the city, are either discharged untreated into wetlands surrounding the city or allowed to infiltrate into ground water, polluting fresh water resources as well as the soil [1]. There are plans to build a sewage treatment plant, but funding remains a major challenge. Kigali lacks financial resources and therefore cannot afford a centralized sanitation system because of the high cost of the associated physical infrastructure which includes a network of pipes, treatment plants and maintenance [1,12].

Although it is known that well-managed systems for piped water, sanitation, drainage, and garbage removal would improve the health of city residents [13], introducing and maintaining centralised systems in developing cities have been hampered by political, economic, ecological and social instabilities. This leads to poor environmental performances and perpetual breakdowns, due to lack of proper maintenance or timely investments [2]. In addition, although Kigali is well-known for the cleanliness of its streets, little is known about the life in informal settlements.

This raises a pressing need to understand the nature and magnitude of the issues affecting sanitation provision in order to find more cost-efficient and sustainable sanitation alternatives to address them. Innovative decentralized sanitation and re-use systems were developed partly in opposition to centralized ones and there have been claims they are more robust, cheaper and better able to deal effectively with environmental challenges [1]. Whichever technologies are used, they must be context appropriate and cost effective to the low-income dwellers [14] of developing cities. However, to the best of our knowledge, no research has been undertaken on how current on-site problems can be solved by the use of other on-site or 'mixed' technologies that match with the context of informal settlements of Kigali.

It is against this background that this article aims to analyse the challenges faced by people using the existing sanitation systems in terms of wastewater treatment, operation and maintenance; and frame sustainable sanitation systems that match with the local conditions of informal settlements of Gatsata and Kimisagara. The findings will contribute to providing the basis for policy makers to make informed decisions on which sanitation systems fit for informal settlements of Kigali, and other major cities.

5.2 METHODS

The study was conducted in two informal settlements in Kigali which were purposely sampled because they have some of the poorest sanitation facilities. The settlements are also characterized by high levels of poverty, high rates of illiteracy, high unemployment, poor housing, and a lack of access to quality health care and transportation, and an unhealthy environment [15]. Other characteristics of these areas include poor drainage systems, poor sanitation facilities, the unauthorized building of extensions, and the high density of settlements with steep slopes and wetland areas.

A mixed method approach was used. This included transect walks through the settlements, a household survey, focus groups discussions and key informant interviews. The study team started an unannounced transect walk with an informal talk with a few community members and then continued the walk to observe the condition of sanitation facilities as well as any evidence of open defecation around the house and backyards.

The household survey, collected quantitative data on sanitation facilities and income using a structured questionnaire. The survey sample was selected through random route sampling techniques in proportion to the population of the study area. It was conducted between May and September 2012. The survey questionnaire was pilot tested before being administered in the communities, and all the staff involved with the survey were trained before use. Out of the 1,883 targeted household, 1,794 households (95%) were interviewed giving a non-response rate of 5%. The head of household or another adult (18 years and over) answered the questionnaire on behalf of the household.

The findings of the survey were supplemented by the qualitative research, undertaken to find out more about the issues and constraints to improving sustainable sanitation. Purposive sampling was used to select informants with the same sampling frame used in both settlements. Focus group discussions and in-depth interviews were used to capture the informants' perspective and allowed for more in-depth information on sanitation and helped us to better understand what was going on and why. This is important because in various fields of science, voices have been raised that research should be done with people and not on or for people [16,17]. Sanitation is an example for such a setting and thus in order to define sanitation technology options with a high chance of long-term success, a thorough understanding of the needs and concerns of residents (own occupiers, tenant, landlords) is essential.

Our qualitative work in Kimisagara and Gatsata yielded a rich data set. This article draws primarily on eight focus group discussions conducted with owner occupiers, eight focus group discussions with tenants (half female and half male) who were the head of households but excluded local leaders and resident landlords; four focus group discussions with landlords (two landlords-residents, two landlords-absent); two focus group discussions with community health workers; two focus group discussions with village leaders; two class discussions with primary six pupils; three individual interviews with people with disability; three individual interviews with old people (above the age of 65); one person with a chronic disease.

Survey data were coded and analysed using the SPSS software version 20. For the statistical analysis, the minimum level of significance accepted was 95% ($p < 0.05$). The data collected through focus groups discussions and in-depth interviews with key informants was transcribed and analysed thematically. In order to improve the validity of the data, a triangulation strategy was used. This strategy involved collecting information from a range of sources (household survey, transect walks, focus group discussions, interviews).This has the advantage of filling weaknesses or gaps in data for one method, which results in strengthening the overall quality of the results. Ethical approval was given by Ethics Committee of University of Surrey. Participation in this study was voluntary and all respondents gave verbal informed consent to their participation in the research.

5.3 RESULTS

5.3.1 SOCIO-ECONOMIC CHARACTERISTICS OF SURVEY RESPONDENTS

The age of respondents ranged from 18 years to 88 with a mean of 33.1 years and a SD of 11.7. The majority of the respondents were between the ages of 25 and 35 years. 62.8% of respondents were married with 23.1% being single and 14.2% being divorced/widowed. Men were more likely to be single (35.5%) than female (16.8%) while the women were more likely to be widowed (19.8%) than men (3.1%). Just over half of the respondents had completed primary school education (53.3%). Women were more likely to have had no education than men (11.7% compared to 6.4%) while men were likely to have higher education than women (3.2% compared to 1.4%).

TABLE 1: Distribution of improved sanitation systems (a) and shared usage according to JMP definition (b) in informal settlements of Gatsata and Kimisagara.

Section (a): Type of Sanitation Systems				Section (b): Sanitation Systems					
				Shared		Not Shared		Grand Total	
		N	%	N	%	N	%	N	%
Improved	Flush	48	2.7	19	1	29	1.6	48	2.6
	Pit latrine with slab	946	52.7	650	36.3	296	16.50	946	52.9
	Other improved categories	15	0.9	11	0.6	4	0.3	15	0.9
	Improved Sub-Total	1,009	56.3 **	680	37.9	329	18.4 *	1,009	56.3
Unim-proved	Open pit latrine without slab	690	38.5	558	31.2	132	7.4	690	38.6
	Other unimproved categories	90	5.0	74	4.1	16	0.9	90	5.0
	Open defecation	5	0.3	-	-	-	-	5	0.3
	Unimproved Sub-Total	785	43.7	632	35.3	148	8.3	785	43.6
	Grand Total of both Improved and Unimproved	1,794	100	1,312	73.2	477	26.7	1,794	100

*Notes: * The proportion improved sanitation (Shared systems excluded); ** the proportion improved sanitation (Shared systems not excluded).*

The level of unemployment was 22.5% while employment was reported to be 77.5%. Of those who said they were employed, the majority were self-employed (66.2%) including farm work, 20.6% were engaged in waged employment, while 13.2% of respondents were dependent family workers. Men were more likely to be in waged employment (25.5%) than women (9.3%) while men were significantly ($p < 0.005$) less likely to be unemployed (11.3%) than women (30.2%) (Cramer's $V < 0.005$). During the focus group discussions with males and females in the two selected settlements, it was consistently revealed that men were financially responsible for the family, although women may partake in small income-generating activities. In terms of household duties, men were mainly responsible for providing food, shelter, clothing, construction of latrine; whereas, women were mainly engaged in domestic work, childcare, raising the children, cooking, cleaning, and collecting water.

5.3.2 EXISTING SANITATION SYSTEMS IN STUDY AREAS OF KIGALI

In the two informal settlements in Kigali, pit latrine facilities (both with a slab and without a slab) were the most common sanitation option (91.2%), as shown in Table 1, Section (a). The table also shows the distribution of improved sanitation systems according to WHO/UNICEF Joint Monitoring Programme (JMP) definition, which excludes shared sanitation. Improved sanitation facilities are defined as the hygienic separation of human excreta from human contact, which includes a flush or pour flush toilet connected to either pipe sewerage, a septic tank or a pit latrine, a Ventilated Improved Pit latrine (VIP), a composting toilet, a pit latrine with a cover slab and other special case (e.g., urine diverting dry toilet) [5]. Thus, when shared sanitation is excluded from improved sanitation, the proportion of residents with improved sanitation in study areas is 18.4% (Table 1), Section (b).

Sharing facilities was common with, on average a latrine, being shared, between four households. At the extreme, in one instance, 15 households were sharing one latrine. Open defecation was not widely reported (0.3%) (Table 1), Section (a) despite being observed during the transect walks. The few households that were not using latrines gave a number of reasons, that the latrine was full; they did not have a latrine; the latrine had collapsed; latrine was under construction.

5.3.3 ISSUES ASSOCIATED WITH EXISTING SANITATION FACILITIES IN INFORMAL SETTLEMENTS OF KIGALI

The survey sought to establish whether respondents think there are problems associated with their sanitation facility. 80% of respondents reported at least one problem, with the most frequently mentioned problem being shared usage (58.5%) followed by smell (38.7%). Other problems frequently reported were difficulty to clean (38%), insect problems (32.4%), safety (30.1%), distance from the dwelling (21.2%), and toilet not always available when needed (20.2%) (Figure 1).

However, while some respondents experienced no problems, others said that they had more than one problem with a majority of respondents having between two and five problems (20.8%, 16.9%, 13.4%, and 8.5% respectively). A few households reported experiencing more than six problems (5.5%). During transect walks, we observed that most of the latrines were open pits that smelled bad because they were very shallow or full; a few had visible breeding areas for flies. The mud floors in traditional pit latrines had dirty floors, preventing the water draining away in a hygienic sanitary way and providing a favourable breeding ground for flies.

The participants in focus group discussions said that most of toilets were dirty and this caused health problems, such as intestinal worms, typhoid and diarrhoea, especially amongst children. One female tenant, for example, explained how diseases can be contracted from using unclean toilets:

> *"The ways of transmission are various. The lack of fresh water and soap in the house can hinder people from hand washing and uncovered pits or stagnant black water can attract flies. Effluent from tanks and pits can pollute surface and ground water used for human consumption with pathogens and pollutants."*

One reason for the unhygienic condition of the toilets was said to be because a large number of households share one latrine. For example, elderly, disabled and sick people reported unclean facilities to be a big challenge for them, especially when shared with many households. This was because they sometimes had to touch the floor or dirty walls to get support while using the facility particularly when squatting in position caused pain in their knees and back. The following quote illustrates their concerns:

> *"Sometimes when I stay too long in the squatting position, to be able to stand up, I find support by touching the ground, where I can easily contaminate with diseases. It is a very embarrassing and stressful experience for us old people"* (Woman with disability).

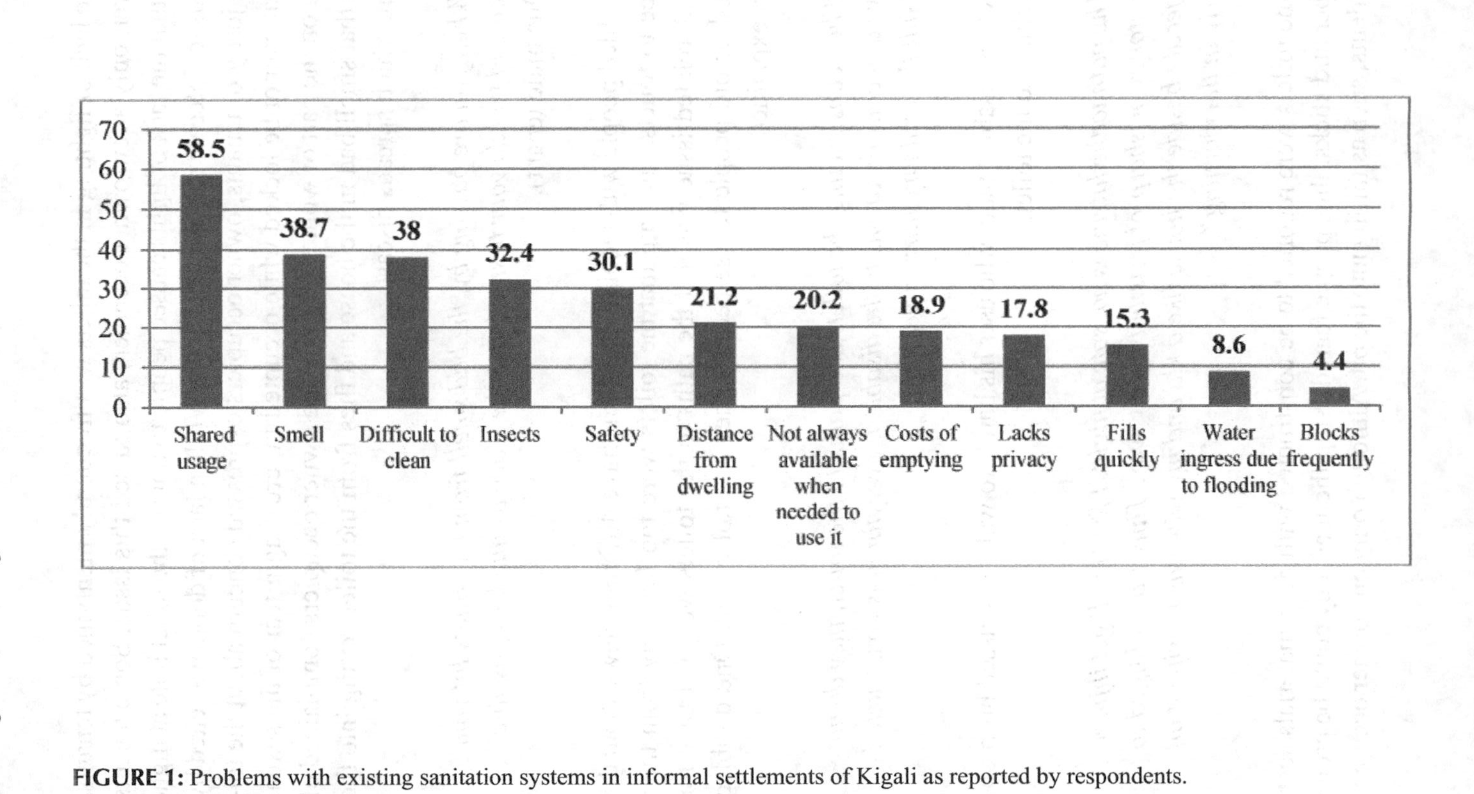

FIGURE 1: Problems with existing sanitation systems in informal settlements of Kigali as reported by respondents.

The lack of lighting in the toilets was frequently mentioned by tenants; however, only a few owner occupiers also raised this issue. Some tenants noted that the lack of lights caused people to not see the toilet hole at night and thus to defecate on top of the toilet. While a lack of doors was cited by the majority of tenants, owner occupiers expressed concern about the toilets that cannot be locked while occupied in, creating a fear of the sexual assault on the part of women. Tenants and owner occupiers complained of toilets that smell bad in the house and flies from the toilet getting into the kitchen, which spreads diseases:

> *"When you are passing by, the bad smell from the toilets hit your face and you do your best to leave the place as quick as possible"* (One male tenant).

Disabled people with mobility problems have different views about the distance to toilets, and said that the toilets were too far away from their house. A related issue was that the paths to the toilets were not easy for disabled or old people to use because they are not paved. One disabled woman explained:

> *"When constructing the toilets and paths to the toilet, they do not give any considerations to the disabled; everything is constructed to suit the ordinary normal person".*

Another disabled man, who uses his hands to walk, discussed his problems getting to the toilet:

> *"The path to the latrine is not safe at night. I fear I will get injured by stones or sharp objects left on my way. Thus I am obliged to defecate beside the house like a kid and my wife has to wipe it off early in the morning".*

Some toilets were reported to be constructed with poor materials such as timber and sticks; thus people feared using them as they could be prone to accidents, such as falling into the pit. Although only one owner occupier mentioned a fear of slipping and falling on the wet toilet floor, the issue of safety was frequently raised by tenants.

> *"I am frightened to use our toilets because I fear that the toilets will collapse at any time and I will fall down the hole and die; so I prefer other alternatives such as using a bucket, or practicing open defecation"* (Male tenant).

They further pointed out that some toilets lacked doors and this compromised privacy:

> *"My toilet is okay, the only problem is that it is not properly constructed and does not promote privacy. If I am inside, people outside can see me while using it"*, complained one female tenant.

Older and disabled people frequently mentioned difficulties they have using a toilet without a seat and explained that their knees and legs get tired easily. To deal with such a situation, they said that they made a kind of stool for them to sit on to go over and around the pit hole. They also said that they used buckets in their homes and their relatives throw out the waste later.

Tenants reported that the size of the holes in some pit latrines are too large for children, thus they defecate on the ground or besides the pit latrine's hole. They also complained about problems with open defecation practiced by people without toilets in their settlement. Similarly, owner occupiers reported the issue of open defecation when their pits fill up:

> *"People in settlements still practiced open defecation in open spaces between houses, in corners, on people's fences and this is mostly done by kids, drunkards, homeless youth or outsiders; the reasons of such malpractice was said to be due to local bars not having a toilet as well as lack of public toilets in the settlement, unavailability of toilet due to a large number of people sharing a facility, poor hygienic conditions of the facility, people without facilities or simply because of poor attitude of some people"* (Female owner occupier).

Although open defecation was frequently cited by tenants and owner occupiers, village leaders went further and said that people defecate in bags and throw the bags on the roofs. In addition, they reported that people empty the toilets into the drainage channels and solid waste dumps at night or in neighbours' toilets. However, observations made during the transect walks found that there was limited open defecation and only few cases were reported in the backyards of house.

The tenants also complained about the pits, which fill quickly. One female tenant stated:

> *"Pit latrines fill up quickly here in informal settlements as a result of the large number of users. Due to the diminishing space available in informal settlements, households must resort to emptying their pit latrine and as constructing a new latrine is impractical, but the issue is the fact we do not even have emptiers around here..."*

While agreeing to some extent with tenants about the pits that filled quickly, landlords provided a slightly different point of view. They stressed that in most cases; pits had shallow depth due to high water table or rocky soil, and blamed tenants for disposing of tins, glasses, plastics and garbage into the facility, thus causing the facility to fill up quickly.

5.3.4 CONSTRAINTS TO SUSTAINABLE SANITATION IN INFORMAL SETTLEMENTS OF KIGALI

Results from the survey showed that the most important constraints to building sanitation facilities were lack of money (68.2%), topography (13.8%), insufficient space (12.3%), difficult in obtaining permit (3%), lack of construction materials (2.4%), lack of specialized equipment (0.2%) and lack of information (0.2%). As far as toilet waste emptying and transport were concerned, only 2% of respondents reported emptying their pit latrines. Yet when a pit fills up, emptying is often the only sustainable option [1].

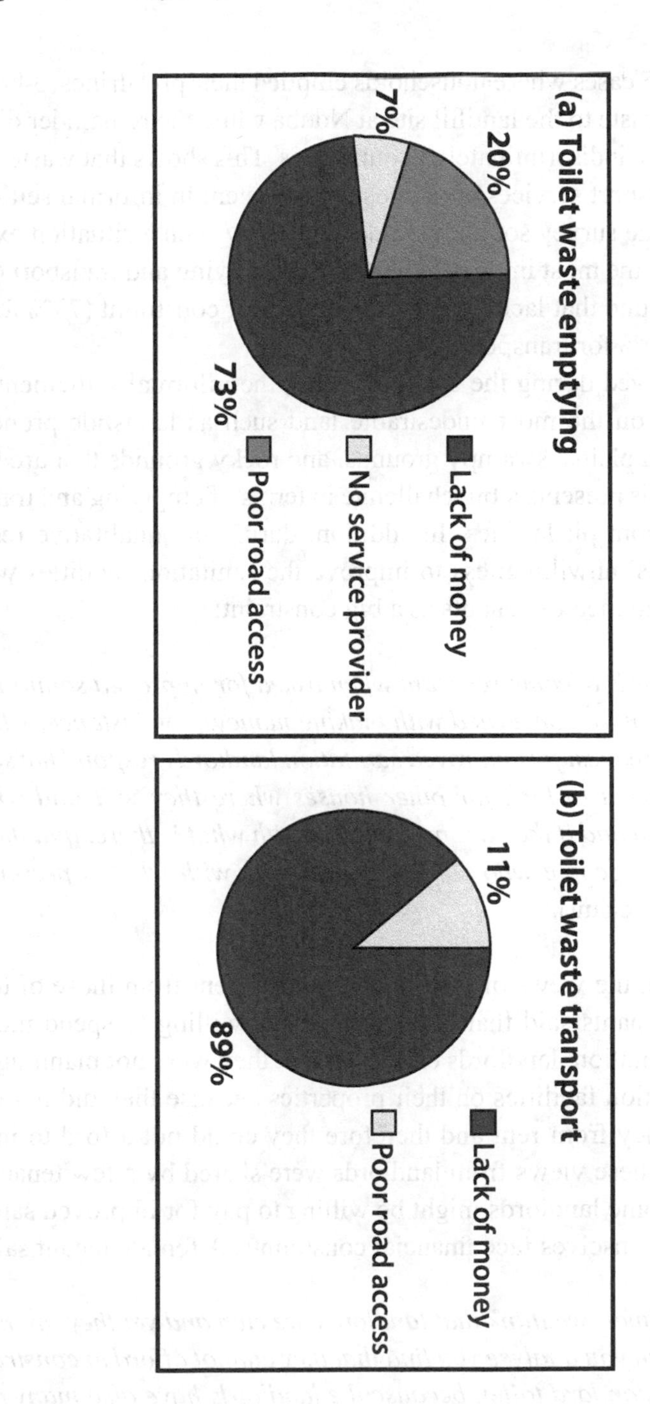

FIGURE 2: The most important constraints in toilet waste (a) emptying and (b) transport in informal settlements of Kigali.

Of the 25 cases where households emptied their pit latrines, 34% transported the waste to the landfill site at Nduba while the remainder disposed of their waste indiscriminately in dumpsites. This shows that waste emptying and transport services are almost non-existent in informal settlements of Kigali. The survey sought to understand why such a situation exists by ascertaining the most important barriers to emptying and transport (Figure 2). It was found that lack of money was a major constraint (73% for emptying and 89% for transport).

As observed during the transect walks, the informal settlements were constructed on the most undesirable land such as landslide-prone steep slopes, flood plains, swampy grounds, and rocky grounds that are hard to build on. This presents a big challenge in terms of emptying and transporting waste from pit latrines. In addition, during the qualitative research, the landlords' unwillingness to improve the sanitation facilities was frequently mentioned by tenants as a big constraint:

> *"Landlords become resistant when asked for improved sanitation and are more concerned with making money. For instance, when tenants request for improved sanitation landlords respond harshly telling them to look for other houses where they will find what they want and if they are not satisfied with what is there, that there are many people who can be comfortable with what is present"* (Female tenant).

However, the views of landlords were different from those of tenants. While the tenants said that landlords were unwilling to spend money to improve sanitation, landlords explained that they were not planning to upgrade sanitation facilities on their properties because they did not receive enough money from rent and therefore they could not afford to improve sanitation. These views from landlords were shared by a few tenants who noted that some landlords might be willing to pay for improved sanitation but might themselves face financial constraints. A female tenant said:

> *"Sometimes, we think that landlords are rich and yet they are not, and when you analyse you find that they cannot afford to construct a good standard toilet, because the landlords have also many de-*

pendents and more pressing needs and priorities than having a proper toilet".

Owner occupiers shared some of views of tenants and landlords and stated that sanitation is not their top priority because they had many other urgent issues to deal with:

"My toilet has no roof because I have no money; when I get money, another urgent problem occurs and then I prefer to solve that one first" (A female owner-occupier).

The participants in the qualitative research also raised the challenges in achieving sustainable sanitation in informal settlements of Kigali. For example, the landlords noted that when an old pit is full, they dig a new hole because emptying services are limited and expensive:

"The major challenge faced is the cost of constructing a toilet which is very expensive. Even if they pay us rent every month, it is too little to construct a toilet" (Male resident- landlord);

"Space is the biggest obstacle of all; it is impossible to construct more than three toilets on my land" (Female absentee-landlord).

Owner occupiers agreed that although the lack of money was one of reasons why emptying services were limited in informal settlements of Kigali, there were other constraints that needed to be taken into consideration:

"Although pit emptying services are too expensive (USD 150) and few can afford to pay this amount of money, they (emptying services) had been banned by the authorities because people had dumped the sewage into the drainage channels; yet those who afford to pay for emptying are limited by the roads which are too narrow in the settlement for pit emptying trucks to reach their toilets, so I think the facilities that are in close proximity to roads could be easily emptied" (A male owner-occupier).

Some key informants went further and pointed out that there was nowhere in Kigali for toilet emptiers to dump the sewage from the toilets:

> *"We really do not have sustainable mechanism to deal with sewage here because there is nowhere to dump the sewage; it is a big problem..."*(KCC official).

Responses for elderly, disabled and sick people showed that while flush toilets suited their needs, the lack of piped water to their compounds makes it impossible for them to have flush toilets:

> *"The facility that I think is suitable for older people is the flush toilet as it gives support but the challenge is there is no water access in the settlement and this would also increase water bills yet money is one of my biggest challenges in accessing a good toilet"* (Old woman).

5.4 DISCUSSION

The aim of this study was to analyse challenges to achieving sustainable sanitation in informal settlements in Kigali and frame appropriate, suitable and sustainable sanitation technologies that match with the local conditions of study settlements of Kigali.

5.4.1 CHALLENGES TO ACHIEVING SUSTAINABLE SANITATION IN INFORMAL SETTLEMENTS OF KIGALI

Regarding the challenges to achieving sustainable sanitation in informal settlements of Kigali, it was found that the high population density and the ensuing congestion of houses contribute to the lack of space for latrines. This leads to pits being dug close to houses, which weaken the foundations of already poorly constructed houses. The majority of the informal low-income communities were located in difficult terrains such as marshy land, swamp, steep slopes, abandoned refuse tips or grave yards, flood plains and rocky areas. Building latrines in these soil conditions can sometimes be challenging due to the instability of the soil, and the difficulty of digging into rocks.

Pit latrines with a slab represent improved sanitation in its most basic form, but once the pit is full it no longer provides this service; and the pit must either be covered over, and a new latrine constructed, or the existing pit emptied [18,19]. However, unlike other developing cities such as Kampala, Nairobi and Dar-es-Salaam, Kigali has no clear strategy for the emptying of pit latrines [1,15] and only 2% of households empty sludge from their pits. Therefore, there is a risk of the full latrine overflowing, contaminating the environment with large quantities of excreta containing harmful pathogens and causing offensive smells [20].

However, smell was not reported as a major problem in some areas of Africa [21,22]. From the findings of this study, a possible explanation for this disparity could be that, a significant proportion of pit latrines in the informal settlements of Kigali might be simple pit latrines without a cover and are thus technologically inferior at reducing smell when compared to VIPs.

Another problem that was noted in relation to full pits is the potential for the pollution of groundwater under or near pit latrines, particularly in areas with high water table [23], which is the case of most of informal settlements of Kigali. This is a serious problem because it affects the quality of drinking-water. For example, due to the high water table, Botswana experienced high groundwater pollution that can be linked to the widespread use of pit latrines as a sanitation option [22,24].

Pits in the informal settlements of Kigali were generally not lined with bricks and vulnerable to collapse. This puts children at risk and most households tend to discourage children from using latrines for fear that they might fall in. This was also reported in Kumasi, Ghana where children were made to defecate in plastic containers, which were later emptied into the latrine [25].

The other issues raised by respondents included closeness of the toilet to the households, accessibility during night, toilet rooms not having light inside, a toilet that is not lockable for privacy when in use, and the propagation of flies. However, although the quantitative results from the survey indicate low percentages for insect nuisance (32.4%), flies and insects are serious issues because they are reported to be responsible for the propagation of faecal-oral diseases, such as diarrhoea or intestinal worms [26,27], children are particularly known to be more at risk [28] because they are used to play in stagnant wastewater. Controlling smells, flies and mosquitoes is, therefore, a high priority for reducing household and environmental health hazards.

A number of factors have constrained progress towards sanitation improvement. In particular, lack of money was reported in survey to be the main reason. The inability to save funds to invest in longer-term sanitation facilities, coupled with a low income, significantly restricts the choices that individuals can make. The situation might be improved if financial support from local and national governments was available. However, unlike other developing cities, there are no Non-Government Organizations (NGOs) working on sanitation issues in the informal settlements of Kigali [15]. In addition, sanitation is not a budgetary priority for the Government [29] and the City of Kigali, which is supposed to support the poor, often lacks financing resources to meet the needs of the population for sanitation and other services [15]. The illegal nature of occupancy and the impact that is likely to have on decisions to invest in sanitation is another challenge.

Research in South Africa supports the argument that poverty is the biggest factor in preventing households from benefiting from improved sanitation [30]. However, there was no agreement amongst the participants of this study as to whether financial problems are a real or rather a perceived constraint that restricted poor urban dwellers of informal settlements from building latrines. Indeed, the results from the questionnaire survey indicated that there were many other factors that hinder the sanitation improvement in informal settlements of Kigali such as the lack of sufficient space to build individual toilets. This is supported by a study conducted in Kibera slum in Nairobi (Kenya) where it was found that it is not feasible to provide individual sanitation facilities in high-density slums with high poverty levels [6].

5.4.2 SUSTAINABLE SANITATION TECHNOLOGIES APPROPRIATE TO INFORMAL SETTLEMENTS OF KIGALI

In developing countries, unimproved sanitation facilities are the prime cause of widespread and serious health problems, but improvements in these services show few health benefits unless they are coupled to improved hygiene behaviour [18]. Overall, 18.4% of households in the informal settlements surveyed had access to some form of improved sanitation system if WHO/UNICEF Joint Monitoring Programme (JMP) definition is used, which excludes shared sanitation. However, this does not genuinely reflect the reality on the ground. This is because access to improved sanitation systems was assessed using physical measures [19] but in most cases this did not capture real access levels since even the flush toilets might not provide adequate sanitation services if they are poorly maintained.

Some respondents argued that the establishment of central sewerage systems can help in addressing some of the challenges and achieving sustainable sanitation in Kigali. However, such systems need high investment and most developing cities, including Kigali, lack the financial resources to pay for centralized sanitation systems [1]. Costly centralized sanitation systems are not only a problem for developing countries. Because of high maintenance cost and little profit returns, centralized or off-site water and sanitation systems have to be directly cross subsidized and the chances to ever become financially sustainable are low even in developed countries [1]. There are other problems caused by centralized sanitation systems associated with over exploitation of natural resources. To transport human waste, networks of sewer pipes consume enormous volumes of water, which is not available in informal settlements.

In comparison with central sewerage systems, decentralized sanitation systems have been promoted by scholars [1,2]. Decentralized sanitation technologies used worldwide include simple pit or traditional latrines, Ventilated Improved Latrines (VIP), Ecological (Eco-san) latrines, pour-flush latrines and Water closet toilets, connected to septic tank. In household survey, it was reported that the majority of respondents reported the use of traditional pit latrines. The preponderance of traditional pit latrines is in line with previous research findings, in which they are reported to be the preferred sanitation option compared to other more advanced technologies such as flush toilets [31] because they are cost-effective and when well designed, built and maintained, they provide adequate sanitary benefits.

However, the problem is that the pit is not lined and thus difficult to empty when they fill up [32]. The unlined pits pose a challenge using emptying trucks because of the excreta mixing up with soil or gravel particles from the pit walls. A common problem with unlined pit latrines is collapse, especially during the rainy season. Excavating new pits within an ever diminishing space due to the high density and close proximity of households is not practicable or sustainable. Other inconvenient drawbacks for traditional pits are soil and ground water contamination with pathogens, bad odours, flies/mosquito breeding, potential pit collapse in cases of heavy rains, the distance from house, especially for women and children during night [33].

On the other side, the flushing toilets connected to septic tanks which constitute the preferred system for most citizens because of their comfort, are also faced with problems. Septic tanks are expensive and therefore not affordable for the majority of the urban poor population and they cease to work properly when they are old and over-utilized, potentially causing serious environmental and public health problems [1]. However, there is no need to start from nothing, since small-scale household composting and other decentralized systems (DeSaR) are widely considered a potential solution for developing countries [2]. Unlike conventional sewers which require a large wastewater system and expensive treatment before there can be any reuse and redistribution, decentralised systems are cost-effective and enable a more efficient separation of liquid and solids for re-use by communities located near the site [1,2].

5.5 CONCLUSIONS

The aim of this study was to analyse challenges to achieving sustainable sanitation in informal settlements of Kigali and propose sustainable sanitation technologies that match with the requirements of study settlements of Kigali. The study used a mixed method approach and this included transect walks, a household survey, focus groups discussions and key informant interviews.

Our findings reveal that the traditional pit latrines are the most common types of excreta management systems that exist in Kigali. However, such systems are not a sustainable sanitation option because they are vulnerable to leakages, collapse during heavy rains and attract flies. In addition, these facilities fill up quickly due to small volumetric capacity for most pits, high number of users (because of sharing), and are not easily or regularly emptied. Also, as a result of the steady increase in the population of the slum dwellers coupled to the construction of unplanned structures, the space available for constructing new traditional pit latrines is continually decreasing [34].

To this end, this implies that dwellers of informal settlements are inclined over time to reject these traditional pit latrines for alternative low-cost more sustainable options, such as innovative decentralized sanitation and reuse (DeSaR) and water serving sanitation technologies, because they can play a part in reduction of over exploitation of natural water sources, which continue to be scarce, as a result of population pressure in the country. DeSaR technologies are appropriate in informal settlements because they occupy less space, do not require emptying by vacuum tankers, pre-treatment/composting, provides opportunity for nutrients re-cycling which is environmentally sustainable and, if well maintained, have minimal harmful effects [35]. However, to be able to provide improved sanitation options for these communities, pilot projects are necessary so as to gauge acceptability [34]. Meanwhile, since the majority of residents do still depend on shared sanitation facilities to reduce the sanitary-related diseases, more emphasis has to be placed on hygiene education practices, focusing on proper use and cleanliness of the facilities.

REFERENCES

1. Sano, J.C. Urban Environmental Infrastructure in Kigali City, Rwanda: Challenges and Opportunities for Modernised Decentralised Sanitation Systems in Poor Neighbourhoods. M. Sc. Thesis, Wageningen University, Wageningen, The Netherlands, August 2007.
2. Oosterveer, P.; Spaargaren, G. Meeting Social Challenges in Developing Sustainable Environmental Infrastructures in East African Cities. In Social Perspectives on the Sanitation Challenge; Springer: Berlin, Germany, 2010; pp. 11–30.
3. Shah, N. Characterizing Slums and Slum-Dwellers: Exploring Household-level Indonesian Data. 2012. Available online: http://storage.globalcitizen.net/data/topic/knowledge/uploads/20120920133613196030_slums_Shah.pdf (accessed on 24 August 2013).
4. Dinye, R.D.; Acheampong, E.O. Challenges of slum dwellers in Ghana: The case study of Ayigya, Kumasi. Mod. Soc. Sci. J. 2013, 2, 228–255.

5. WHO/UNICEF. Progress on Sanitation and Drinking-Water: 2010 Update; WHO Joint Monitoring Programe: Paris, France, 2010.
6. Schouten, M.; Mathenge, R. Communal sanitation alternatives for slums: A case study of Kibera, Kenya. Phys. Chem. Earth. 2010, 35, 815–822.
7. National Institute of Statistics for Rwanda. 2012 Population and Housing Census: Provisional Results. Republic of Rwanda National Institute of Statistics for Rwanda: Kigali, Rwanda, 2012.
8. National Institute of Statistics of Rwanda. Rwanda Demographic and Health Survey 2010: Final Report; National Institute of Statistics for Rwanda Republic of Rwanda: Kigali, Rwanda, 2010.
9. Hohne, A. State and Drivers of Change of Kigali's Sanitation—A Demand Perspective. Proceedings of the East Africa Practitioners Workshop on Pro Poor Urban Sanitation and Hygiene, Kigali, Rwanda, 2011; Available online: http://www.google.ca/url?sa=t&rct=j&q=&esrc=s&source=web&cd=1&ved=0CC0QFjAA&url=http%3A%2F%2Fwww.irc.nl%2Fcontent%2Fdownload%2F162378%2F590890%2Ffile%2F&ei=YjWgUotLYjs0gXQ9oG4Aw&usg=AFQjCNFHdLyZFqAHU0kaWajsyyA5A3oaAA&bvm=bv.57155469,d.d2k (accessed on 20August 2013).
10. Parkinson, J. Improving servicing of on-site sanitation-a neglected issue for the UN Year of Sanitation. Water 2008, 21, 40–42.
11. Lubaale, G.N.; Musyoki, S.M. Pro-poor Sanitation and Hygiene in East Africa: Turning Challenges to Opportunities. Proceedings of the East Africa Practitioners Workshop on Pro Poor Urban Sanitation and Hygiene, Kigali, Rwanda, 2011; Available online: http://www.irc.nl/page/64589 (accessed on 12 September 2011).
12. Scheinberg, A.; Spies, S.; Simpson, M.H.; Mol, A.P.J. Assessing urban recycling in low-and middle-income countries: Building on modernised mixtures. Habitat Int. 2011, 35, 188–198.
13. Satterthwaite, D. The Under-Estimation of Urban Poverty in Low and Middle-Income Nations; IIED: London, UK, 2004.
14. Tilley, E.; Morel, A.; Zurbru, C.; Schertenleib, R. Compendium of Sanitation Systems and Technologies; Swiss Federal Institute of Aquatic Science and Technology (Eawag): Geneva, Switzerland, 2008.
15. Tsinda, A.; Abbott, P. A Review and Analysis of the Situation Pertaining to the Provision of Sanitation to Low-Income Settlements in Kigali City (Rwanda); Diagnostic Report, Institute of Policy Analysis and Research. IPAR-Rwanda: Kigali, Rwanda, 2012.
16. Chevalier, J.M.; Buckles, D.J. Participatory Action Research. Theory and Methods for Engaged Inquiry; Routledge: London, UK, 2013.
17. Lüthi, C.; Panesar, A.; Schütze, T. Sustainable Sanitation in Cities: A Framework for Action, Sustainable Sanitation Alliance (SuSanA) & International Forum on Urbanism (IFoU); Papiroz Publishing House: Rijswijk, The Netherlands, 2011.
18. Grimason, A.M.; Davison, K.; Tembo, K.C.; Jabu, G.C.; Jackson, M.H. Problems associated with the use of pit latrines in Blantyre, Republic of Malawi. J. R. Soc. Promote. Health 2000, 120, 175–182.
19. Mtungila, J.; Chipofya, V. Issues and challenges of providing adequate sanitation to people living on the shore of lake Malawi: Case of Monkey Bay, Malawi. Desalination 2009, 248, 338–343.

20. Moe, C.L.; Rheingans, R.D. Global challenges in water, sanitation and health. J. Water Health 2006, 4, 41–57.
21. Saywell, D.; Shaw, R. On-plot Sanitation in Urban Areas; Water and Environmental Health at London and Loughborough (WELL): London, UK, 2005.
22. Bolaane, B.; Ikgopoleng, H. Towards improved sanitation: Constraints and opportunities in accessing waterborne sewerage in major villages of Botswana. Habitat Int. 2011, 35, 486–493.
23. Cross, P.; Morel, A. Pro-poor strategies for urban water supply and sanitation services delivery in Africa. Water Sci. Technol.: J. Int. Assoc. Water Pollut. Res. 2005, 51, 51–57.
24. UN-HABITAT. UN-Habitat Annual Report 2009; United Nations Human Settlements Programme: Nairobi, Kenya, 2010.
25. Adubofour, K.; Obiri-Danso, K.; Quansah, C. Sanitation survey of two urban slum Muslim communities in the Kumasi metropolis, Ghana. Environ. Urban. 2013, 25, 189–207.
26. Prüss, A.; Mariotti, S.P. Preventing trachoma through environmental sanitation: A review of the evidence base. Bull. World Health Organ. 2000, 78, 267–273.
27. Graczyk, T.K.; Knight, R.; Tamang, L. Mechanical transmission of human protozoan parasites by insects. Clin. Microbiol. Rev. 2005, 18, 128–132.
28. Thye, Y.P.; Templeton, M.R.; Ali, M. A critical review of technologies for pit latrine emptying in developing countries. Crit. Rev. Environ. Sci. Technol. 2011, 41, 1793–1819.
29. Tremolet, S.; Kolsky, P.; Perez, E. Financing On-Site Sanitation for the Poor: A Six Country Comparative Review and Analysis. Water and sanitation Programme. 2010. Available online: http://www.wsp.org/sites/wsp.org/files/publications/financing_analysis.pdf (accessed on 23 June 2012).
30. Chinyama, A.; Chipato, P.; Mangore, E. Sustainable sanitation systems for low income urban areas-A case of the city of Bulawayo, Zimbabwe. Phys. Chem. Earth. 2012, 50, 233–238.
31. Kulabako, R.N.; Nalubega, M.; Wozei, E.; Thunvik, R. Environmental health practices, constraints and possible interventions in peri-urban settlements in developing countries—A review of Kampala, Uganda. Int. J. Environ. Health Res. 2010, 20, 231–257.
32. Isunju, J.B.; Schwartz, K.; Schouten, M.A.; Johnson, W.P.; van Dijk, M.P. Socioeconomic aspects of improved sanitation in slums: A review. Public Health 2011, 125, 368–376.
33. Mara, D.; Alabaster, G. A new paradigm for low-cost urban water supplies and sanitation in developing countries. Water Policy 2008, 10, 119–129.
34. Thye, Y.P.; Templeton, M.R.; Ali, M. Pit Latrine Emptying: Technologies, Challenges and Solutions. London, UK, 2009. Available online: http://www.ewb-uk.org/system/files/Yoke+Thye+report.pdf (accessed on 23 August 2013).
35. Jha, P. Health and social benefits from improving community hygiene and sanitation: An Indian experience. Int. J. Environ. Health Res. 2003, 13, S133–S140.

PART II

SUSTAINABLE TRANSPORTATION

CHAPTER 6

Transport Infrastructure and the Environment: Sustainable Mobility and Urbanism

ROBERT CERVERO

6.1 INTRODUCTION

Urban areas, home to more than half of the world's population, face unprecedented transport and mobility challenges. With rapid population and economic growth, demands for urban mobility are steadily rising. Globally, some 8 billion trips are made every day in cities of which nearly half (47%) are by private motorized modes, almost all of which are propelled by fossil fuels (Pourbaix, 2011). In 2050, there may be 3 to 4 times as many passenger-kilometers travelled as half a century earlier, infrastructure and energy prices permitting (International Transportation Forum, 2011).

Concurrent to rapid rates of motorization, more sustainable forms of transport like public and non-motorized transport face mounting challenges, especially in developing countries. Public transport and non-motorized modes, despite being the chief way many poor people get around, are fast

Transport Infrastructure and the Environment: Sustainable Mobility and Urbanism. Cervero R. Paper prepared for the 2nd Planocosmo International Conference Bandung Institute of Technology October 2013. Reprinted with permission from the author.

losing customers to private cars in much of the world (Gakenheimer and Dimitriou, 2011). In 2005, walking and cycling accounted for only 37 percent and public transport 16 percent globally (Pourbaix, 2011). At the same time, informal modes of transport are proliferating to fill the gaps left by inadequate or non-existent public transport services.

The transportation sector is also inextricably linked to the climate-change challenge since it is currently responsible for 13 percent of Greenhouse Gas (GHG) emissions worldwide and 23 percent of total energy-related GHG emissions (UN Habitat, 2011). If recent trends hold, the sector's share of global GHG emissions could reach 40 percent by 2050 (International Energy Agency, 2011). Fueling this increase is the growing demand for urban mobility. In the hundredplus years of motor vehicles relying on gasoline as a fuel, the world has used approximately 1 trillion barrels of petroleum to move people, materials, and goods (Black, 2010). The transport sector's share of global oil demand grew from 33 percent in 1971 to 47 percent in 2002 and by one account could reach 54 percent by 2030 if past trends hold (IFP Energy Nouvelles, 2012). With increasing motorization and investments in roads and highways, cities find themselves in a vicious cycle—reliance on the private car unleashes more sprawl and road building further increases reliance on the private car.

It is widely accepted that cities of the future must become more sustainable, and that the transportation sector has a major role to play in this regard. The idea of a paradigm shift in urban transport is gaining currency in many parts of the world, not only to de-carbonize its fuel supply but also to create cleaner, economically viable, and socially just cities of the future. In particular, a shift towards the design of more compact cities based on the inter-mixing of land uses that prioritize sustainable forms of mobility such as public transport and non-motorized movement is broadly advocated. The post-oil city of tomorrow will need to be one that allows people to easily get around by foot, two-wheelers, buses, and trains. It is also recognized that urban transportation systems needs to be inclusive, providing mobility opportunities for all. In a car-dependent city, those without access to a private vehicle—often the poor, physically disabled, youth, elderly, or those forsaking car ownership out of choice—are unable to access opportunities and services. It will also be essential to enhance the pivotal role of transportation in the shaping the economic future of cities, in recognition that it is the transport sector that connects workers to jobs, raw materials to plants, produce and goods to markets, and people to retail shops and places of entertainment and recreation.

This paper discusses key challenges in advancing sustainable urban mobility in the 21st century, particularly in a developing cities context. Issues facing different modal options discussed, particularly with regard to public transport. Reforms needed to achieve sustainable urban mobility on multiple fronts—environmentally but also socially and economically—are also reviewed. The paper then shifts to a particularly important transport-infrastructure challenge: investment in BRT systems that not only enhance mobility but also promote more efficient, sustainable, and socially just urban forms. Experiences in three global cities are reviewed in this regard. The paper closes with commentary on the institutional challenges and social equity considerations of advancing the sustainable mobility agenda.

6.2 URBANIZATION AND MOTORIZATION

Since the mid-half of the last century, rapid urbanization has been accompanied by urban sprawl. Spread-out patterns of growth carry high costs. It not only increase automobile dependence but also consume farmland and open space, threaten estuaries and natural habitats, and burden municipal treasuries with the high costs of expanding urban infrastructure and services.

From 1995 to 2005, 85 percent of the 78 largest cities in the developed world experienced a faster growth in their suburban belts than their urban cores (UN Habitat, 2011). In Bangkok and Jakarta, 53 and 77 percent of urban growth by 2025, respectively, is expected to be in peri-urban regions (Angel, 2011). In Greater Cairo and Mexico City, sprawl is fueled mostly by informal housing settlements while on the outskirts of Mumbai and Delhi new towns and employment sub-centers have been the largest consumers of once exurban land. Sprawl in China is partly induced by local government policy wherein municipalities buy agriculture land at low prices, add infrastructure and services, and then lease to developers at much higher prices—effectively practicing value capture as a revenue generating tool.

Urbanization has both encouraged and been shaped by the growth in motorized movements in cities. The global count of motorized vehicles has been increasing at unprecedented rates. In 2010, there were nearly 1.2 billion passenger vehicles worldwide (UN Habitat, 2011; Wright and Fulton, 2005). Based on data from five years earlier, nearly half of all urban trips were by private motorized modes, a figure that continues to climb (Pourbaix, 2011). A key factor contributing to rising motorization in both developed and developing countries is the availability of fairly cheap oil, which has literally and figuratively fueled low-density development. In China, urban growth is occurring as far as 150km to 300km from the core of cities. A recent study of Shanghai residents who were relocated from the compact, mixed-use, highly walkable urban core to isolated residential towers on the periphery found dramatic shifts from non-motorized to motorized modes, accompanied by substantial increases in travel duration and vehiclekilometers- traveled (VKT) (Cervero and Day, 2008). Economic growth and rising incomes have also triggered motorization. From 2002 to 2007, China's per capita incomes almost doubled and car ownership nearly tripled. Societal values also play a role given that for many who join the ranks of the middle class in rapidly emerging economies, owning a car is a rite of passage.

Rapid motorization unavoidably shifts future travel from the most sustainable modes—public transport and non-motorized ones (walking and cycling)—to private vehicles. Daily trips in urban areas by private cars are projected to jump from 3.5 billion in 2005 to 6.2 billion in 2025, an 80 percent rise (Pourbaix, 2011). Much of this growth will be in developing countries. If past trends continue, petroleum consumption and greenhouse gas emissions are projected to increase by 30 percent, matched by a similar growth in traffic fatalities. While they provide tremendous mobility benefits to those who cannot afford a car, motorcycles, which are the dominant mode of transport in many Asian countries, come at a high cost. Besides congesting city streets, they can be exceedingly loud, contribute to traffic accidents, and when powered by two-stroke engines, spew dirty tailpipe emissions. A poorly tuned two-stroke engine, for example, can emit 10 times as many hydro-carbons and particulate matter as a four-stroke engine or private car (Badami, 1998; World Bank, 2002).

Motorization is also marked by environmental justice concerns given the growing international trade of old second-hand vehicles from high-income to low-income countries. Over 80 percent of the vehicle stock in

Peru was originally imported as used vehicles from the United States or Japan (Davis and Kahn, 2011). In many African countries, import liberalization policies from the 1990s made it easier and cheaper for households to buy second-hand vehicles shipped across the Mediterranean Sea from Europe.

6.3 MOBILITY AND MODALITY

Challenges faced by the two most resourceful forms of mobility—public transport and nonmotorized transport—are reviewed in this section. Being pro-transit, pro-walking, and procycling means not only enhancing the service quality of these options but also removing the many built-in subsidies and incentives that promote auto-mobility.

6.3.1 PUBLIC TRANSPORT

In 2005, 16 percent of the roughly 7.5 billion trips made in urban areas worldwide were by some form of public transport (i.e., formal, institutionally recognized services, such as local buses and rail transit) (Pourbaix, 2011). Public transport's mobility role varies widely, accounting for 45 percent of urban trips in Eastern Europe and Asia, 10 to 20 percent in much of Western Europe and Latin America, and less than 5 percent in North America and Sub-Saharan Africa (where informal services dominate the mass transit sector) (UITP, 2006).

In cities of the developing world, the mobility role of public transport also varies markedly, particularly among African cities. Only a handful of Sub-Saharan Africa cities, such as Addis Ababa, Abidjan, and Ouagadougou, have reasonably well-developed, institutionalized forms of local bus services that are of a high enough quality to capture 25 to 35 percent of motorized trips. In most other parts of Sub-Saharan Africa, private paratransit and informal operators dominate, with local buses serving but a small fraction of trips, if any. In Sub-Saharan Africa as well as poorer parts of South and Southeast Asia, government-sponsored transit is either inadequate or non-existent, mainly because governments are too cash-strapped and understaffed to mount and sustain effective and reliable mass transit services.

In Southeast Asia, conventional 50-passenger buses are the workhorse of the public transport networks of most cities. In Bangkok, 50 percent of passenger trips are by bus, rising to 75 percent during peak hours. In East Asia, buses serve slightly larger shares of mechanized trips than metrorail in Taipei (14.4 versus 12.9 percent) and Shanghai (12.9 percent versus 5.7 percent) whereas metrorail is more dominant in Hong Kong (35.5 percent of mechanized trips), Seoul (34.8 percent), and greater Tokyo (57 percent). Buses similarly predominate throughout Latin America, even in rail-served cities like São Paulo, Santiago, and Buenos Aires. When buses operate on exclusive dedicated lanes, they tend to gain even more popularity by mimicking the speed advantages of metros, however, usually at a fraction of the construction cost. As discussed later, the most extensive Bus Rapid Transit (BRT) networks are today found in Latin America.

In many parts of Asia, Africa, and Latin America, the informal transport sector serves the mobility needs of most people. The lack of affordable and accessible public transport systems in developing countries has led to the proliferation of informal operators, such as private microbus and minibus services. These modes help fill service gaps but can also worsen traffic congestion and air quality. In some settings, informal carriers are the only forms of mass transport available. In India, for example, only about 100 of the more than 5,000 cities and towns have formal public transport. Everything from hand-pushed rickshaws to private minibuses have stepped in to fill the gap.

6.3.2 NON-MOTORIZED TRANSPORT

Walking and bicycling are the healthiest, least intrusive, and most affordable forms of movement. In 2005, 37 percent of urban trips worldwide were made by foot or bicycle, the two predominant forms of non-motorized transport (NMT). In African cities, 30 to 35 percent of all trips are by walking but in some cities, like Dakar and Douala, the share is much higher, over 60 percent (Montgomery and Roberts, 2008). In general, the poorer and smaller the city, the more important NMT becomes, capturing as many as 90 percent of total person trips. In densely packed urban cores,

NMT provides access to places that motorized modes cannot reach and are often the fastest means of getting around. Among South Asia's densest, most congested cities, more than half of all passenger and goods trips are by foot, bicycles, and rickshaw.

Walking is often the only form of transport for the very poor. Many people from the developing world are "captive walkers", meaning that they cannot afford an alternative. For them, having a well-connected and safe pedestrian environment is critical to meeting their daily needs. As the least expensive form of mobility, walking allows the very poor to allocate income for other purposes, thus helping to reduce poverty. It also promotes physical fitness, provides feeder access to bus and rail stops, and enhances security by providing "eyes on the street".

Cycling's mobility role contrasts sharply among the world cities. In general, the lower the per capita income, the bigger the mobility role played by bicycles however when high-quality cycling infrastructure is provided, bicycles can be a prevalent mode in even well-to-do cities. Today, bicycles are used for more than 40 percent of trips in some Dutch and Danish cities. Historically bicycles have also played a prominent mobility role in Chinese cities but today their use is in rapid decline, partly due to motorization but also government policies. In Beijing, for example, it is still illegal to park bicycles in front of many modern office buildings yet cars can be parked nearby. Bicycle lanes have been taken away in cities like Guangzhou and Shenzhen to make way for motorists. Shanghai and Nanjing officials recently announced the goal of cutting bicycle trips in half.

In some of the poorest cities of the world, bicycles serve as "mass transport", in the form of rickshaws. Cycle rickshaws are found all over Bangladesh, India, Pakistan, and Sri Lanka. They are particularly important modes for women and children. In Dhaka, around 40 percent of school trips are by rickshaw (Jain, 2011). Rickshaw pulling is often the first job for many rural migrants in cities of South Asia. In Dhaka, 20 percent of the population, or 2.5 million people, rely on rickshaw pulling for their livelihood, directly or indirectly (Jain, 2011). Still, the vehicles are being banned for slowing motorized traffic and a belief that they detract from the city's image as a modern metropolis.

6.4 IMMOBILITY: TRAFFIC CONGESTION

Traffic congestion is an unwanted by-product of widespread, or what some might call "excess", mobility in cities around the world. A recent study in 20 cities across six continents revealed that traffic congestion levels markedly worsened during the 2007-2010 period (IBM, 2010). Moscow motorists reported the worst commute, with an average daily delay of two and a half hours. With a 24 percent annual growth rate in registered vehicles, traffic conditions are deteriorating most rapidly in Beijing according to 95 percent of surveyed residents.

Congestion has widespread impacts on urban quality of life, consumption of fossil fuels, air pollution and economic growth and prosperity. World Bank (1994) studies from the 1990s estimated that traffic congestion lowered GDP of cities in the range of 3 to 6 percent, with the higher value applying mostly to rapidly growing cities (e.g., places with busy port traffic, reliance on just-in-time inventorying and manufacturing, and other time-sensitive activities). Time losses from traffic congestion are estimated to comprise 2 percent of GDP in Europe and 2 to 5 percent in Asia. The hidden external costs of traffic congestion in Metro Manila, Dakar, and Abidjan have been pegged at nearly 5 percent of those cities' GDPs (Chin, 2011). Such costs not only exact a burden on the present generation but also commit future generations to long-term debts, which can eventually slow global growth.

Limited road capacity in the face of growing demand for motorized mobility partly explains deteriorating traffic conditions. The nature of the problem, however, varies markedly across the globe. Less than 10 percent of land area is devoted to roads in many developing country cities (e.g., Kolkata, Jakarta, Nairobi) (Vasconcellos,1999). This contrasts with 15 to 20 percent in many rapidly emerging economies (e.g., Seoul, São Paulo), 20 to 25 percent in much of continental Europe (e.g., London, Paris), and 35 percent or more in America's largest automobile-oriented cities (e.g., Houston, Atlanta) (Vasconcellos, 2001). In India, the annual growth rate in traffic during the 1990s was around 5 percent in Mumbai, 7 percent in Chennai, and 10 percent in Delhi. However none of these cities have expanded their road supply by even one percent annually (Pucher et al., 2005).

In the developing world, buses are most vulnerable to the speed-eroding effects of traffic congestion. Because many are long, lumbering vehicles with slow acceleration and deceleration, restricted turning radii, and limited maneuverability to switch lanes, buses move the slowest in highly congested conditions. Average peak-period bus speeds in Bangkok are 11 km/hr, for example, compared to 20 km/hr in Curitiba, Brazil, one of the first cities to provide exclusive bus-lanes (Cervero, 2000). Stop-and-go traffic causes buses to over-heat and break down. Unreliable services in turn chase away choice consumers who have the option of driving a car instead.

6.5 TOWARD SUSTAINABLE TRANSPORT

It is increasingly recognized that sustainability in the urban transportation realm must be pursued and achieved on multiple fronts—environmentally, socially, and economically. This section addresses these challenges.

6.5.1 ENVIRONMENTAL SUSTAINABILITY

The urban transport sector's ecological footprint is enormous and expanding. Many environmental problems in the urban transport sector are rooted in its reliance on petroleum, the automotive fuel source of choice, to propel motor vehicles, increasingly ones that are privately owned and used. The share of the world's oil consumption accounted for by transportation rose from 45.2 percent in 1973 to 61.7 percent in 2009, and the sector is expected to continue to drive the growth in oil demand (IEA, 2011). World reserves of conventional oil exceed what has been used to date, but with rapid motorization and thus increasing demands for oil, many observe believe it is unlikely that this energy source will last beyond the mid-century mark. Rising GHG emissions and global temperatures as well as levels of photochemical smog and particulates in urban air basins further underscore the urgency of weaning the sector from its dependency on oil and more generally auto-mobility. A combination of technological advances, demand management, and externality-based pricing will be critical in charting an environmentally sustainable future in the urban transport sector. On the technological front, clean-fuel vehicles and information systems that enable innovations like dynamic ridesharing and carsharing, will have pivotal roles to play. Reducing the demand for indiscriminant automobility, such as by designing compact, mixed-use cities that shorten trips and encourage NMT, will also be important. Setting price signals so that polluters and those driving in rush hours internalize costs are similarly part of the environmental sustainability equation.

Environmental sustainability will depend on good economics (e.g., congestion pricing) but also the presence of the other two pillars of sustainability—institutional capacities and social equality. Setting maximum air and noise pollution standards will be useless unless there is the political will and regulatory resources in place to enforce them. Nor will the premature introduction of costly low-carbon fuel alternatives aid the poor if bus fares increase as a consequence.

6.5.2 SOCIAL SUSTAINABILITY

Urban transport is socially sustainable when mobility benefits are equally and fairly distributed, with few if any inequalities in access to transport infrastructure and services based on income, social, and physical differences (including gender, ethnicity, age, or disabilities). Social sustainability is rooted in the principle of accessibility wherein equality exists among groups in accessing opportunities for employment, housing, retail markets, and other essential urban services. It recognizes mobility and accessibility as human rights, not privileges. Cities that ensure accessibility for all are socially inclusive and ones that do not are socially exclusive.

One important aspect of accessibility is the affordability of transport modes. By affordability is meant the financial capacity to pay for the ability to reach destinations for everyday needs, such as work, education, and shopping, without undue economic hardships. For many urban dwellers in developing countries, the availability of reliable and affordable bus and rail services can be the difference between being integrated into the economic and social life of a city or not. The share of marginalized city-dwellers with poor access to essential facilities and services, including public transport but also clean water and sanitation, is increasing worldwide. In the poor informal housing settlements on the outskirts of Mexico City, beyond the service jurisdiction of the city's 201-km metro, residents sometimes must take 2 to 3 separate collectivos (shared-ride taxis and microbuses) to reach a metro terminal which provides low-cost connections to the core city and job opportunities (Cervero, 1998). Travel can consume 25 percent or more of daily wages. Time costs can also be exorbitant: 20 percent of workers in Mexico City spend more than 3 hours traveling to and from work each day. Studies show that taking a series of informal minibuses and motorized tricycles to and from work can cost 20 to 25 percent of daily wages in rapidly growing cities like Delhi, Buenos Aires, and Manila and as high as 30 percent in Nairobi, Pretoria and Dar Es Salaam (Vasconcellos, 2001; Kaltheier, 2002; Ferrarazzo and Arauz, 2000; Carruthers, et al., 2000).

6.5.3 ECONOMIC SUSTAINABILITY

The urban transport sector is economically sustainable when resources are efficiently used and distributed to maximize the benefits and minimize the external costs of mobility, and investments in and maintenance of transport infrastructure and assets can be sustained. The translation of investments in walkways, bikeways, transitways, and roadways into jobs, business expansion, and increased economic output means that the urban transport sector is on an economically sustainable pathway. Increasingly, the litmus test of cost-effective transport infrastructure is whether the project is "bankable"—capable of attracting loans and private investors. Urban transport infrastructure is expensive. It can consume a large share of the public largesse in emerging economies. In Ho Chi Minh City, a US$5 billion subway is currently under construction and in Jakarta a new ring road is expected to cost about the same amount. Crafting reliable and equitable funding programs for transport infrastructure that reward efficient and sustainable behavior remains a formidable challenge.

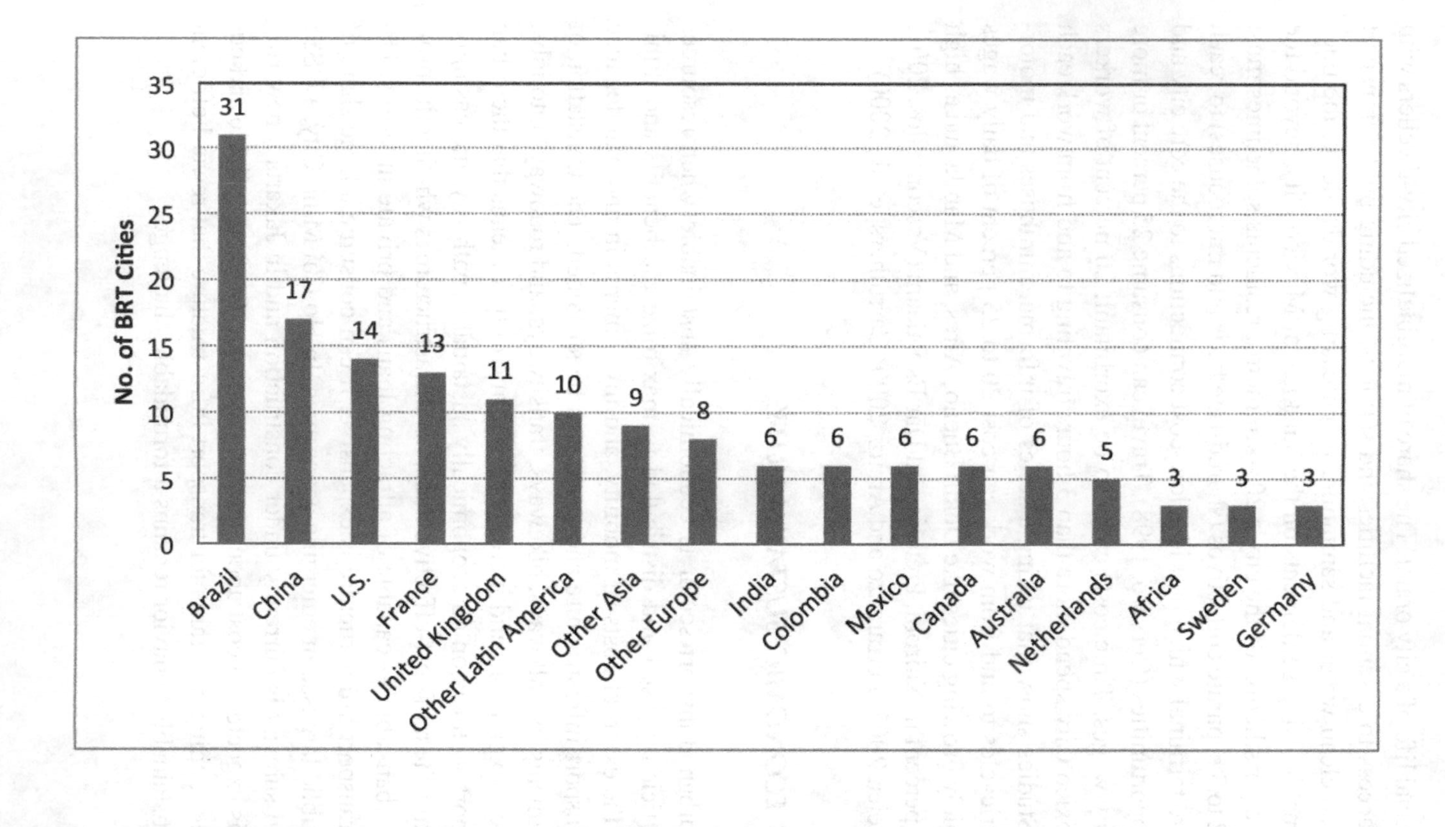

FIGURE 1: Number of Cities with BRT Systems, by National and Regional Settings, 2013. Source: BRTDATA.ORG.

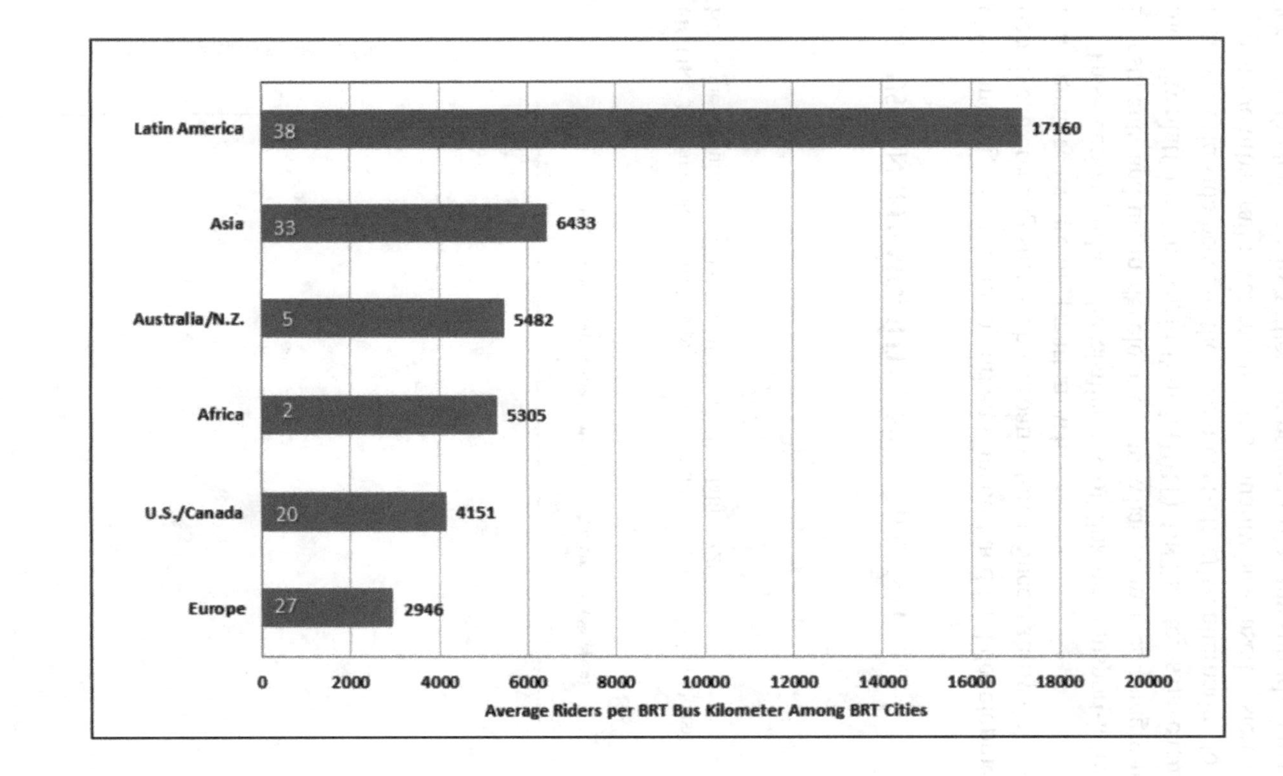

FIGURE 2: Average Weekday Riders per BRT Kilometer Among BRT Cities, by Continent-Region. Numbers in bars denote number of BRT cities in region that are included in the analysis. Source: BRTDATA.ORG.

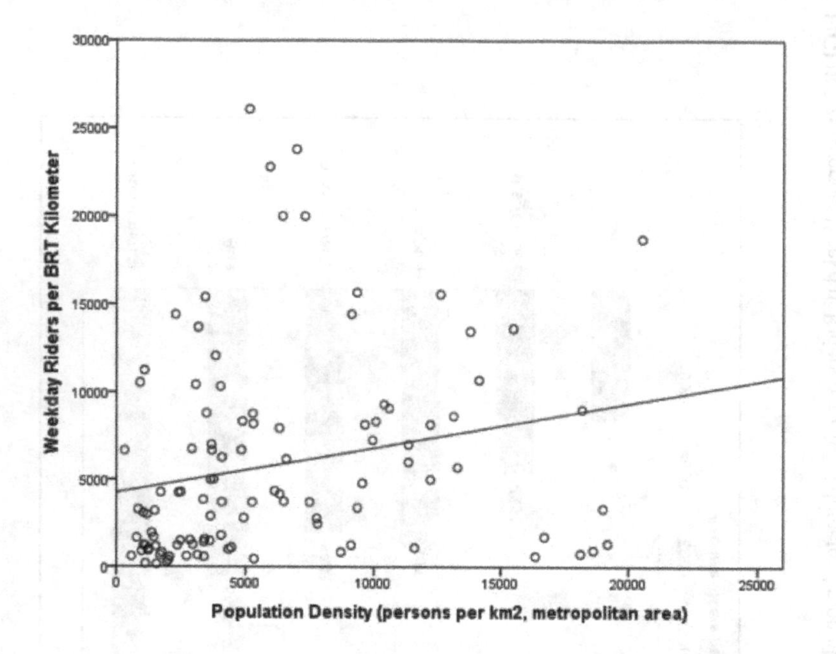

FIGURE 3: Scatterplot of Riders per BRT Kilometer and Population Density Among 105 BRT Cities. Sources: BRTDATA.ORG and UN Habitat (2012).

6.6 SUSTAINABLE MOBILITY AND URBANISM

Coordinating and integrating urban transport and land development is imperative to creating sustainable urban futures. Successfully linking the two is a signature feature of "smart growth".

This section probes the challenges of linking transport-infrastructure investments and urban development in what is an increasingly important mobility platform: Bus Rapid Transit (BRT). BRT systems have gained popularity worldwide because they are a cost-effective alternative to far more expensive urban rail investments. High-quality bus-based systems also better serve the low-density settlement patterns of many suburban markets and small-to-medium size cities due to the inherent flexibility advantages of rubber-tire systems—the same vehicle that provides speedy line-haul services on a dedicated bus lane or busway can morph into a feeder vehicle, collecting and distributing customers on local streets. To date, more than 150 cities have implemented some form of BRT system worldwide, carrying around estimated 28 million passengers each weekday (BRTDATA.ORG).

6.6.1 BUS RAPID TRANSIT AND URBAN DEVELOPMENT

New kilometers of BRT lines are today being added at a rapid-fire pace, gaining particular favor in the developing world, following on the heels of widely publicized BRT successes in Curitiba, Bogotá, Mexico City, Istanbul, Ahmedabad, and Guangzhou. These developing cities show that high-performance BRT systems that yield appreciable mobility and environmental benefits can be built at an affordable price. Metrorail systems, studies show, can cost 10 times as much a BRT system of similar length (Suzuki et al., 2013). Light Rail Transit (LRT) can be more than four times as expensive. Besides cost-savings, highly congested mega-cities of the world, like Jakarta, Delhi, Sao Paulo, and Lagos have been drawn to BRT because high-capacity transit can be built and expanded quickly during periods of rapid motorization and ever-worsening traffic congestion. The ability to open segments before an entire system is in place is particularly attractive to politicians and taxpayers who want quick results. Politicians are also drawn to the economic development potential of BRT. In its Liveanomics series, the Economist Intelligence Unit (2011) found that 61 percent of surveyed mayors reported that "improving public transport/roads" was the most important thing that could done to make their city more competitive for business on the global stage. This was nearly twice the share that felt investing in schooling and education was the key to being economically competitive.

BRT will no doubt play an increasingly prominent role in the global campaign to achieve more sustainable urban and mobility futures. This is partly because the bulk of future population growth will be in intermediate-size cities, the very places where BRT is often more cost-effective than its pricier alternative, metrorail transit (UN Habitat, 2011). Future growth of not only population but also economic outputs is also projected for intermediate-size cities (Glaeser and Joshi-Ghani, 2012).

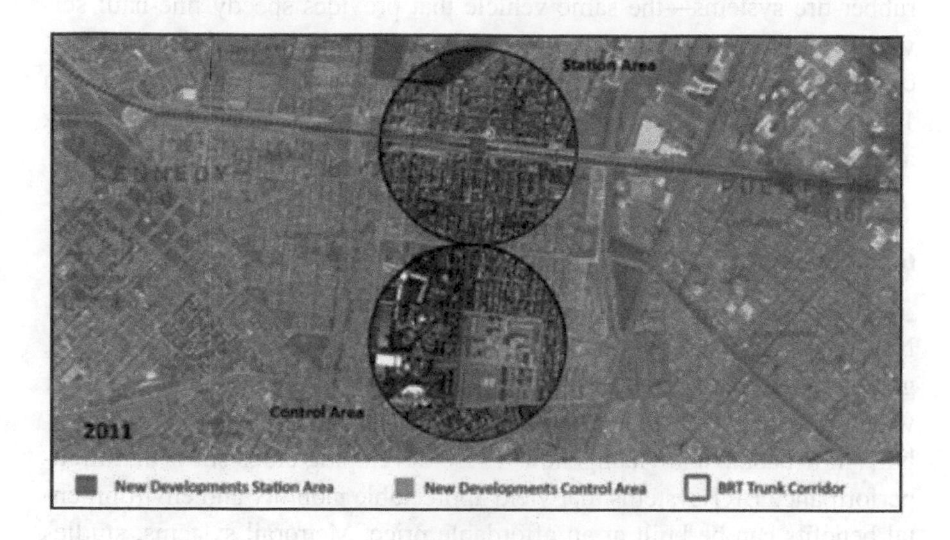

FIGURE 4: Footprints of new developments in Station Area and Control Area for an Intermediate Station, 1998 to 2011. Source: Suzuki, Cervero, and Iuchi, 2013.

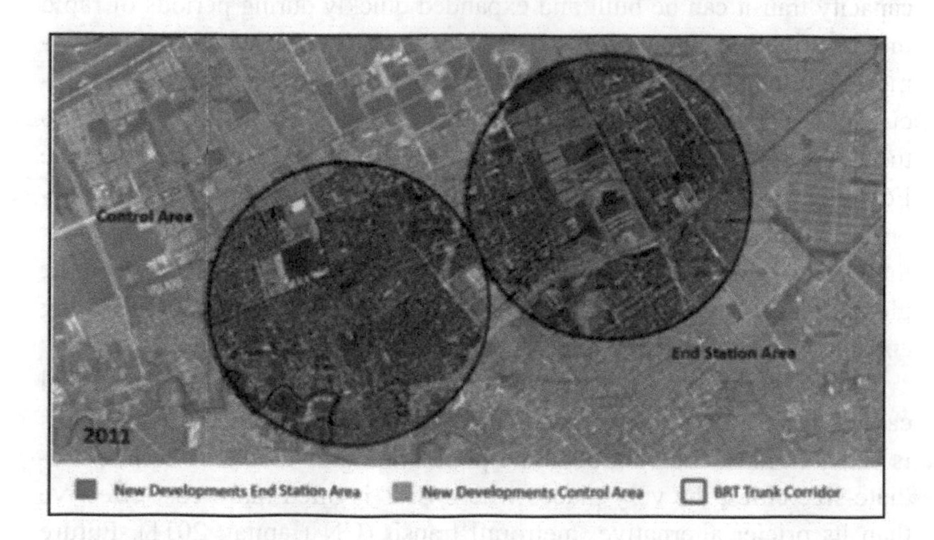

FIGURE 5: Footprints of new developments in Station Area and Control Area for an End-ofthe-Line Station, 1998 to 2011. Source: Suzuki, Cervero, and Iuchi, 2013.

Figure 1 rank-orders countries or regions based on the number of cities with BRT systems as of mid-2013. The vast majority of these systems have been built in the last 15 years. Brazil has emerged as the global leader in building BRT systems, extending the success of Curitiba's pioneering system to 30 other cities. Other Latin American countries, notably Colombia and Mexico but also Chile, Peru, and Ecuador, have since followed Brazil's lead. Latin America is today the epicenter of the global BRT movement. A third of BRT route kilometers and nearly two thirds (63%) of ridership are in Latin America (BRTDATA.ORG, 2013). Among 38 Latin American BRT cities with reliable data from BRTDATA.ORG, average weekday ridership is more 10 times greater than averages for BRT cities of the U.S. and Europe. Latin American BRT systems are also considerably more productive than systems elsewhere. Figure 2 shows that they averaged more than 2 ½ times as many weekday riders per BRT kilometer as Asian It is widely accepted that for public transit systems to be successful, they must be accompanied by high densities (Pushkarev and Zupan, 1977; Cervero, 1998; Newman and Kenworthy, 1999). Mass transit, as the saying goes, needs "mass". For 105 BRT cities for which reliable data could be obtained, Figure 3 suggests a moderately positive relationship between BRT ridership and urban density. The presence of outliers weakens the simple correlation (.225) and as the scatterplot reveals, the number of riders per BRT kilometer tends to vary more as urban densities increase. Regardless, the positive association between urban densities and ridership productivity argues in favor of BRTOD—Bus Rapid Transit-Oriented Development.

The challenges of leveraging TOD with BRT investment is probed in the next three subsections. The struggles faced by two of the world's most extensive and highly regarded BRT cities—Bogotá and Ahmedabad—are contrasted with what remains the world's best-case example of BRT-land-use integration—Curitiba, Brazil.

6.6.2 THE CHALLENGES OF LEVERAGING TOD IN BOGOTБ

Bogotá, the capital of Colombia and home to 7.6 million inhabitants, has gained a reputation as one of the world's most progressive cities, underscored by the 2000 opening of what has been called the gold standard of BRT, the 110-km TransMilenio system. Delegations of officials and dignitaries from around the world visit Bogotá to marvel at the system. Operating on a two-lane dedicated carriageway, TransMilenio carries upwards of 40,000 passengers per hour per direction, which matches the passenger-throughputs of most metros. The system also boasts enhanced stations (accessible by networks of skyways), smart card-based fare collection, advanced control systems, distinctive images, and affordable fares. TransMilenio's patronage is growing at a healthy pace of around 10 percent annually, from 800,000 daily riders when it opened in 2001 to around 1.7 million today, accounting for 74 percent of public transit trips in the city. Finance policy has played a role in TransMilenio's success. In 2000, a 20 percent surcharge was tacked onto all gasoline sales in Bogotá, with half the revenues earmarked for TransMilenio infrastructure. As a cross-subsidy from the 19 percent of Bogotá's population that owned cars to transit-dependents, the policy promoted social as well as environmental sustainability.

While Bogotá's TransMilenio is a substantial, widely celebrated BRT investment, able to carry some 45,000 passengers per direction per hour, reshaping urban form and land-use patterns was not a primary objective in its design. Building the system quickly and enhancing affordable transport for the poor was. Placement of BRT lines in mostly economically stagnant zones that were largely built out has suppressed land development. So has the siting of BRT stations in busy roadway medians, which limits joint development opportunities and creates unattractive pedestrian environs around stations. Minimal pro-active station-area planning and a dearth of incentives for private property-owners to redevelop parcels have also tempered TOD activities.

Since TransMilenio's 2000 opening, Bogotá's population has grown by 21 percent. Building densities have increased throughout the city, but mostly in areas away from TransMilenio corridors. The initial TransMilenio lines were built quickly in response to worsening traffic congestion but also to build political momentum and curry political favor for future expansions. Aligning corridors in mostly economically stagnant zones that were largely built out has suppressed land development. So has the siting BRT in busy roadway medians, which limited land supplies for leveraging TOD and resulted in mostly unattractive pedestrian environment immediate to stations.

Minimal pro-active station area planning or incentives for private property-owners to redevelop parcels also tempered TOD activities.

Cadastral data obtained from the city of Bogotá reveals the degree to which urban growth turned its back on TransMilineo. Between 2004 and 2010, the mean floor-area ratio (FAR) of residential and commercial development increased by 7 percent throughout the city of Bogotá versus 5 percent within 1000 meters of stations along the initial 42-kilometer system (Suzuki, Cervero, and Iuchi, 2013). In fact, more densification occurred along surface bus routes that feed into suburban TransMilineo stations than around BRT stops. Matched pair comparisons of changes in building footprints between 1998 and 2011 for 1-km radii around BRT stations and otherwise similar control areas further revealed weak effects on urban growth. For all but endof- line stations, more new construction occurred beyond than within 1000 meters of stations.

Figure 4 shows one paired comparison for an intermediate station on a Phase II line toward the southwest of the city, near the low-income neighborhood of Kennedy. Far less new development occurred within 1000 meters of the BRT station than the control area off the line. For terminal stations, however, there tended to be relatively more new building activities than in control areas, as revealed by one of the matched-pair comparisons shown in Figure 5, for the Americas terminal station. Other researchers have similarly found more land-use densification near TransMilenio's terminal stations than control areas (Bocharejo, Portilla, and Perez, 2013). This higher degree of station-area activities was largely due to the commercial opportunities at terminals, representing busy transfer points between feeder buses and trunkline BRT services.

Findings from Bogotá square with earlier assessments of transit investments and urban development (Knight and Trygg, 1977; Cervero and Seskin, 1995; Cervero and Landis, 1997), namely that transit cannot overcome weak local real estate markets. Station siting also matters. Placing stops in the medians of active roadways inevitably means a poor-quality pedestrianaccess environment and thus little commercial development near the stations themselves. TransMilenio's design gave little weight to the pedestrian experience. The visually prominent skywalks that connect to BRT stops create lengthy, circuitous walks, can be noisy (resonating like steel drums during peak traffic conditions, by some accounts), and are difficult for the elderly, disabled, and semi-ambulatory individuals to negotiate. Bogotá's experiences further show that planning matters. Neither the city nor neighborhood districts (where detailed land use planning is regulated and implemented) prepared station-area plans to orchestrate private development, change zoning (including increasing permissible densities), introduce complementary improvements (like streetscape enhancements) to entice private investments, or take any other pro-active steps to leverage new development.

The one area for which local leaders win kudos has been the bundling of transit investments and the provision of affordable social housing for the poor. In 1999, at the time Bogotá's successful Transmilenio BRT system was being built, an innovative land-banking/poverty-alleviation program, called Metrovivienda, was launched (Cervero, 2005). Under Metrovivienda, transportation and housing are treated as bundled goods. The city acquires plots when they are in open agricultural uses at relatively cheap prices and proceeds to plat and title the land and provide public utilities, roads and open space. Property is sold to developers at higher prices to help cover infrastructure costs with the proviso that average prices be kept under US$8,500 per unit and are affordable to families with incomes of US$200 per month.

To date, four Metrovivienda sites have been created near one of Transmilenio's terminuses, each between 100 and 120 hectares in size and housing some 8,000 families. At build out, the program aims to construct 440,000 new housing units. Putting housing near stations helps the city's poor by "killing two birds with one stone"—i.e., providing improved housing and public transport services. Those moving from peripheral illegal settlements into transit-served Metrovivienda projects enjoy both "sites and serviced" housing and material improvements in access to major economic centers in the city. It is estimated that job-accessibility levels via transit within one-hour travel times increased by a factor of three for those moving from illegal housing to legal Metrovivienda projects (Cervero, 2005).

An important aspect of the program is the acquisition of land well in advance of BRT services. Because Metrovivienda officials serve on the Board of Transmilenio, they are aware of strategic plans and timelines for extending BRT. This has enabled the organization to acquire land before prices are inflated by the arrival of Transmilenio. Acquiring land in advance has enabled Metrovivienda to keep prices affordable for households relocated from peripheral "clandestine" housing projects. Transmilenio also makes commuting more affordable. When living in the hillsides, most residents used two different public transit services (a feeder and a mainline), paying on average US$1.40 a day to leave and return home (Cervero, 2005). With Transmilenio, feeder buses are free, resulting in an average of US$0.80 in daily travel costs. Metrovivienda serves as a model of multi-sectoral and accessibility-based planning in a developing country. By coupling affordable housing with affordable transport, Bogotá leaders have improved access to jobs, shops, and services while reducing the joint costs of what often consumes two-thirds of the poor's income: housing and transport. Whether Metrovivienda makes a serious dent in the city's housing shortages and traffic woes remains to be seen, however most observers agree that it is a significant and positive step forward.

6.6.3 THE CHALLENGES OF LEVERAGE TOD IN AHMEDABAD

In the 2009, Ahmedabad opened India's first and what today remains the country's largest BRT network. Called Janmarg ("People's Way), the current 45 km system was built to relieve mounting traffic congestion in India's fifth largest city. With some 5.5 million inhabitants, Ahmedabad is today one of the world's fastest growing cities (Forbes, 2010). The ingredients are thus there for BRT to shape future urban growth: rapid growth and motorization coupled with worsening traffic congestion. To date, however, few notable changes have occurred near Janmarg stations.

As in Bogotá, Janmarg was envisaged and design as a mobility investment, not a city-shaping one. Janmarg lines were and are being selected to serve the city's fastest growing areas, more so than in the case of Bogotá, however little attention has been given to the physical integration of BRT stops with surrounding neighborhoods or increasing the share of future populations and workers near BRT. Janmarg, slated to span some 220 kilometers at build-out, which would make it one of the most extensive BRT systems anywhere, was designed mainly to keep costs low. Little thought was given to urban development possibilities. So far, no land-use or TOD plans have been prepared for any Janmarg stations. What land development is occurring has been left solely to private market forces.

So far, Ahmedabad officials have opted to maintain uniform densities throughout the city, regardless of how close parcels might be to transit corridors. This has been done to disperse trips and thus decongest the city. It has also been done for socio-cultural reasons, namely to avoid creating a privileged class of land owners whose new-found wealth is create through government fiat. However keeping densities uniform also shifts growth to the periphery, in a more autooriented configuration. In the near term, the city may experience less traffic congestion due to density caps however over the long term, the resulting auto-oriented urban form could backfire, creating more traffic congestion and air pollution for the region as a whole.

Several design shortcomings also need to be overcome if Ahmedabad is to spawn TOD. Janmarg was and is being designed as a closed system, requiring users to access stations sited in the medians of roadways by foot, bicycle, car, two-wheeler, three-wheelers, or surface-street buses. Little attention, however, has been given to perpendicular connectors to BRT stops. No secondary feeder systems provide safe and efficient pedestrian, bikeway, and transit connections to mainline services. While a substantial network of cycletracks was built in conjunction Janmarg, for the most part bike-paths run parallel rather than perpendicular to the busway, thus functioning more as competitive than complementary systems. Moreover, there is no bicycle parking at stations. What few pedestrian-ways exist near Janmarg stops are often occupied by motorcycles and fast-moving three-wheel vehicles.

6.6.4 BRT AND URBANISM IN CURITIBA

A counterpoint to failures in coordinating BRT and urban development is the well-chronicled experiences of Curitiba, Brazil. Guided by a cogent long-term vision of the future city, the municipal government mandated that all medium- and large-scale urban development be sited along a BRT corridor. Orchestrating regional growth has been the Institute for Research and Urban Planning (IPPUC), an independent entity charged with ensuring integration of all elements of urban growth.

A design element used to enhance transit accessibility in Curitiba is the "trinary"—three parallel roadways with compatible land uses and building heights that taper with distance from the BRT corridor. The first two floors of the busway, which do not count against permissible plot ratios (building height/land area), are slated for retail uses. Above the second floor, buildings must be set back at least five meters from the property line, to allow sun to cast on the transitway. The inclusion of upper-level housing entitles property owners to density bonuses, which has led to vertical mixing of uses within buildings. An important benefit of mixed land uses and transit service levels along these corridors, in addition to extraordinarily high ridership rates, has been balanced bidirectional flows, ensuring efficient use of bus capacity. The higher densities produced by the trinary design have translated directly into higher ridership. Concentrated commercial development has also channeled trips from residences beyond BRT terminuses to the trinary corridors. In 2009, for example, 78.4 percent of trips boarding at the terminus of Curitiba's north-south trinary corridor were destined to a bus stop on the same corridor (Duarte and Ultramari 2012). Today, Curitiba's share of motorized trips by transit (45 percent) is the highest in Latin America (Santos, 2011). High transit use has appreciably shrunk the city's environmental footprint. Curitiba's annual congestion cost per capita of $0.67 (in US$2008) is a fraction of São Paulo's (Suzuki et al., 2010). The city also boasts the cleanest air of any Brazilian city with more than 1 million inhabitants, despite having a sizable industrial sector. The strong, workable nexus that exists between Curitiba's bus-based transit system and its mixed-use linear settlement pattern deserves most of the credit.

Sustained political commitment has been pivotal to Curitiba's success. The harmonization of transit and land use took place over 40 years of political continuity, marked by a progression of forward-looking, like-minded mayors who built on the work of their predecessors. A wellarticulated long-term vision and the presence of a politically insulated regional planning organization, the IPUCC, to implement the vision have been crucial in allowing the city to chart a sustainable urban pathway.

One area where Curitiba's BRT investment has fallen short is the provision of housing for the poor. Most social housing built in the last 40 years for Curitiba's poor has been far from main transit axes and transport corridors (Duarte and Ultramari, 2012). The availability of cheaper land

and laxer environmental regulations on floodplain development prompted Curitiba's authorities to put the most disadvantaged households in the least transit-accessible locations.

6.7 CLOSE

The best ideas for advancing sustainable urbanism and mobility will go nowhere unless there is the political will and institutional capacity to embrace and move forward with them. The ability to manage and respond to escalating demands for urban travel is often limited in developing cities. Institutional shortcomings—such as an insufficiently trained and educated civil-service talent pool or absence of a transparent and corruption-free procurement process for providing transport infrastructure—abound. Limited experience with urban management, budgeting and accounting, urban planning, finance, and project supervision have thwarted Indonesia's decentralization of infrastructure programs from the central to local governments over the past decade.

Sustainable mobility futures will depend upon a re-ordering of priorities, a paradigm shift if you will, that promotes inherently resourceful forms of mobility, frames investments in more holistic (and less mobility-focused) terms, and importantly seizes opportunities to integrate transport infrastructure and urban development when and where they avail themselves. As more and more growth shifts to cities of the Global South, opportunities for linking land development and transport infrastructure should not be squandered. Given that a large share of future urban growth is projected for small-to-medium size cities, bus-based forms of smaller scale transitoriented development interlaced by high-quality infrastructure for pedestrians and cyclists holds promise in many global settings. Many developing cities have the kinds of pre-requisites needed if BRT investments are to trigger meaningful land-use changes, including rapid growth, rising real incomes, and increased motorization and congestion levels. This, of course, assumes there is supportive planning and zoning, public-sector leveraging and risk-sharing, a commitment to travel demand management to remove many built-in incentives to car use, and the capacity to manage the land-use shifts that are put into motion by transportation infrastructure investments.

While integrated transport and land development can relieve congestion, cleanse the air, and conserve energy, its potential to reduce what remains the gravest problem facing the Global South—extreme and persistent poverty—is every bit if not more important. All that is done in the developing world must pass the litmus test of helping to alleviate poverty. Designing cities and transport systems to enhance accessibility and affordability is pro-poor. So are initiatives that strengthen non-motorized and public transport, keep fares affordable, and protect vulnerable populations from the hazards of motorized travel. Mass transit needs to be pro-poor across the board. In many developing countries, this means investing in busways over metros to keep fares affordable and targeting affordable housing to transit-served corridors. In Brazil, transit is kept affordable via national legislation, called Vale Transport, that requires employers to provide bus passes for commuting expenses that exceed 6 percent of workers' earnings. In Cairo and Bogotá, tens of thousands of low-income households have been relocated to more transitaccessible sites.

Being pro-poor also means designing high-quality and safe walking and cycling environments. Mixed land-use patterns and walking/cycling friendly environments allow the very poor to allocate income for other urgent purposes and thus helps reduce poverty. In the very poorest cities, small interventions—e.g. siting basic services such as schools, health centres, markets, and water standpipes to reduce travel distances—can make a big difference in the amount of time and energy devoted to transport. The time freed up allows women to achieve gainful employment and children to attend schools. What are cardinal features of integrated and sustainable transport and urbanism everywhere -- accessible urban activities and safe, attractive walking and cycling environs—are particularly vital to the welfare of the neediest members of the world's poorest countries.

REFERENCES

1. Angel, S., 2011. Making room for a planet of cities. Cambridge, Massachusetts: Lincoln Institute of Land Policy.
2. Badami, M., 1998. Improving air quality in Delhi. Habitat Debate 4(2), pp. 22-23.
3. Black, W., 2010. Sustainable transportation: problems and solutions. New York: Guilford Press.

4. Bocharejo, J., Portilla, I., and Perez, M., 2013. Impacts of TransMilenio on density, land use, and land value in Bogota. Research in Transportation Economics, 40(1), pp. 78-86.
5. BRTDATA.ORG. Accessed July 24-25, 2013.
6. Carruthers, R., Dick, M. and Saurkar, A. 2005. Affordability of public transport in developing countries. Washington: World Bank, World Bank Transport Paper, TP-3.
7. Cervero, R. 1998. The transit metropolis: a global inquiry. Washington, D.C.: Island Press.
8. Cervero, R. 2000. Informal transport in the developing world. Nairobi: UN Habitat .
9. Cervero, R. 2005. Progressive transport and the poor: Bogotá's bold steps forward. Access, 27, pp. 24-30.
10. Cervero, R. and Landis, J., 1997. Twenty years of BART: land use and development impacts, Transportation Research A 31(4), pp. 309–333.
11. Cervero, R. and Seskin, S., 1995. An evaluation of the relationship between transit and urban form. Washington: Transit Cooperative Research Program, National Research Council, Research Results Digest 7.
12. Cervero, R. and Day, J. 2008. Suburbanization and transit-oriented development in China, Transport Policy, 15, pp. 315-323.
13. Chin, H.C. 2011. Sustainable urban mobility in South-Eastern Asia and the Pacific. Nairobi: UN Habitat. http://www.unhabitat.org/grhs/2013
14. Davis, L. and Kahn, M. 2011. Cash for clunkers? The environmental impact of Mexico's demand for used vehicles. Access, 38, pp. 15-21.
15. Duarte, F. and Ultramari, C. 2012. Making public transport and housing match: accomplishments and failures of Curitiba's BRT. Journal of Urban Planning and Development 138(2), pp. 183-194.
16. Economist Intelligence Unit. 2011. Lievanomics: urban liveability and economic growth. London: The Economist.
17. Ferrarazzo, A. and Arauz, M. 2000. Pobreza y transporte, consultación con grupos de foco en Buenos Aires, Informe Final. Washington: World Bank, WB UTS Review, Santiago Conference.
18. Forbes. 2010. In pictures: the next decade's fastest-growing cities. http://www.forbes.com/2010/10/07/cities-china-chicago-opinions-columnists-joelkotkin_ slide4.html.
19. Gakenheimer, R. and Dimitriou, H. 2011. Introduction. In: Urban Transport in the Developing World: A Handbook of Policy and Practice, H. Dimitriou and R. Gakenheimer, eds. Cheltenham UK: Edward Elgar, pp. 3-7.
20. Glaeser, E. and Joshi-Ghani, A. 2012. Overview, Rethinking cities. Washington: World Bank.
21. IBM (International Business Machines) Corporation, 2010. The globalization of traffic congestion: IBM 2010 commuter pain survey. Armonk, New York: IBM Corporation.
22. IEA (International Energy Agency), 2011. Key world energy statistics. Brussels. IEA. http://www.iea.org/textbase/nppdf/free/2011/key_world_energy_stats.pdf
23. IFG Energies Nouvelles, 2012. IFG Energies Nouvelles activity report 2012. Lyon: IFG Energies Nouvelles.

24. International Transport Forum, 2011. Transport outlook: meeting the needs of 9 billion people. Paris: Organization for Economic Development/International Transport Forum.
25. Jain, A.K., 2011. Sustainable urban mobility in Southern Asia. Nairobi: UN Habitat. http://www.unhabitat.org/grhs/2013
26. Kaltheier, R., 2002. Urban transport and poverty in developing countries: analysis and options for transport policy and planning. Eschborn, Germany: Deutsche Gesellschaft für Technische Zusammenarbeit (GTZ) GmbH.
27. Knight, R. and Trygg, L. 1977. Evidence of land use impacts of rapid transit systems, Transportation, 6(3), pp. 231-247.
28. Montgomery, B. and Roberts, P. 2008. Walk urban: demand, constraints and measurement of the urban pedestrian environment. Washington: World Bank, Transport Papers, TP-18.
29. Newman, P. and Kenworthy, J. 1999. Sustainability and cities: overcoming automobile dependence. Washington: Island Press.
30. Pourbaix, J. 2011. Towards a smart future for cities: urban transport scenarios for 2025. Public Transport International, 60(3), pp. 8-10.
31. Pucher, J., Korattyswaropam, N., Mittal, N. and Ittyerah, N. 2005. Urban transport crisis in India. Transport Policy 12: 185-198.
32. Pushkarev, B. and Zupan, J. 1977. Public transportation and land use policy. Bloomington: Indiana University Press.
33. Santos, E. 2011. Pioneer in BRT and urban planning. Saarbrücken, Germany: Lambert Academic Press.
34. Suzuki, H., Cervero, R. and Iuchi, K. 2013. Transforming cities with transit: transit and land-use integration for sustainable urban development. Washington: World Bank.
35. UITP (International Association of Public Transport), 2006. Mobility in cities database, CD Rom. Brussels: UITP.
36. UN Habitat, 2011. Global report on human settlements 2011: cities and climate change. Nairobi: UN Habitat.
37. UN Habitat. 2012. Database on urban densities among global cities. Nairobi: UN Habitat. Proprietary data provided to the author.
38. Vasconcellos, E. (1999) 'Urban transportation and traffic policies: The challenge of coexistence in developing countries', Transportation Quarterly 54(1): 91-101
39. Vasconcellos, E., 2001. Urban transport, environment and equity: The case for developing countries. London: Earthscan.
40. World Bank, 1994. World development report. Washington: World Bank.
41. World Bank. 2002. Cities on the move: a World Bank transport strategy review. Washington: World Bank.
42. Wright, L. and Fulton, l, 2005. Climate change mitigation and transport in developing nations', Transport Reviews, 25(6): 691-717.

CHAPTER 7

Personalized Routing for Multitudes in Smart Cities

MANLIO DE DOMENICO, ANTONIO LIMA,
MARTA C. GONZÁLEZ, AND ALEX ARENAS

7.1 INTRODUCTION

Rapid development of wireless communication and mobile computing technologies call new research that explores the responses of urban systems to the flow of instant information. Thus, the analysis of spatial signals becomes an increasingly important research theme.

The required four steps to model trips consist of calculating trip generation, trip distribution, modal split and route assignments. The sources to inform these steps traditionally have come from travel diaries and census data [1]. However, the presence of new information and communication technologies (ICT) provide big data sources that are allowing novel research and applications related to human mobility. Recent studies have advanced the knowledge on trip generation by studying the number of different locations visited by individuals through mobile phones and quantifying their frequent return to previously visited locations. These have demonstrated that the majority of travels occur between a limited number of places, with less frequent trips to new places outside an individual radius [2], [3]. In the domain of trip distributions, new models have helped us to predict number of commuting trips when lacking data for calibration [4].

An important topic is to explore route assignments in the context of smart multimodal systems [5], [6], where individual daily trips follow recommendations based on personal and global constraints. This is of special interest towards efficient cities, where individuals could be automatically routed reducing the probability of traffic congestion and at the same time reducing the environmental impact. From the individual's point of view, for instance, one might want to choose a trip which minimizes the amount of traffic along the route, or to avoid routes across areas with high criminality level, or to favorite routes across more touristic areas, etc. On the other hand, the choices of certain routes at individual level, without accounting for the state of the system, often leads to traffic congestion [7], [8] which, in turn, is responsible for increasing pollution while decreasing the quality of the environment, with evident impact on the community.

In this work we model the trips in an urban system as interacting particles with data-driven origin-destination pairs that can be routed in their trips. Their route choices are based in a time-varying potential energy landscape that seeks to satisfy individual's and community's requirements simultaneously. Main streams methods for distributed routing seek to avoid congestion by global travel time reduction based on optimization methods [7], [9]. More recently, adaptive path optimization on networks (London underground network and global airport network) related the problem to physics of interacting polymers [10]. In this work we go one step forward in that direction and use a framework based on potential energy landscapes to integrate diverse layers of constraints to favor certain routes and to study the effects of the level of adoption of the proposed recommendations. In this work our main focus is to explore a new framework of analysis to study routing strategies for urban mobility, while the road network constrains are left to further studies.

7.2 DATA-DRIVEN ROUTING OF HUMAN MOBILITY

We consider a geographic area of interest (e.g., a city, a district, etc.) and we discretize it into a grid $\mathcal{G}$ with size $L \times L'$. In the following, for sake of simplicity, we will consider squared grids with size L.

We model individuals moving within the grid as a complex system of interacting sentient particles whose goal is to move between two geographic points according to certain criteria. Each criterion is encoded by a matrix C, with the same dimension of the grid, where each entry indicates the state of the corresponding cell in $\mathcal{G}$. In the same spirit of physical models of an electromagnetic surface, we use the convention that $C_{ij}>0$ indicates a repelling cell, i.e., a geographic area that should be avoided. Similarly, $C_{ij}<0$ indicates an attracting cell, i.e., a geographic area that should be involved for routing. Areas where $C_{ij}=0$ are considered as neutral.

The origin of a constraint can be of different nature. In fact, there are constraints at individual level, i.e., the ones corresponding to requirements of the single user (e.g., avoid areas with high criminality level), and at global level, i.e., the ones corresponding to the requirement of the whole community (e.g., keep minimum the pollution level). Moreover, there are static (or quasi-static) constraints corresponding to restrictions that do not change over time or change over large temporal scales, and dynamic constraints corresponding to rapid changes within the system itself, like the traffic flow or the weather. On one hand we should account for individuals' goals and requirements, while on the other hand it is crucial to satisfy constraints imposed for the wealth of the community.

In the following, we consider the set of all constraints, static and dynamic at individual and collective level, and we assign to each of them a time-varying matrix $C^{(\alpha)}(t)$, where $\alpha=1,2,\ldots,M$ and M is the total number of constraints. In the case of static constraints, the matrix is considered constant over time. Moreover, the entries of each matrix are rescaled to the range $0 \leq C^{(\alpha)}_{ij} \leq 1$, for all values of i, j and α to assign a relative importance to each constraint and to settle on a common scale. Finally, the total constraint matrix is defined by the linear combination of such constraints at each time step:

$$C(t) = \sum_{\alpha=1}^{M} w_\alpha(t) C^\alpha(t), \quad \sum_{\alpha=1}^{M} w_\alpha(t) = 1 \tag{1}$$

where the coefficients $w_\alpha(t)$ are empirical and define a trade-off between individual's and global constraints. It is worth remarking that these coefficient might vary over time because, depending on the circumstances (special events, incidents, etc.), it could be necessary to change their value to satisfy different priorities.

We define another matrix, $\mathbf{D}_\ell(t)$ ($\ell=1,2,\ldots,N(t)$), encoding the starting and destination cells of each individual in the system, where the starting point is considered to be a repelling or neutral area and the destination point is an attractor. The number of individuals N(t) is allowed to change over time. The matrix $\mathbf{D}_\ell(t)$ might change over time because, in principle, the individual might change destination during his or her travel, and for simplicity we assume that $-1 \leq D_{ij} \leq 0$ for each individual. It is worth remarking that attracting cells are in general associated to destinations and should be encoded in the set of matrices $D_\ell(t)$, whereas repelling cells are associated to constraints and should be encoded only in the set of matrices $C^\alpha(t)$.

We interpret the set of matrices $\mathbf{C}(t)$ and $\mathbf{D}_\ell(t)$ as potential energy landscapes and the routing of individuals is performed by means of a gradient descent, where each user moves along geodesics while reducing his or her potential energy until he or she reaches the destination. For simplicity, we assume no dependence on time for matrices $\mathbf{D}_\ell(t)$. We consider the case of a gravitational field in two dimensions permeating the areas encoded by $\mathbf{D}_\ell(t)$. More specifically, let (i_ℓ, j_ℓ) and $(i_\ell^{(d)}, j_\ell^{(d)})$ denote the cells of the underlying grid and destination point of the journey of individual ℓ, respectively, and let the following formula indicate their distance:

$$r = \sqrt{\left(i_\ell^{(d)} - i_\ell\right)^2 + \left(j_\ell^{(d)} - j_\ell\right)^2}$$

The potential energy landscape is defined by

$$D_\ell(r) = \begin{cases} -\dfrac{\Omega}{\sqrt{r}} & (i_\ell, j_\ell) \neq \left(i_\ell^{(d)}, j_\ell^{(d)}\right) \\ -\dfrac{\Omega}{\sqrt{1-10^{-4}}} & (i_\ell, j_\ell) = \left(i_\ell^{(d)}, j_\ell^{(d)}\right) \end{cases} \tag{2}$$

where Ω is a constant factor, defining the scale of the potential which should guarantee that the potential is strong enough in each cell. In our simulations, we considered $\Omega = 30L\sqrt{2}$.

The choice of the value of the potential at the destination is somehow arbitrary and, as a rule of thumb, it should be a number smaller than the potential of the neighbors (whose distance is $r=1$ or $r=\sqrt{2}$, the latter if movements along diagonals are allowed), but not so small to avoid a potential well so deep that the rest of the landscape is almost flat.

To guarantee the convergence of the gradient descent even in presence of constraints or noise resulting in potential wells, we weight the overall landscape for each particle by

$$V_\ell(r,t) = \gamma(t) D_\ell(r) + (1-\gamma(t))\, C(r,t) \tag{3}$$

where $C(r,t)$ is the potential energy landscape corresponding to constraints encoded by matrix $\mathbf{C}(t)$. The weighting factor $\gamma(t)$ should be a function ranging between 0 and 1 accounting for the importance given to the constraints with respect to the destination. The key to ensure the convergence of the gradient descent, while accounting at the same time for the constraints, is to make this function changing over time from an initial value up to 1. A candidate function is given by

$$\gamma(t;a,b) = 1 - (1-b)e^{-at} \tag{4}$$

where a is a non-negative number whose inverse $\tau = a^{-1}$ defines the time scale for convergence to 1 and b is the relative importance to be assigned at time t=0 to constraints and destination. A reasonable choice is to balance the two potential energy landscapes to allow the particles to be routed according to the constraints and the destination up to a time scale τ, above which the influence of the destination becomes more important. Small values of b might give more importance to the constraints rather than destination, leading to a routing less oriented to the final destination during the first time steps. Therefore, we require $\gamma(0) \geq 1 - \gamma(0)$ leading to $b \geq 0.5$.

We rewrite Eq. (3) to put in evidence the terms corresponding to different constraints. Let $C_\sigma(r)$ and $C_\delta(r,t)$ denote the potential due to all static and dynamical constraints, respectively, which are not related to the state of the other particles of the system. For instance, $C_\sigma(r)$ might encode the landscape corresponding to crimes, supposed to change over very long time scales, while $C_\delta(r,t)$ might encode the areas where it is raining, snowing or being affected by other meteorological events. On the other hand, we make the realistic assumption that not all individuals follows the routing provided by the smart system. While the information about the traffic of all individuals can be available by sensors properly disseminated across the grid, it is not possible to predict the behavior of a certain fraction p of individuals. To account for such a fraction p of individuals, we consider a set of $N(1-p)$ individuals moving along shortest paths between pairs of origin and destination, sampled from real data as discussed further in the text, and a set Np of individuals moving randomly in the city, i.e., following random walks instead of shortest paths. We indicate by $F_{in}(r,t)$ the potential corresponding to the flow of individuals within the system, i.e., those ones following suggestions from the smart system, and by $F_{out}(r,t)$ the potential corresponding to the flow of individuals out of the system. The latter is modeled by a noisy flow in terms of random walking individuals, although other mobility models can be used. In order to preserve conservation of the flow, we rescale each term by the number of particles in the most visited cell, i.e., by a weight $m(t)=\max[\mathbf{F}(t)]$, being $\mathbf{F}(t)$ the matrix accounting for the flow of individuals in the city at time t, with $\sum_{cell \in \mathcal{G}} \mathbf{F}(t) = N(t)$. The matrix $\mathbf{F}(t)$ is not weighted by the factor $[1-\gamma(t)]$ as in the case of $C_\sigma(r)$ and $C_\delta(r,t)$, because it would wash out the contribution of

F(t) to the potential landscape for increasing time. This choice makes our model more realistic: in fact, while it is possible to decide to traverse an undesirable area to balance the time spent looking for alternatives, it is not possible to traverse those areas which are congested or overcrowded. Therefore, the potential energy landscape accounting for the traffic flow should not be weighted by the function $1-\gamma(t)$, whose existence is justified only to introduce a trade-off between the needing to reach the destination and the time spent to achieve this goal while accounting for personalized constraints. Finally, Eq. (3) maps to

$$V_\ell(r,t) = \gamma(t)D_\ell(r) + (1-\gamma(t))\ [C_\sigma(r) + C_\delta(r,t)] + m(t)[(1-p)F_{in}(r,t) + pF_{out}(r,t)]. \tag{5}$$

This model is rather general, accounting for the presence of traffic and, simultaneously, for personalized and collective, static and dynamic, constraints. However, in this study we focused only on static constraints and we aggregated time-varying constraints for simplicity. It is worth remarking here that the potential landscape $V_\ell(r,t)$ experienced by individual ℓ still changes over time, because of the traffic flow term. Moreover, if agents are distributed in the grid according to the underlying population distribution and they move along shortest-path adapting over time in the evolving potential landscape, it is not possible to perform quantitative predictions about the state of the full system at a given time without numerical simulations.

7.3 OVERVIEW OF THE DATASET

Most of the datasets used in this work were acquired as part of the Telecom 'Big Data Challenge' and all of them are related to the city of Milan, Italy (see Figure 1).

The constraints encoded by matrices $\mathbf{C}^\alpha(t)$ can be represented as different 'layers' of the city, as shown in Figure 2. The weighted combination of such layers, as in Eq. (1), allows to build the potential energy landscapes

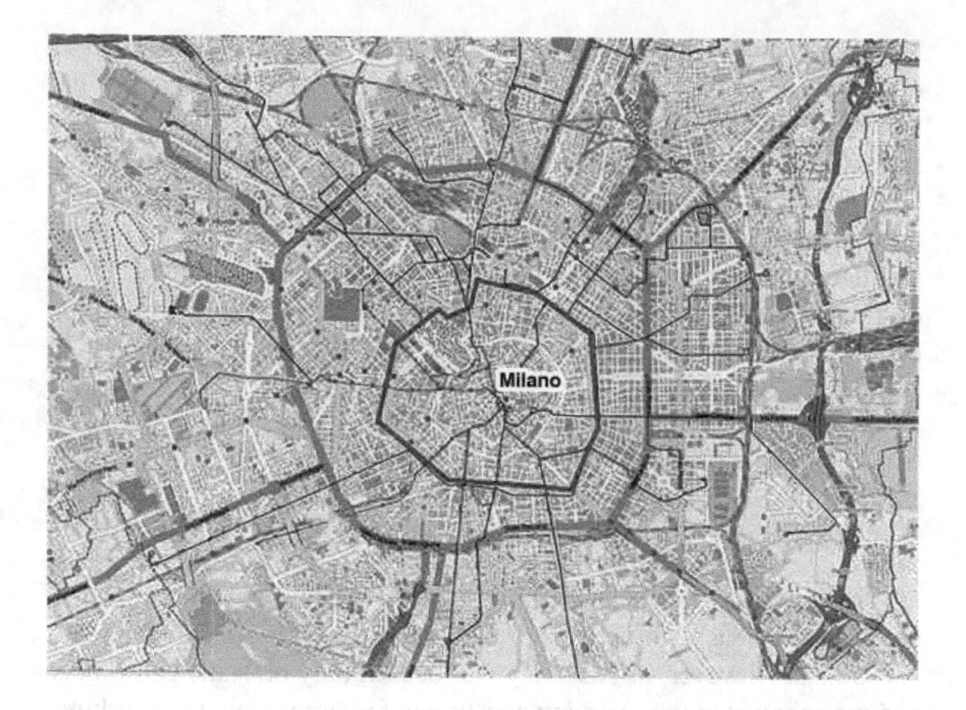

FIGURE 1: City of Milan. Area of Milan (Italy) considered in the present study. The area is divided into a squared grid with 10,000 cells of size 55,225 m^2.

$C_\sigma(r)$, $C_\delta(r,t)$, $F_{in}(r,t)$ and $F_{out}(r,t)$ influencing the overall landscape defined by Eq. (5).

For simplicity, we considered four static layers obtained from the provided datasets and here we explain how the layers were generated. The 'pollution' layer was generated from readings of 7 sensors scattered around the city, taken hourly over the course of 2 months. Because these sensors are very sparse in space, we smoothed their readings conveniently. The 'events' layer was generated by looking at the number of tweets coming from each grid of the city. It contains 100,000 geolocated tweets generated over a 30-day period. Lastly, the 'crime' layer was generated from a list of crimes, manually curated, and sourced from newspaper articles. It

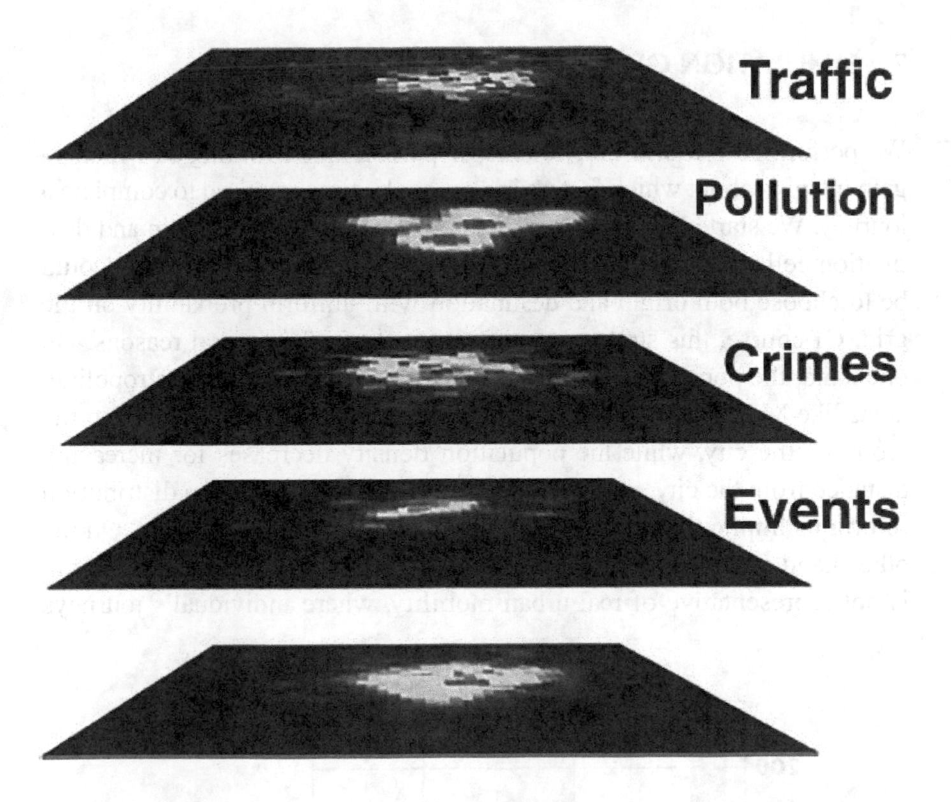

FIGURE 2: Layered potential energy landscapes. Each layer represents a potential energy landscape corresponding to a specific individual or collective constraint. The potential energy landscape in the bottom of the figure represents an example of weighted combination of such potentials.

contains 1276 crimes happened during the course of 12 months in Milan and reported by newspapers and local media.

Finally, we used data about the total number of calls and texts generated in Milan by all users of a mobile carrier, over a period of two months. We used the aggregated fraction of calls and texts between areas of the city, aggregated over the whole 2-month period, to determine the distribution of trip origin and destination, as detailed in the next session.

7.4 SIMULATION OF PERSONALIZED ROUTING

We performed massive simulations of personalized routing in Milan to gain insights about which factors influence the time required to complete a journey. We started by exploring different ways to sample origin and destination cells for each individual in the city. The simplest strategy would be to choose both origin and destination with uniform probability on the grid. Of course, this strategy can not be realistic for several reasons. On one hand, the population is never uniformly distributed over metropolitan areas like Milan, where there is a high concentration of individuals in the 'core' of the city, while the population density decreases for increasing distance from the city centre [11]. In fact, assuming a uniform distribution of origins implicitly considers a population uniformly distributed. On the other hand, the choice of a random destination, regardless of the origin, is not representative of real urban mobility, where individual's journeys

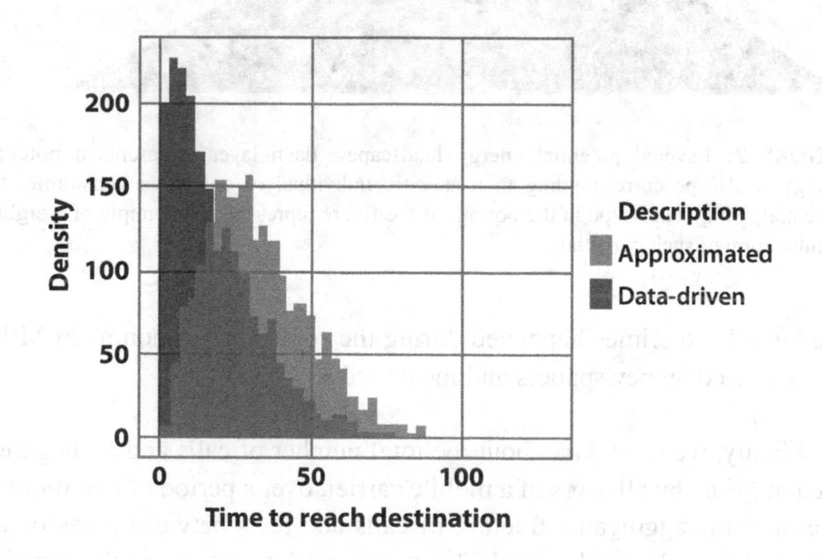

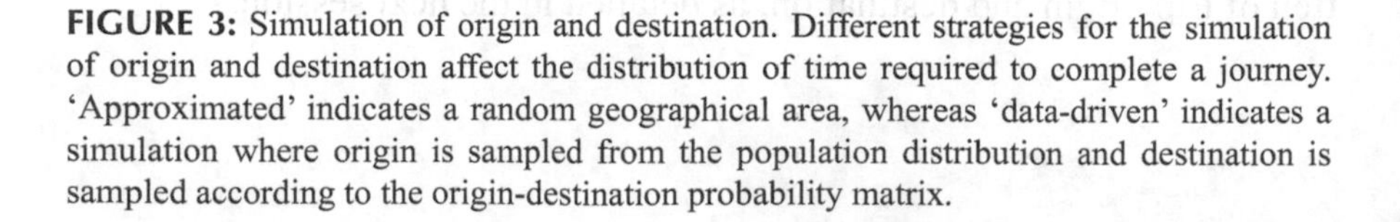
FIGURE 3: Simulation of origin and destination. Different strategies for the simulation of origin and destination affect the distribution of time required to complete a journey. 'Approximated' indicates a random geographical area, whereas 'data-driven' indicates a simulation where origin is sampled from the population distribution and destination is sampled according to the origin-destination probability matrix.

show a high degree of spatio-temporal regularity, with a few highly frequented locations [12]–[14] and high predictability of the underlying trajectories [3], [15], [16].

For this reason, we employed a data-driven approach accounting for intrinsic correlations in human mobility and leading to a more realistic distribution of origin-destination pairs. As a proxy for the population distribution, we have used the human activity measured by calls and texts generated by mobile phones. The calls dataset also provided information about the distribution of calls across all the pairs of grids; we exploited this information to sample a realistic ensemble of origin-destination pairs and to build an origin-destination probability matrix. Although this is a strong assumption, recent works [17]–[19] show how one of these quantities can be used to measure the other. Our simulations, summarized in Figure 3,

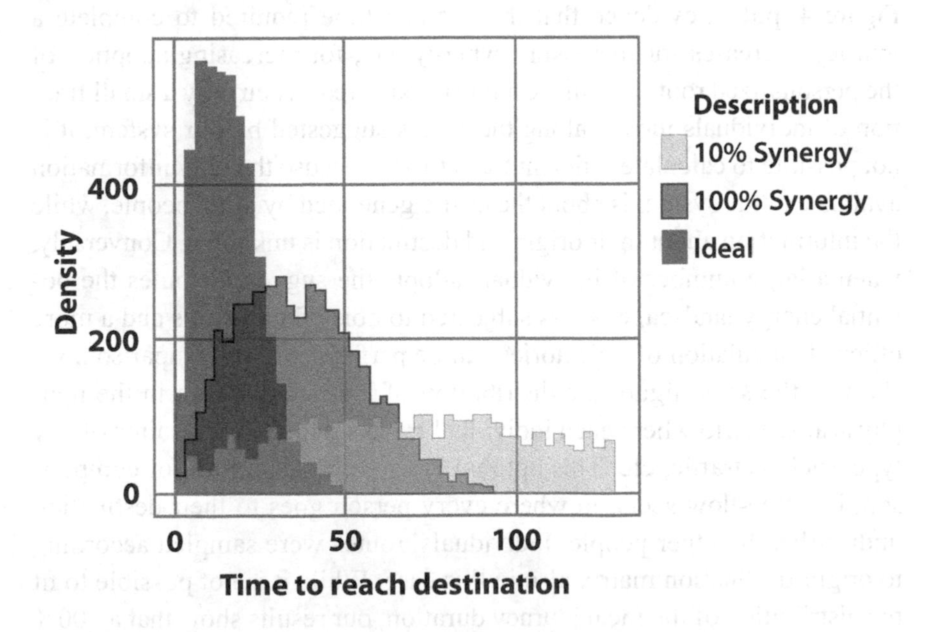

FIGURE 4: Simulation of personalized routing in Milan. Data-driven simulations where a certain fraction of individuals traveling in the city are routed by our system. The average time required to complete a journey decreases for increasing synergy, i.e., for increasing adoption of the personalized routing. The distribution of journey duration is shown in the ideal case, i.e., the non-physical scenario where each individual travels without constraints of any type, such as traffic, etc.

show that the time required to complete a journey is, on average, faster when a data-driven strategy is employed vs. the one approximated by random origin and destinations.

We capitalized on this result to perform data-driven simulations by varying the fractions of individuals traveling by adopting our routing system. For each individual, we calculated again the time required to complete his or her journey, sampled according to the origin-destination probability matrix. To understand how the efficiency of our re-routing algorithm is affected by the fraction of individuals adopting the recommended routes, we define this fraction (1−p) as the synergy of the system and we calculate the time required to reach the destination for each individual. The remaining fraction p of individuals does not follow the recommended routes.

We found that the underlying synergy has a non-negligible effect on the way individuals experience mobility the city. Our results, shown in Figure 4, put in evidence that the average time required to complete a journey decreases for increasing synergy, i.e., for increasing adoption of the personalized routing. This result was expected: when only a small fraction of individuals moves along the routes suggested by our system, it is not possible to calculate efficient trajectories because the only information available to the system is about the traffic generated by other people, while the information about their origin and destination is unknown. Conversely, when a large number of individuals adopts the suggested routes the potential energy landscape is less subjected to noisy fluctuations and a more efficient calculation of trajectories can be performed. For comparison, we show in the same figure the distribution of journey duration in the non-physical scenario where each individual travels without constraints of any type, such as traffic, etc. This optimal case, shown in figure for comparison, is a free-flow scenario where every person goes to their destination undisturbed by other people. Individuals' routes were sampled according to origin-destination matrix also in this case. While it is not possible to fit the distribution of the ideal journey duration, our results show that a 100% synergy produces a distribution close to the ideal one. It is worth remarking that this analysis would be able to quantify the benefits of synergy for urban traffic if information on the individual adoption of routing technology could be available to researchers.

Our routing system also allows to monitor mobility of the city from a new point of view. Interpreting individuals as particles moving in a thermodynamical system, it is possible to calculate the 'temperature' of the city. For each particle ℓ we calculate the mean speed at time t by

$$V_\ell(t; t_\ell^{(0)}) = \frac{\sqrt{\left[i_\ell(t) - i_\ell\left(t_\ell^{(0)}\right)\right]^2 + \left[j_\ell(t) - j_\ell\left(t_\ell^{(0)}\right)\right]^2}}{t - t_\ell^{(0)}} \tag{6}$$

i.e., as the ratio between the distance travelled up to time t and the time required to travel. Here $t_\ell^{(0)}$ indicates the time at which the particle has been injected into the system, i.e., the time at which the individuals leaves the origin of his or her route. The temperature of this system can be defined as the mean squared speed $\langle v_\ell^2 \rangle_\ell$. This measure is better understood in terms of permeability (or connectivity) of the city, as defined in urban studies allowing us to quantify how fast individuals flow through the city. Therefore, we define the permeability by

$$\mathcal{P}(t) = \langle v_\ell^2(t) \rangle_\ell = \frac{1}{N_{in}(t)} \sum_{\ell=1}^{N_{in}(t)} v_\ell^2(t) \tag{7}$$

where the sum and the average are limited to individuals adopting the routing system, because of the lack of information about origin-destination of the others. Nevertheless, $\mathcal{P}(t)$ is indirectly affected by the traffic generated by $N_{out}(t)$ individuals, therefore it is a robust measure of permeability. Higher the value of $\mathcal{P}(t)$ faster the flow of individuals trough the city and, conversely, lower the value of $\mathcal{P}(t)$ and slower the movements in the city, i.e., higher the probability that there are congested areas or, in the worst case, 'frozen' cells in the grid. In the upper panel of Figure 5 we show how the permeability changes over time for a data-driven simulation with N=100 individuals, a=0.1, b=0.5 and p=0, i.e., for 100% synergy.

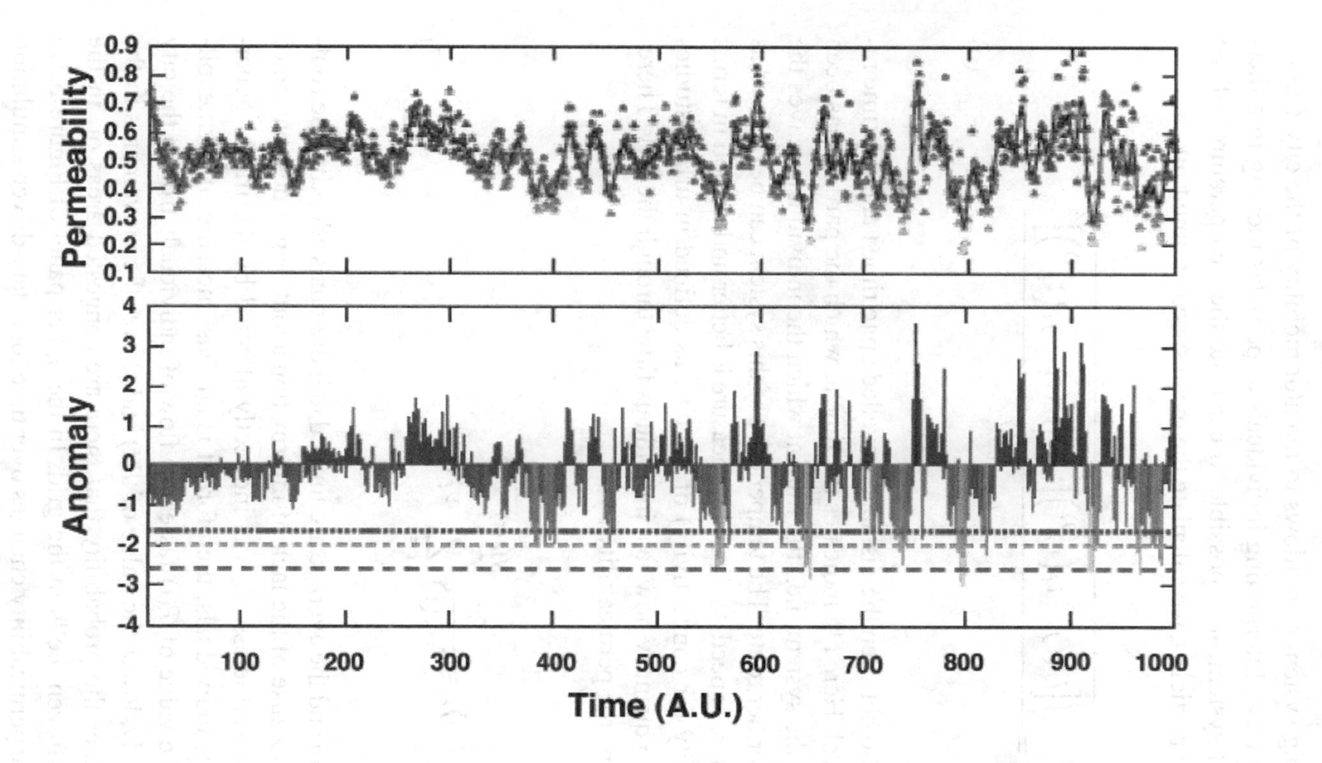

FIGURE 5: Monitoring traffic congestion in the city. Permeability of the city and corresponding anomaly versus time in the case of a data-driven simulation with N=100 individuals, a=0.1, b=0.5 and p=0. The horizontal lines in the bottom panel corresponds to different possible levels of 'traffic congestion alert'. The color gradient codes the status of the city with respect to its historical permeability (see the text for further detail).

The color gradient codes the status of the city with respect to its historical permeability. The existence of congested areas is more evident when the time series of anomaly $\mathcal{A}(t)$ is observed. The anomaly is defined as the departure of $\mathcal{P}(t)$ from the historical average $\mu\mathcal{P}(t)$ with respect to the historical standard deviation $\sigma\mathcal{P}(t)$

$$\mathcal{A}(t) = \frac{\mathcal{P}(t) - \mu\mathcal{P}(t)}{\sigma\mathcal{P}(t)} \tag{8}$$

where

$$\mu\mathcal{P}(t) = \frac{1}{t}\sum_{\tau=1}^{t} \mathcal{P}(\tau) \tag{9}$$

$$\sigma\mathcal{P}(t) = \sqrt{\frac{1}{t}\sum_{\tau=1}^{t} \mathcal{P}(\tau)^2 - \left[\frac{1}{2}\sum_{\tau=1}^{t} \mathcal{P}(\tau)\right]^2} \tag{10}$$

In the bottom panel of Figure 5 we show the anomaly changing over time. The traffic experiences large fluctuations for large values of t, positive and negative ones, alternating periods of high permeability with a few periods of low permeability. This is due to a few overcrowded cells that are quickly and automatically uncrowded by the system itself. Therefore, it is possible to monitor the traffic of the city by looking at the permeability and its anomaly over time, programming different alert levels such as low ($-2 \leq \mathcal{A}(t) < -1.7$), medium ($-2.6 \leq \mathcal{A}(t) < -2$) or critical $\mathcal{A}(t) < -2.6$.

7.5 DISCUSSION AND CONCLUSIONS

We have presented a strategy to route individuals between pairs of points of interest according to constraints of different type. Our method accounts for the simultaneous inter-playing between personalized constraints, as avoiding specific areas of the city because of personal choices, and collective constraints, from pollution reduction in certain areas of the city to the presence of adverse atmospherical conditions requiring targeted intervention. We have shown that the synergy plays a fundamental role in designing a smart city: only when all individuals take part in the routing system and move according to the recommended routes, the overall traffic in the city is closer to the most ideal mobility scenario. In the presence of real time information, our method allows to monitor the state of the city in real time, automatically identifying areas that are experiencing a temporary congestion and giving authorities the possibility to intervene timely.

Finally, the potential applications of our routing strategy are multiple. For instance, for certain values of the parameters (i.e., $a=b=0$, leading to $\gamma(t)=0$), we obtain a routing strategy from an origin and without a fixed destination, while accounting for specified constraints. This case could be useful to perform automated routing of objects or individuals through the city. For instance, it would be possible to route cars or drones which are collecting data about the city (as Google cars) and to route people in charge of social services like cleaning the streets or performing targeted intervention, as disseminating salt in areas with snow. An additional application could be in the field of social security, to route police cars in areas with high crimes rate. Finally, our framework can help decision-makers to real-time application of urban mobility policies in responses to crisis, e.g. the emergence of hotspots of infection in specific areas of the city (or a larger area) can be incorporated into the model to avoid people passing through dangerous areas before physical quarantine is employed.

REFERENCES

1. Hazelton ML: Statistical inference for time varying origin-destination matrices. Transp Res, Part B, Methodol 2008, 42(6):542-552.
2. Schneider CM, Belik V, Couronné T, Smoreda Z, González MC: Unravelling daily human mobility motifs. J R Soc Interface 2013., 10(84)
3. Song C, Qu Z, Blumm N, Barabási A-L: Limits of predictability in human mobility. Science 2010, 327(5968):1018-1021.
4. Simini F, González MC, Maritan A, Barabási A-L: A universal model for mobility and migration patterns. Nature 2012, 484:96-100.

5. De Domenico M, Solé-Ribalta A, Gómez S, Arenas A: Navigability of interconnected networks under random failures. Proc Natl Acad Sci USA 2014, 111(23):8351-8356.
6. Gallotti R, Barthelemy M: Anatomy and efficiency of urban multimodal mobility. Sci Rep 2014., 4
7. Youn H, Gastner MT, Jeong H: Price of anarchy in transportation networks: efficiency and optimality control. Phys Rev Lett 2008., 101(12)
8. Wang P, Hunter T, Bayen AM, Schechtner K, González MC: Understanding road usage patterns in urban areas. Sci Rep 2012., 2
9. Delling D, Goldberg AV, Pajor T, Werneck RF: Customizable route planning. In Experimental algorithms. Springer, Berlin; 2011:376-387.
10. Yeung CH, Saad D, Wong KM: From the physics of interacting polymers to optimizing routes on the London underground. Proc Natl Acad Sci USA 2013, 110(34):13717-13722.
11. Makse HA, Havlin H, Stanley H: Modelling urban growth. Nature 1995., 377
12. Gonzalez MC, Hidalgo CA, Barabasi A-L: Understanding individual human mobility patterns. Nature 2008, 453(7196):779-782.
13. Lima A, De Domenico M, Pejovic V, Musolesi M (2013) Exploiting cellular data for disease containment and information campaigns strategies in country-wide epidemics. arXiv:1306.4534
14. Salnikov V, Schien D, Youn H, Lambiotte R, Gastner M: The geography and carbon footprint of mobile phone use in cote d'ivoire. EPJ Data Sci 2014., 3(1)
15. Song C, Koren T, Wang P, Barabási A-L: Modelling the scaling properties of human mobility. Nat Phys 2010, 6(10):818-823.
16. De Domenico M, Lima A, Musolesi M: Interdependence and predictability of human mobility and social interactions. Pervasive Mob Comput 2013, 9(6):798-807.
17. Crandall DJ, Backstrom L, Cosley D, Suri S, Huttenlocher D, Kleinberg J: Inferring social ties from geographic coincidences. Proc Natl Acad Sci USA 2010, 107(52):22436-22441.
18. Farrahi K, Emonet R, Cebrian M: Epidemic contact tracing via communication traces. PLoS ONE 2014., 9(5)
19. Palchykov V, Mitrović M, Jo H-H, Saramäki J, Pan RK: Inferring human mobility using communication patterns. Sci Rep 2014., 4

CHAPTER 8

Transport Accessibility Analysis Using GIS: Assessing Sustainable Transport in London

ALISTAIR C. FORD, STUART L. BARR, RICHARD J. DAWSON, AND PHILIP JAMES

8.1 INTRODUCTION

Accessibility to jobs, services or other destinations of interest, has long been recognized as key to the development of sustainable transport, land use and spatial planning strategies [1,2,3,4,5]. Increased emphasis on sustainable urban development has underlined the importance of accessibility for (i) economic development as it enables the movement of people and goods to support the functioning of the economy [6]; (ii) environmental objectives such as reducing greenhouse gas emissions and pollutants that result from different transport modes and how they are used [7]; and (iii) equitable access for all socio-economic groups to core services such as healthcare [8].

Transport systems are key mediators of sustainability in urban areas, as they influence the way people and goods move through a city, and hence the energy (and thus carbon emissions) required to ensure a city functions successfully [9]. Well-designed low-carbon transport systems can encourage transitions from high emission to low emission forms of transport (e.g., from private car to public transport or bicycles). Spatial planning can complement this by encouraging development to occur in areas of good connectivity, ensuring the provision of low-carbon transport options in city plans, or prioritizing mixed-use developments where walking and cycling are more attractive options to travel [9].

Models of transport accessibility, and its interaction with sustainability, have been developed over many years. Hansen [10] showed a strong correlation between the accessibility of an area within a city and its ability to attract new urban development or investment. This concept was extended by Lowry [11] as a land-use transport model where the spatial separation of (or the ease of travel between) population and employment is a key determinant of land-use (see Levinson [12] for a more recent application of such a model). These models used measurements of time or distance in their assessment of sustainability and thus the carbon emissions of a journey were not considered. Further developments in urban land-use transport models [4,13,14,15,16,17,18] use a range of approaches, often with more sophisticated representations of urban processes, but central to these models is the characterization of transport accessibility [19]. Consequently, whilst accessibility calculations have often been wrapped within land-use transport models, the importance of understanding accessibility in its own right has more recently led to the development of a number of specific accessibility tools (see [5] for a review of such tools employed in Europe), some of which are standalone tools whereas others have been developed in a GIS environment.

Many of these tools have been developed for specific cities (e.g., CAPITAL is a London-specific tool [20]) or their access is restricted to specific audiences (e.g., the UK Department for Transport Accession GIS tool [21]). The data and computational overheads with these, and many other models, can be a barrier to adoption by non-specialist decision-makers (see Te Brömmelstroet et al. [22] for a discussion on usability), reduce their utility for rapidly exploring a wide range of options and policies, and limit the feasibility of their incorporation into a broader assessment of non-transport urban sustainability issues (e.g., which might consider a diverse range of issues such as flood risk and energy consumption for a number of different scenarios).

GIS platforms have, for some time, supported transport planning by analysis of spatial patterns, such as calculating the shortest path between two points on a network [23]. Tools commonly included in commercial GIS software offer interactive and rapid calculations on simple networks but are insufficient for a rich understanding of urban accessibility (e.g., inclusion of scheduling information or the monetary costs of journeys). Improvements to these basic GIS functions have been presented in recent years; Liu and Zhu [2] developed an accessibility toolkit for ArcView GIS, measuring accessibility by different modes and to various destinations, Lei and Church [24] included walking times and transit frequencies in their assessment of accessibility, Benenson et al. [25] introduce an ArcGIS-based toolkit to calculate service areas and travel times including transfers and timetable information for public transport, and Mavoa et al. [26] calculated different accessibility scored by public transport for 17 different destination types. In addition, more advanced analysis of accessibility using graph-theoretical approaches have been developed [27,28].

Building on these advances, this paper introduces a model designed to exploit the capabilities of GIS in order to support strategic planning for sustainable urban transport by enabling rapid appraisal of city-wide transport options, here using generalized cost of travel as a comparative metric. This methodology addresses some of the limitations of standard desktop GIS packages to provide a more sophisticated understanding of accessibility (e.g., following some of the methodologies highlighted above to include such improvements as generalized cost of travel and access times to services as opposed to standard shortest pathway analysis), whilst taking advantage of the strengths of GIS (over standalone tools) as an interactive data manipulation, spatial analysis, network analysis and visualization tool. In particular, a GIS framework enables non-specialists and decision-makers to interact directly with the model and the input data in order to rapidly test scenarios on their desktop and to explore the results and data which show the accessibility landscape for a city in a familiar environment. Such results can be further examined in the GIS software using in-built geo-statistical analysis tools.

An additional advantage is that GIS enables the combination of other model outputs and spatial datasets of interest in order to examine the transport appraisal results in context, enabling for example direct comparison between low-carbon and carbon-intensive forms of transport, thus giving insights into areas where sustainable transport investment is needed. Moreover, this has enabled the rapid accessibility analysis tool to be widely employed within an "Urban Integrated Assessment Facility" (described in Hall et al. [29] and Walsh et al. [30]) which is used for the analysis of climate change policy in cities. In this paper, we analyze the potential benefits of a number of infrastructure investment scenarios for Greater London in the UK, including future investment scenarios proposed by Transport for London, which demonstrate how London might increase urban accessibility with more sustainable modes of transport.

8.2 MEASURING TRANSPORT COSTS AND ACCESSIBILITY

There are many definitions of accessibility in the literature; however, a general definition by Wachs and Kumagi [31] is that accessibility is the ease (or difficulty) that opportunities (e.g., employment) or services can be reached from a location. Accessibility captures the effort required to overcome the spatial separation of two locations, and usually reflects the utility (e.g., travelling from home to a job) associated with travelling between these locations [5,32,33]. Geurs and Van Wee [3] provide a full review of accessibility measures, but the most generalized formulation of the accessibility, A, of location, i, is from Koenig [34]:

$$A_i = \sum_j O_j f(C_{ij}) \tag{1}$$

where O_j are the opportunities (utility or activity) to be gained from travelling to location j, C_{ij} is the distance, time, or cost of travelling from i to j, and $f(C_{ij})$ is a function which ensures that the accessibility increases as the cost of travel between two locations decreases [35]. Thus, fundamental to understanding accessibility is the cost of travel between an origin and a destination. This cost, or impedance, can be measured in a number of ways. The simplest measure is the Euclidian distance between two points which was typically used in earlier analyses [10,11]. Availability of transport network information has enabled more realistic network path analysis of distance; the shorter the distance, the higher the accessibility [36]. However, this does not take into account the physical network structure of transport modes or their different financial overheads, speeds, frequencies, levels of effort required in their use, interchange times and capacities [37,38]. Therefore, in this work the cost of travel is expressed as a generalized cost, taking into account both time and monetary components of any journey along the transport network in one unified value [39,40]. This makes the assumption that the cost of travel between an origin and destination can be generally expressed in the form:

$$G = (C_1 + C_2 + \ldots + C_n) \tag{2}$$

where generalized cost G is a function of a set of cost components of a journey, C_1 to C_n, which may be expressed in terms of time or money [39]. These cost components may be physically measurable (such as the cost of a ticket or the time taken to walk to a railway station) or more subjective (such as a user's preference for a particular type of transport or the relative comfort of a mode). The use of generalized cost provides a simple unit of comparison for the cost of journeys which includes as many of the influences which affect the choice of transport as possible. The UK Department for Transport's "Transport Analysis Guidance" [41] defines generalized cost as "the sum of both the time and money cost" for a journey, or:

$$C = aD + bT \tag{3}$$

where D is the distance (km) between origin and destination, T is the time in hours taken to complete the journey, a is a distance coefficient based on vehicle operating costs per km, and b is the value of time coefficient. Costs are incurred differently by private, C_{PVT}, public, C_{PUB}, or cycling, C_{CYC}, transport, expressed in units of time by WEBTAG [41] as follows:

$$C_{PVT} = (V_{wk} * A) + T + D * \frac{VOC}{occ * VOT} + \frac{PC}{occ * VOT} \tag{4a}$$

$$C_{PUB} = (V_{wk} * A) + (V_{wt} * W) + T + \frac{F}{VOT} + I \tag{4b}$$

$$C_{CYC} = T + T\left(V_{Topo} + V_{Safe}\right) \tag{4c}$$

where A is the access time to the network (walk time to car, bus stop, light rail station or railway station); V_{wk} is the disincentive weight for walking; T is the journey time of transport; VOC is the vehicle operating cost per km; D is the distance in km; PC are parking and other costs; occ is the number of vehicle occupants; V_{wt} is the disincentive weight for waiting; W is the total waiting time for the journey; I is the interchange penalty (if applicable); F is the fare paid for a journey; VOT is the value of time coefficient; V_{Topo} is the disincentive weight for cycling up slopes; and V_{Safe} is the disincentive weight to represent safety concerns for cyclists on busy roads. Further explanation, and the values used for the modeling study presented in Section 4, are given in Section 3.

It can be seen from Equations (4a)–(4c) that the in-vehicle time of any given journey is thus only one of several factors that contribute to the overall cost of travel [2]. Furthermore, time for waiting or walking to access a particular mode of transport captures the disincentive of certain options, and is sometimes referred to as the system accessibility [21]. VOT defines the value that the average person places on a unit of their time, or their willingness to pay for a service, in monetary terms [42]. The value of time varies for different socio-economic groups and other factors such as the urgency of a given journey and differs depending on whether the time is working or non-working time (which is of concern here), with higher values for the former. However, it enables direct comparison and combination of both the monetary and time components of journey cost [42,43].

A transformative shift in improving the sustainability of transport is switching journeys from carbon-intensive modes (such as private cars) to low-carbon or zero-carbon modes (i.e., walking and cycling) [9]. As such, it was considered important to include a representation of the cycling mode in this generalized cost framework. Cycling routes can be considered to predominately follow the road network in most cities, with the addition of a limited number of paths and routes exclusively for cycle use. The generalized cost calculation for a cycle journey is thus a simplified version of the method for private vehicles (Equation (4a)) since there are fewer monetary costs associated with trips. There are, however, additional factors to consider in the cost of cycling. Hopkinson and Wardman [44] highlighted that the gradient of routes along which cyclists may travel is one of the factors influencing cyclists' route choices, whilst Noland and Kunreuther [45] indicated that the perception of risk (i.e., the safety) of a journey is an important consideration. Rodriguez and Joo [46] also allude to this in their study, showing that the physical environment (e.g., the provision of pedestrian or cycling infrastructure) has an effect on the perceived cost of non-motorized journeys. These factors are therefore included as additional weights in the computation generalized costs for cycling journeys.

8.3 IMPLEMENTATION OF GIS-BASED MODEL

To develop a tool that is generic and transferable, able to accept standard GIS input in the form of spatial data, at any given scale, and perform the generalized cost computation to produce a set of accessibility measures, an add-in was developed in VBA for ESRI's ArcGIS©. The utility of this add-in was demonstrated using a case study in the Greater London Authority (GLA) area in the UK (see Section 4 for details description of the implementation and results of this case study). Figure 1 shows the computational process for calculating generalized cost matrices for a given mode.

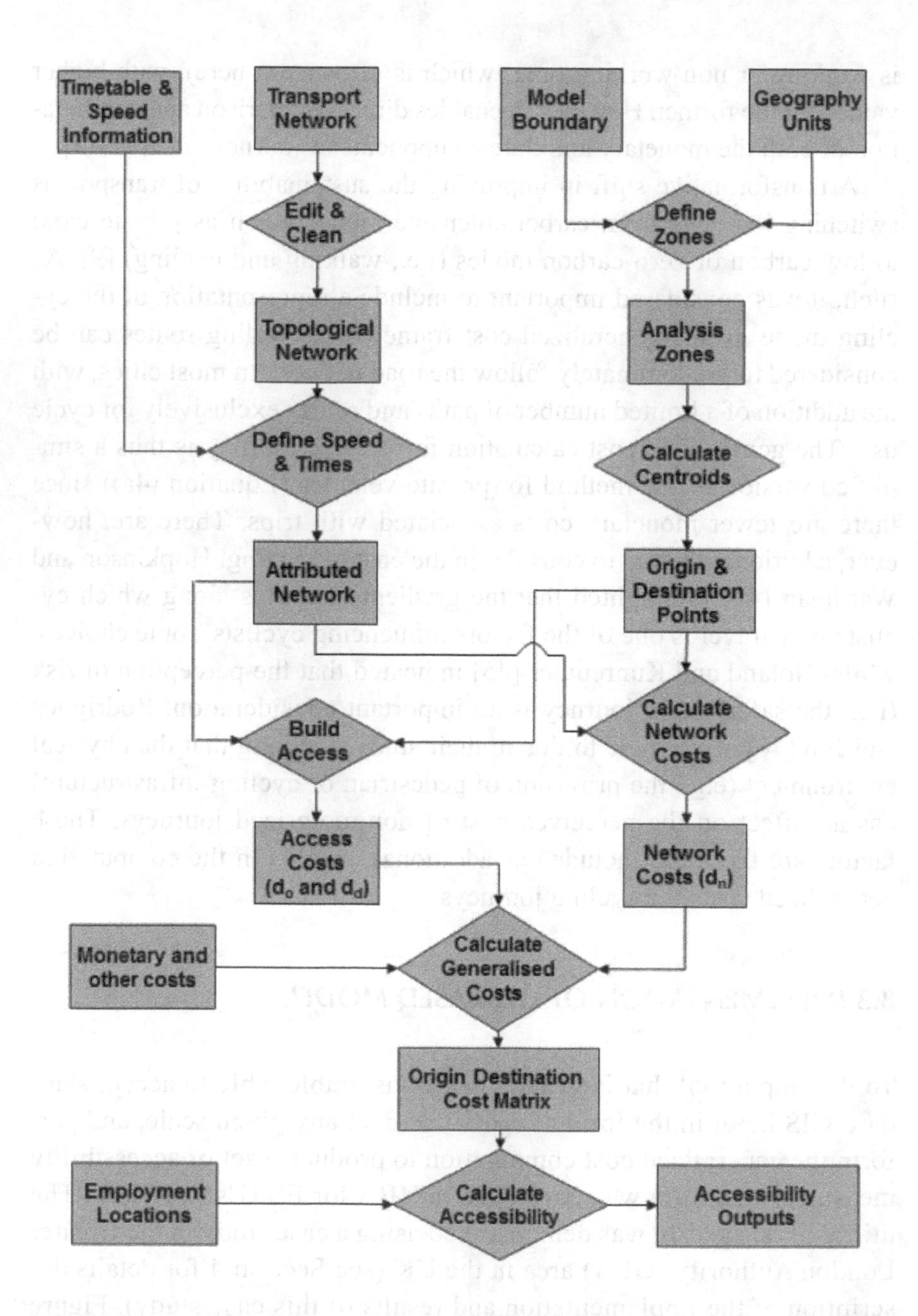

FIGURE 1: Computation framework for matrices of generalized cost, and thus accessibility measures. Employment locations could be replaced by other information to calculate accessibility to other facilities.

The generalized cost model (Figure 1) was implemented as follows using values in Table 1 for the different terms of Equations (4a)–(4c):

1. First, the spatial geography of the model is defined in terms of zonal geography and the spatial boundary of analysis (for example for the London case study presented below, these are UK Census Area Statistics Wards, i.e., zones on which 2001 UK census outputs are reported).
2. Build M transport networks:

    - Analyze, and clean if necessary, the data to ensure the correct topological structure for creation of a spatial network model.
    - Build spatial networks within GIS software.
    - Calculate the length of each network link from geometry.
    - Multiply each network link length by the relevant travel speed to obtain the travel time for each link.

3. For N units of spatial geography, create an N x N matrix of generalized costs for each of the M transport modes (similar to the approach outlined by Benenson et al. [24]):

    - Calculate the location of the centroid and use it to define the Origin, and Destination of zone i.
    - Calculate access distance, do, and associated travel time from centroid to nearest network access location (private and cycling modes), or boarding point (public transport modes) (Figure 2)
    - Calculate shortest (network) path, dn, between origin and destination centroids using Dijkstra's algorithm [47] (Figure 2). Shortest routes based on time, rather than distance, are computed in order to take into account speeds of travel. This is facilitated by ESRI's Network Analyst extension [48].
    - Calculate access distance, dd, and associated travel time from destination station or stop (public transport) or road access location (private and cycling modes) (alighting point) to the centroid of the destination zone (Figure 2).
    - Eliminate nonsensical journeys (e.g., where nearest station is shared between the origin and destination) and return a no-data value.
    - Add on other costs, including non-monetary and monetary components in Equations (4a)–(4c) such as fuel or perception weights, converted to time.

- Sum all journey components to calculate the generalized cost of travel, C_{ij}, between two zones.
- For situations where i = j and the preceding steps calculate C_{ij} = 0 then assume that $C_{ij} = 2/3\min(C_{ij(i \neq j)})$ in line with the approach used by Feldman et al. [49].

4. Use computed generalized costs to determine accessibility to destinations of interest (e.g., employment locations) and determine the proportion of employment which is accessible by a given mode in a given cost of travel.

Figure 3 shows the geographic extent of the road-based (private vehicle travel and Bus) networks as implemented in the London case study, highlighting the size of these networks (in the order of 65,000 links). Figure 4 shows a sample of the OpenStreetMap cycle network constructed for this study, demonstrating the density of the network in the urban area and the number of possible travel routes to be assessed. The calculation of network distance, dn, used the above algorithm for all modes, with appropriate values (i.e., speeds) from Table 2 to establish a UK context. However the calculation of network access distance do and dd for public and private networks differs as public transport networks must be accessed from stops where passengers can board and alight from services, with these stops being connected by routes. Unlike the public transport modes, the road and cycle networks can be accessed at any point along a link. Therefore d_o and d_d are calculated as the Euclidean distance from the centroid to the nearest stop or station for public transport and nearest road link for private transport. This is undertaken using a spatial join in the GIS software.

It is possible that origin or destination zones may contain more than one station or stop (for example, in London 10% of zones have more than one heavy rail station and over half have no station at all, whilst the numbers are one in three and one in five for light rail). If there are fewer than two stations in the zone the nearest is used, whether it falls within the zone or not (Figure 2), to define d_o and d_d. If both the origin and destination zones have fewer than two stations then there is only one possible route between the zones. However, if either the origin or destination contain more than one station, then for a total of S_{ij} stations (e.g., maximum S_{ij} for light rail in the London is 16) the average distance over all route combinations is calculated:

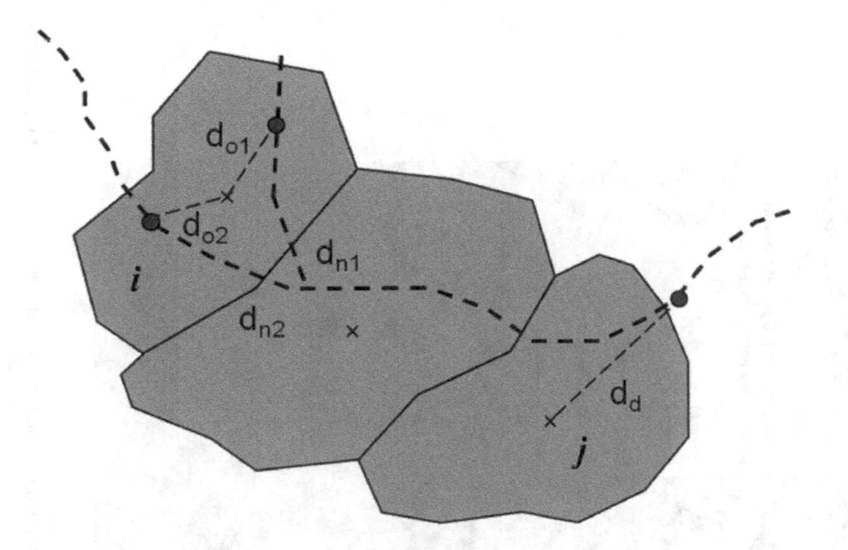

FIGURE 2: Schematic depiction of journey from origin i to destination j.

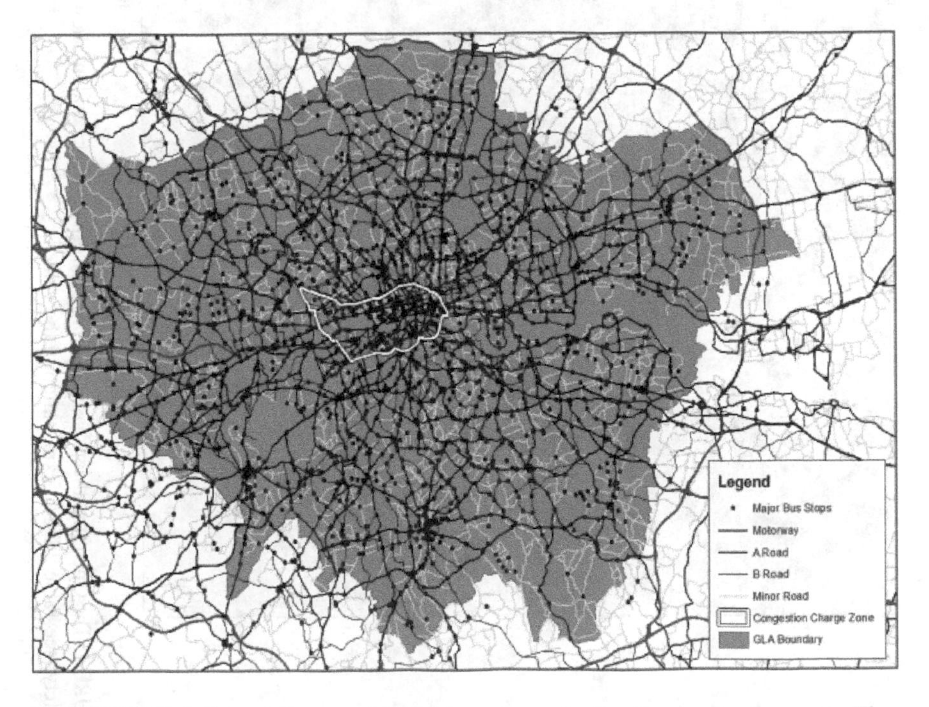

FIGURE 3: Baseline road network, and bus stops, in the Greater London area. The Congestion Charge Zone is also depicted.

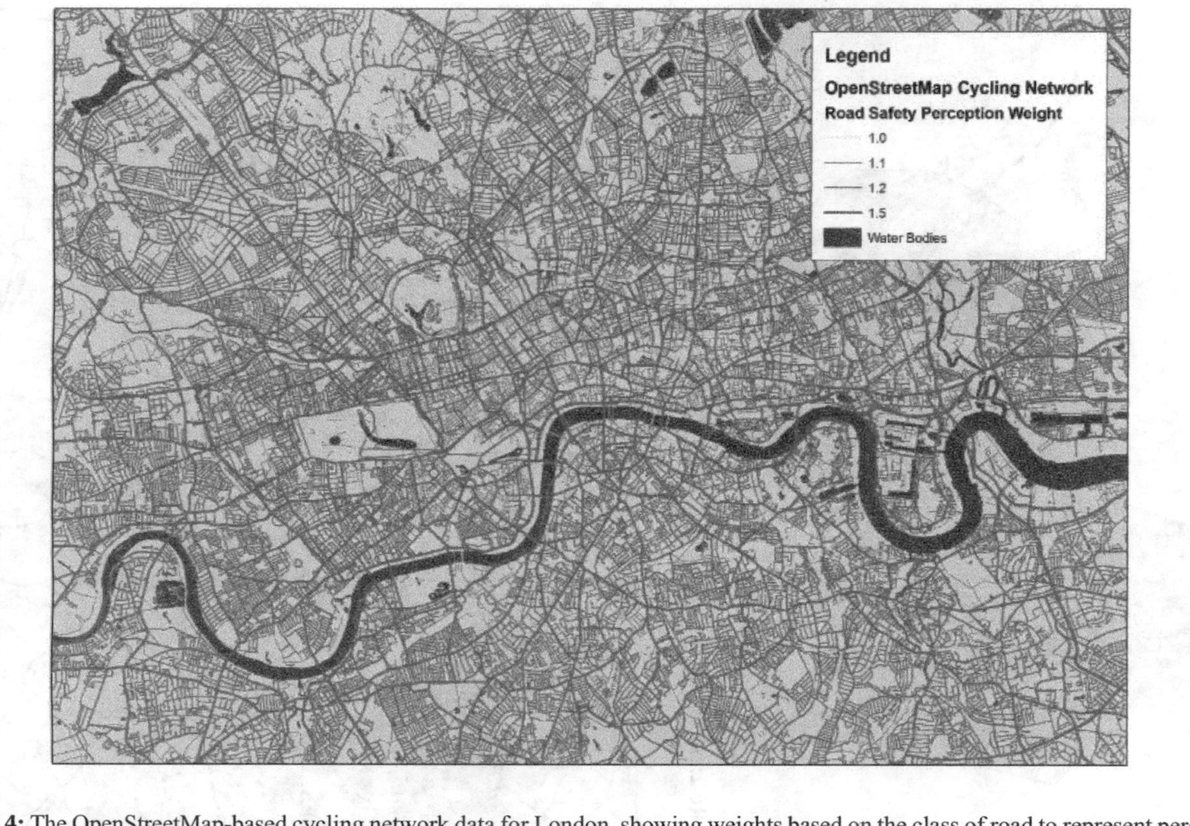

FIGURE 4: The OpenStreetMap-based cycling network data for London, showing weights based on the class of road to represent perceptions of safety (see Table 2 for an explanation of the values).

$$D_{ij} = \sum_{k=1}^{S_{ij}} \frac{d_o^k + d_n^k d_d^k}{S_{ij}} \tag{5}$$

A number of additional assumptions have been made to ensure that the generalized cost model is flexible, can be constructed on simple, publicly-available data, and runs quickly. For public transport networks (bus, heavy rail, and light rail), it is assumed that all portions of the heavy and light rail networks are passenger-carrying (although pre-processing can be undertaken to only include rail lines with stations). There is no limit imposed on access times to the transport networks (e.g., walking times to stations); a maximum walking time to a public transport stop or station could be implemented to reflect the low likelihood of someone walking further without using an alternative mode of transport. For example, NHTSA [50] show that many people will only walk 0.5 miles. To enable direct comparison between modes, for journeys between all zones, this was not implemented here. However, the disincentive of long walking times is still captured in the generalized cost calculation. Interchanges within the same transport mode are calculated by adding walking costs between stations to the cost of travel.

In the case of private transport networks (road and cycling) most cost elements are included on the links themselves, since non-time components are often accrued in a distance-based manner. Thus, for example, the impact of topography on cycling is incurred on a link-by-link basis. Congestion charging or other road user charging schemes must be included as on-link costs (i.e., by means of increasing travel times along a link through Value of Time conversion). See Section 4 for a discussion of the London Congestion Charge and its implications.

All network elements (e.g., roads, cycle routes, or railway lines) are bi-directional. Some road network data (e.g., Ordnance Survey ITN in the UK) includes driving direction which means that it is possible to trace routes through one-way systems, however for large analyses (e.g., the 1500 km^2 of London) such local-scale factors are not significant for global accessibility. On the same basis, the effect of gradient is assumed to impact on cycle journeys equally in both directions.

TABLE 1: Infrastructure improvements implemented in Baseline, Low and High investment scenarios (based on Transport 2025 report [51]).

T2025 Scenario	Road	Bus	Rail	Light Rail
Baseline			Crossrail	
				High Speed 1
			Heathrow Express to Terminal 5	Heathrow Terminal 5 extension
Low	Thames Gateway Bridge	20% increase in bus supply (and thus frequency)	Reduce journey time by 4.5%	DLR extensions, Greenwich and East London transit systems
High	Silvertown Link Bridge National Road User-charging scheme	40% increase in bus supply.	Crossrail 2, East London line extension (Overground).	Tramlink extensions, DLR extension to Dagenham Dock

8.4 RESULTS AND DISCUSSION OF LONDON ACCESSIBILITY STUDY

The above computation framework was used to compute origin-destination matrices of generalized cost of travel under a number of current and future transport infrastructure scenarios for an application in the Greater London Authority (GLA) area in the UK. The future scenarios which were examined are based on the Transport for London "Transport 2025" study [51] options for future infrastructure developments. This study sets out a number of strategies to achieve the aim of ensuring London becomes a "sustainable world city", including supporting sustainable economic development, improving social inclusion, and tackling climate change and enhancing the environment [51]. Part of this strategy is a desire to improve the use of low-carbon transport modes through new infrastructure provision and encouraging behavioral change.

These scenarios were generated in the form of spatial network representations and parameterizations within the methodology outlined in the previous section, thus allowing the testing of their impacts on accessibility in the Greater London area. To facilitate model set-up and to demonstrate ease of use, accessibility is characterized using nationally-available datasets in the UK. Datasets of this type are often available in other countries and increasingly collected according to the same data standards, which limits the time required for data preparation. Improved accessibility to areas of high economic activity by lower-carbon modes of transport can therefore be provided as evidence that such infrastructure plans are meeting the city's sustainability goals. Table 1 summarizes the future infrastructure scenarios examined in this study. Specific improvements for cycling network infrastructure were difficult to obtain (none are mentioned in the Transport 2025 report beyond the development on on-road cycle lanes) so are not included in this table.

The model boundary for this study was defined as the edge of the Greater London Authority for reasons of political jurisdiction and data access, as this area is the defined administrative boundary that is most pertinent to decision-makers. Transport network representations were also restricted to this boundary, although this introduced a limitation in that some possible radial routes outside the administrative boundary of the city are not considered in the generalized cost calculation. Only journeys internal to this study area were considered, since the consideration of this study was accessibility between locations within the GLA region. Zonal units for origins and destinations in this study were the UK Census Area Statistics Wards (i.e., zones on which 2001 UK census outputs are reported) of which there are 633 within the GLA area ranging in population from 106 to 17,000, and in size from 0.13 km^2 to 29 km^2 (mean size of 2.5 km^2) of UK Census Area Statistics Wards were used as they allowed a direct comparison between the computed transport accessibility measures and the socio-economic variables of the zones.

One feature of London's transport policy of particular interest is the central London "Congestion Charge zone", which levies a fee on vehicles entering between 07:00 to 18:00 (see Table 2). Residents commencing their journeys within this zone receive a 90% discount. As mentioned above, these charges were added to the spatial road network representation directly to enable it to be factored into the calculation of the least cost network route, d_n in the generalized cost computation. This ensures that journeys will be routed around the congestion zone if the overall cost is lower than traversing the center of London. Table 2 lists the complete parameterisation and data inputs used in the London case study.

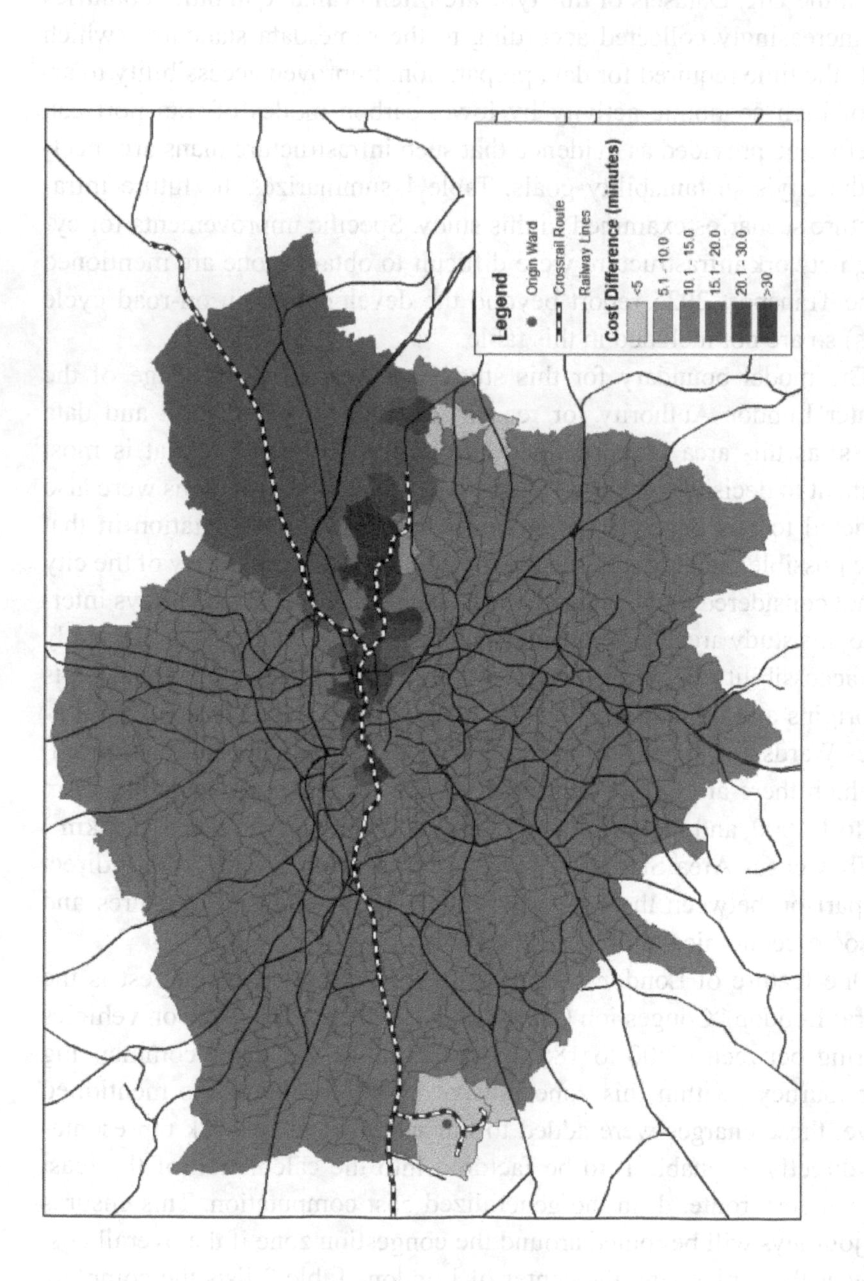

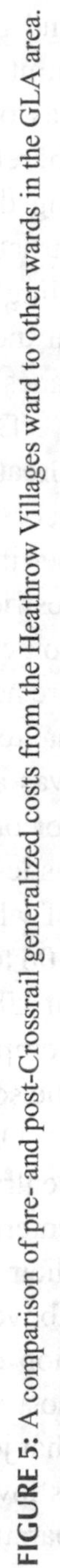
FIGURE 5: A comparison of pre-and post-Crossrail generalized costs from the Heathrow Villages ward to other wards in the GLA area.

8.4.1 TESTING NEW INFRASTRUCTURE INVESTMENT

A key purpose of this GIS tool is to rapidly assess the effects of new infrastructure developments on accessibility patterns. Here, some options from the aforementioned T2025 scenarios of infrastructure development ([51], summarized in Table 1) are considered individually. The changes in generalized costs resulting from such developments are first shown spatially in this section, then their impact on accessibility to employment and wider sustainability impact discussed in later sections.

Crossrail is a program to construct a new East-West Heavy Rail link in London connecting Paddington to the Liverpool Street and Canary Wharf areas. This includes a new section of railway line in a tunnel beneath central London and improvements to existing lines either end of the tunnel to allow higher speeds and greater capacity [55]. These improvements were included in the network model by means of new links and stations to reflect the new tunnel route and increased speeds on the existing lines to reflect greater frequency and shorter travel times. The fare was assumed to be consistent with other rail travel.

The difference between the generalized cost of travel from Heathrow Villages to other census wards pre- and post-Crossrail is shown in Figure 5. This reveals a mean generalized cost benefit of 12 min to census wards across London, but the zones to the immediate east of the Crossrail tunnel show a benefit of up to an hour. Such patterns of improvement demonstrate the increased accessibility to areas to the east of central London (such as Canary Wharf and the former docklands regions) where there are a high number of jobs (around 132,000 people employed in 15 CAS wards in 2005) and a large amount of development. This demonstrates that the construction of Crossrail may improve the use of sustainable modes of transport to such areas.

Another proposed infrastructure upgrade is the Thames Gateway Bridge, a new road crossing of the River Thames to the east of the City of London. Figure 6 shows the generalized cost benefit by road from the Plumstead ward. It can be seen that most reductions in generalized cost are in cross-river journeys (as would intuitively be expected) although there are also reductions in journey times to more distant wards in north London. Wards immediately north of the river experience the greatest reduction in cost (up to 30 min) as it is possible to cross the River Thames at a much closer point than in the baseline network. However, if a proposed £2 bridge crossing toll is added, then the generalized cost benefit is negated (since £2 equates to approximately 24 min of journey time). Whilst some of these reductions are modest, they highlight the wider spatial effects of localized infrastructure improvement. Both projects provide regional benefits, but the structure of the rail network leads to the greatest benefits from Crossrail being focused in fewer wards than for the bridge. This highlights the need to examine the system improvements at city-scale from improvements which could be considered local.

Whilst there are no specific infrastructure investments for cycling outlined in the Transport 2025 report, there is mention of an aspiration to further develop the London Cycle Network Plus (LCN+), a network of up to 900 km cycle routes. Much of this network is in the form of on-road cycle lanes but, more recently, Cycle Superhighway routes have been proposed. One of these is the 10 km-long East-West Cycle Superhighway, running along the River Thames through central London on a dedicated piece of infrastructure [56]. This is a relatively small piece of infrastructure, but its impact on cycling journey costs through this part of central London was analyzed to examine the local improvements which could be seen from constructing such dedicated infrastructure.

Figure 7 shows the reduction in travel cost by the cycling mode between the Plaistow South ward in east London and the other wards in the local area. Improvements of up to 14 min of generalized cost can be observed, due to a combination of shortened journey times (due to higher speeds on dedicated cycling infrastructure) and a reduced perception of risk from the provision of segregated cycle lanes. Such improvements increase the competitiveness of cycling journey costs in comparison with other modes (for example, cycle costs from Plaistow South to Golborne in west London are 91 min after the construction of the East-West Cycle Superhighway. This compares favorably with the travel cost by private car (162 min, since this journey travels through the Congestion Charge zone), light rail (105 min), and bus (also 105 min). If other such infrastructure was provided in a wider context across London, the reductions in travel cost could be large enough to ensure that zero-carbon modes of transport are a competitive alternative for short distance journeys.

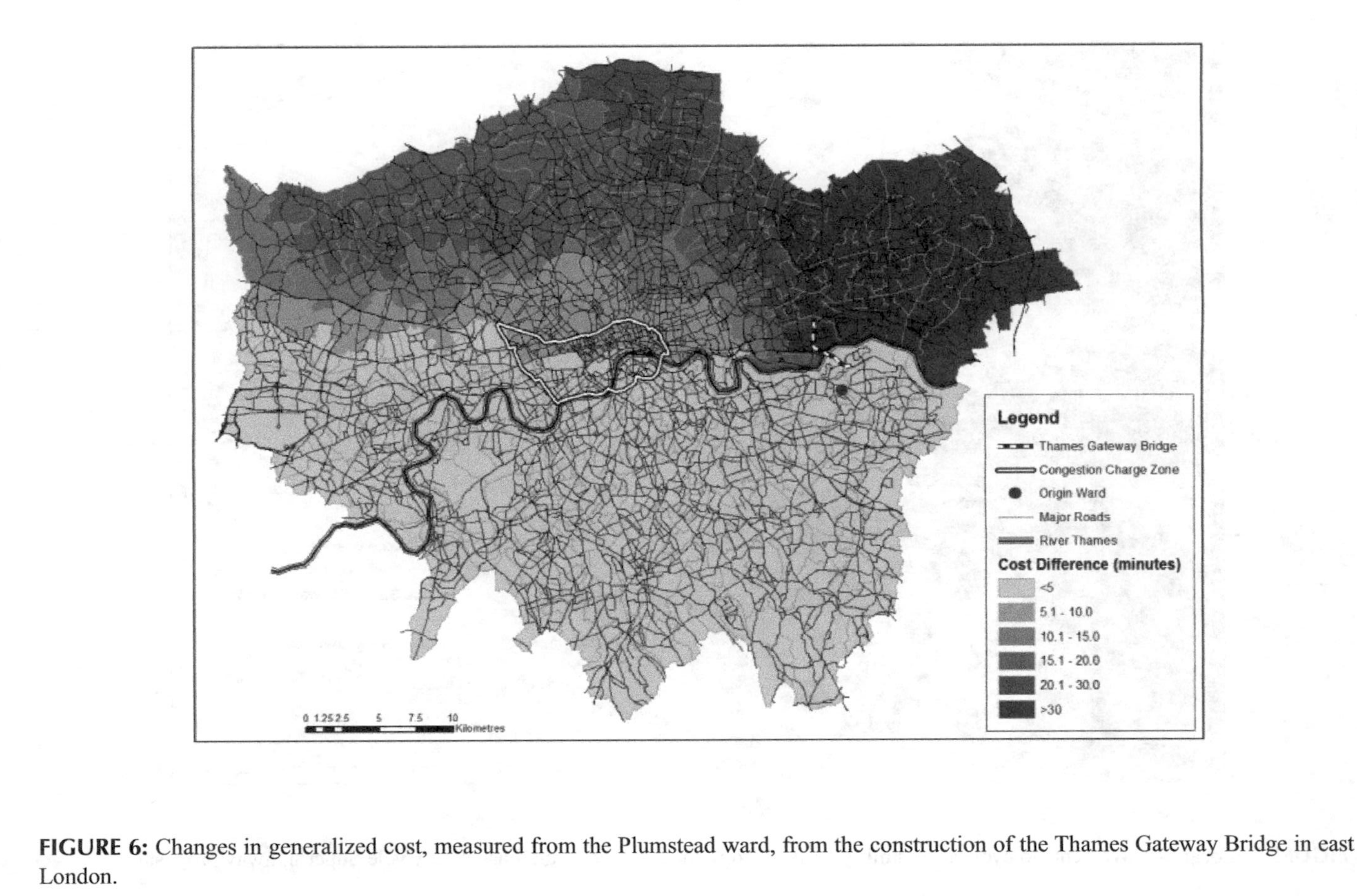

FIGURE 6: Changes in generalized cost, measured from the Plumstead ward, from the construction of the Thames Gateway Bridge in east London.

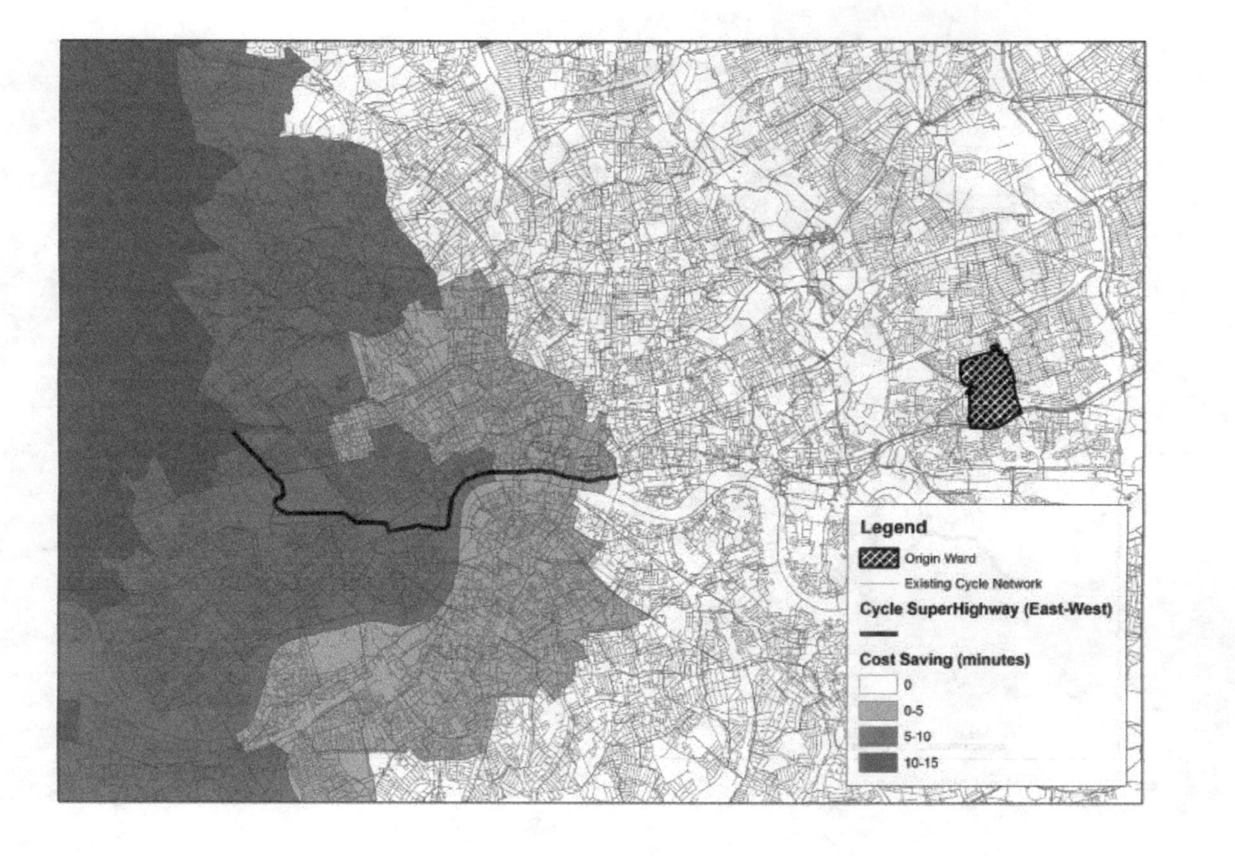

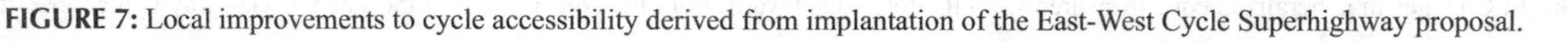

FIGURE 7: Local improvements to cycle accessibility derived from implantation of the East-West Cycle Superhighway proposal.

8.4.2 ACCESSIBILITY TO EMPLOYMENT WITHIN LONDON

Of the 30 million daily reported transport journeys in the 2012 London Travel Report [57], 40% were work-related journeys. Accessibility to employment is therefore an important driver of demand on transport networks and a large contributor to greenhouse gas emissions from transport. Since the zones used in the generalized cost calculation are also zones where population and employment records are available, it is possible to examine the accessibility to employment within London—and thus assess the number of people for whom accessibility is improved. 44% of the journey stages documented in the 2012 London Travel Report were by public transport, with 33% by car. Therefore, whilst patronage of public transport modes is significant in London, there is room for improvement. Providing lower travel costs between home and work locations by more sustainable transport modes could encourage fewer journeys by carbon-intensive modes, and this is a key driver in the London Plan [58].

The 2008 Annual Business Inquiry data [59] provided total employment in each of the zones utilized in the study and used to construct functions of the proportion of London's employment (~3.5 million jobs) that is accessible for a given generalized cost of travel from each of the 633 census wards. Figure 8 shows these employment accessibility functions for the Heathrow Villages ward (Figure 8a) and Aldersgate ward (Figure 8b) for the road, rail, light rail, and bus modes for current transport infrastructure. Travel by cycling does not feature in these accessibility functions as it was important to be able to examine accessibility to all employment in the study area. The average cycling trip length recorded in the UK National Travel Survey in 2012 [60] was 3.2 miles, suggesting that cycles are used mainly for short-distance trips. Since these curves examine accessibility to employment across the whole Greater London area, cycling costs were not included.

The shape of these functions is determined by the characteristics of the mode of travel (e.g., the access time, the frequency of service, the travel speeds) as well as the spatial location of the origin ward and the spatial distribution of employment. Wards, such as Aldersgate, situated nearest to centers of employment (e.g., the City of London or Canary Wharf) have more employment accessible at a lower generalized cost of travel than wards situated in the suburbs or in areas of low accessibility.

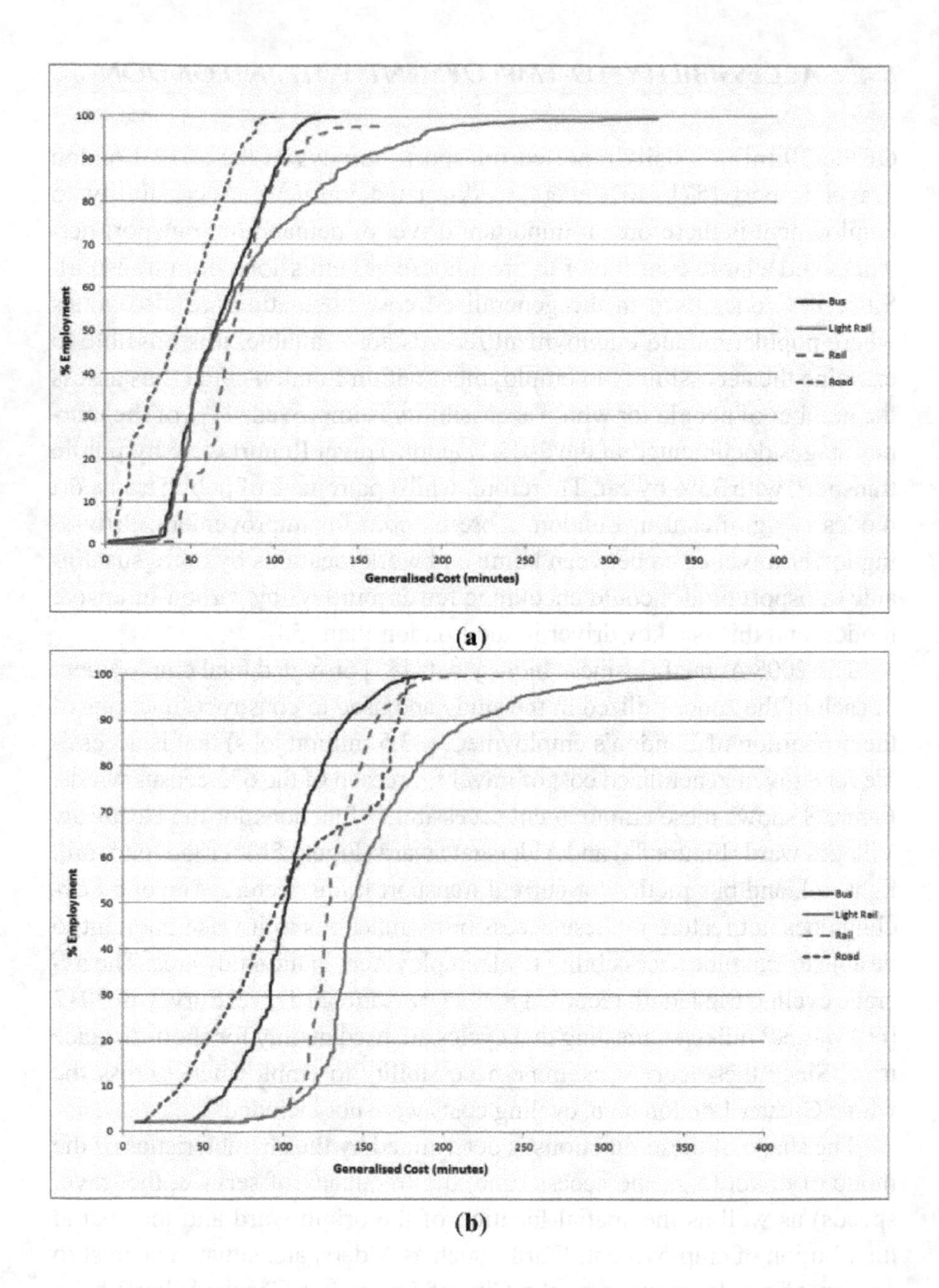

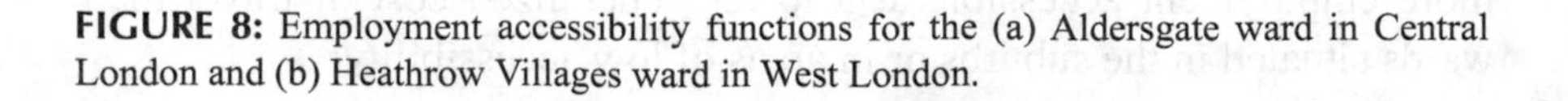

FIGURE 8: Employment accessibility functions for the (a) Aldersgate ward in Central London and (b) Heathrow Villages ward in West London.

Aldersgate (Figure 8a) is located near a number of transport hubs within the Congestion Charge zone. The road mode gives access to a large proportion of jobs for comparatively low generalized cost (approximately 50% of all of London's jobs within 40 min) since this is an area of dense employment and the Congestion Charge is reduced by 90% for journeys originating within the zone. However, this does not include factors such as insufficient parking spaces which may preclude many journeys by this mode. The public transport modes all exhibit higher costs to access a similar level of employment but there is some variation across the modes. Heavy rail has a larger generalized cost than other modes to access employment initially (with a lag of 75 min), highlighting the discentive required to walk to the nearest heavy railway station, but is a lower-cost option for accessing more distant employment than light rail (100% of employment is reached in around 200 min, whilst for light rail the cost is around 350 min). The generalized cost to reach all employment by rail and light rail is higher than that by bus, showing the effect of low bus fares on the generalized cost of travel.

Figure 8b shows comparable functions for journeys originating in the Heathrow Villages ward. It is immediately noticeable that the travel cost to access employment is higher for all destination wards, reflecting the spatial separation of this origin from the main employment centers in central London (100 min of travel cost to reach 50% of jobs by private transport in this instance). It is interesting to note the steepening of the function for the Road mode at 160 min, which is due to the Congestion Charge reducing the accessibility of jobs by road within the charging zone. The initial lag in the Rail and Light Rail modes are due to the long access walks to those modes from this ward (since the ward centroid is 800 m from the nearest station, all of which are located within the airport). The spatial coverage of the Light Rail network, requires some long walks from the destination station to the workplace census ward, leading to a very long tail for the final 5% of jobs.

Both accessibility curves for the current day transport networks highlight a number of interesting patterns. Journeys by private car are, in the most part, the lowest cost means of accessing employment in Greater London (consistently up to 30 min lower in cost than other modes from Aldersgate). The Congestion Charge does, however, reduce the accessibility by this mode to the jobs in central London by adding 95 min to the cost of those journeys. Accessibility by more sustainable public transport modes is initially curtailed by the additional cost of accessing these modes initially (a lag of between 30 and 75 min), but then over short distances in central London the light rail mode is competitive and over longer distances to the wider area the rail mode more attractive. The low cost of bus tickets in London ensures that, despite its relatively slow journey speeds, the overall generalized cost to access employment by this mode is low. Such comparisons suggest that in order to encourage uptake of lower-carbon forms of transport, either the costs of these transport option to access employment must be reduced (through investment or lower ticket prices) or the cost of private car travel must be increased (e.g., through ensuring the cost of car journeys reflects the true cost, for instance through carbon pricing). This paper highlights accessibility functions for only two wards, however such curves are produced for all wards by the generalized cost tool (i.e., 633 such curves for the GLA area).

Such employment accessibility functions can be used to examine benefits to accessibility from infrastructure improvement, as an alternative to examining individual generalized cost improvements as shown in Figure 5, Figure 6 and Figure 7. Figure 9 shows the accessibility change from the Heathrow Villages ward as a result of different infrastructure scenarios which include investments from the Transport 2025 report (see Table 1). These curves show that accessibility as measured by generalized cost improves most notably in the bus mode (due to an increase in the frequency of bus services in the area). Other low-carbon modes, however, exhibit little overall change as a result of the proposed infrastructure improvements in the report. The largest difference in these scenarios is between road scenarios with and without the inclusion of the Congestion Charge. Since a large proportion of jobs (approximately 25%) are located within this part of central London, the Congestion Charge can be seen as a large incentive for people to use lower-carbon forms of transport to access this employment.

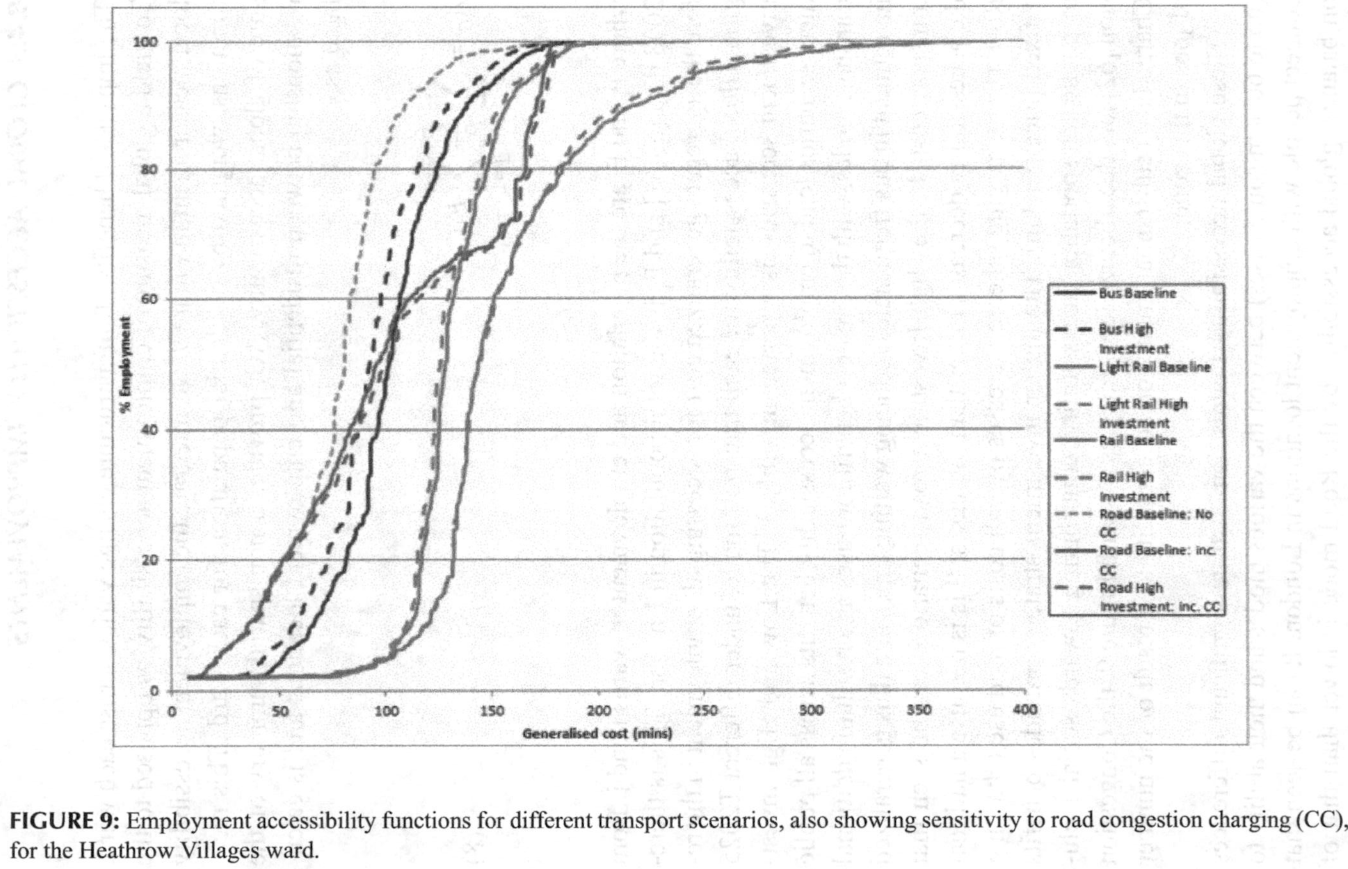

FIGURE 9: Employment accessibility functions for different transport scenarios, also showing sensitivity to road congestion charging (CC), for the Heathrow Villages ward.

8.4.3 GLOBAL ACCESSIBILITY IMPROVEMENTS

The impact and benefits of infrastructure change varies across each ward. To yield a global measure of employment accessibility, as opposed to the local measurement in Figure 9, we integrate the employment accessibility functions, weighted by residential population, for each ward. This measure of global accessibility, AG, provides a non-spatial summary of the relationship between residential and employment locations and is calculated as:

$$A_G = \sum_{i=1}^{N} \frac{P_i}{P_T} \sum_{j=1}^{N} E_j C_{ij} \tag{8}$$

where P_i and E_j are the population and employment in wards i and j, from 2008 data [59,61] and P_T is the total population in London. This is therefore the aggregate generalized cost for accessing all employment in the region. Global accessibility for each transport mode, under different T2025 investment scenarios, is reported in Table 3. The Low and High investment scenarios lead to employment accessibility benefits across all public transport modes, with the greatest benefit for buses as a result of additional investment in bus frequencies reducing waiting times and thus generalized transport cost. In the High Investment road scenario it can be seen that the accessibility decreases rather than increases; this is because a national Road User Charging scheme increases overall costs of road use (charging a fixed price per km). The higher investment strategy provides benefits to public transportation accessibility, contributing towards social inclusion [62] and decarbonization strategies. As with Figure 9, the Congestion Charge has a substantial impact on accessibility as a result of the number of jobs in this zone.

These global accessibility figures demonstrate significant differences (up to 60% in some cases) between the various modes and their ability to connect people with employment locations in London. It can be seen that on balance, global accessibility by the Road mode is lower than that of other public transport modes. This is in part due to some missing factors in the calculation of road generalized cost values (most significantly parking charges) but generally represents the lower cost of travel by car when taking into account all (monetary and temporal) factors. This demonstrates that encouraging the use of sustainable travel options must overcome the overall lower costs of private car travel in Greater London. However this global measure does not capture the local variability of accessibility across the area of interest, only providing a city-wide overview of the various modes. It is useful for comparison of various future investment scenarios and gives an understanding of the implications of policy decisions. Table 3 shows accessibility to employment locations only, whilst access to other services (e.g., shops, parks, hospitals) would have different patterns and totals.

TABLE 3: Global accessibility to employment, AG, for each transport mode under the T2025 transport infrastructure investment scenarios (a lower value means lower cost to access employment and therefore better employment accessibility), and sensitivity of the road mode to congestion charging (CC).

	Bus	Light Rail	Rail	Road:With CC	Road: No CC
Baseline	9193	13868	9767	8548	6027
Low investment	9063 (−1.4%)	13,605 (−1.9%)	9575 (−2.0%)	8520 (−0.3%)	5997 (−0.5%)
High investment	8285 (−9.9%)	13,325 (−3.9%)	9490 (−2.8%)	8698 (+1.7%)	6176 (−2.5%)

8.5 CONCLUSIONS

This paper has presented a GIS-based tool that has been developed to enable rapid characterisation of transport costs and accessibility over large spatial domains using readily available data. The results of this tool can be presented in a variety of ways, including as plots or graphs, and as spatial maps of generalized cost or accessibility. The key benefit of this approach is to enable rapid appraisal of the effects of new infrastructure developments on accessibility at a local to city-wide scale, and the comparison of accessibility patterns between competing modes. Te Brömmelstroet et al. [22] indicated that decision-makers find "visual map-based media to be a very useful tool for communicating accessibility" and that "planners prefer maps, while transport planners are more at ease when presented with quantitative outputs". The tool presented here has been designed to provide both such outputs using widely-available GIS software. Furthermore, this tool is provided as a free add-on for standard GIS and can use publicly-available data thereby addressing concerns of a significant proportion of local government organizations that lack the time, money, data, and computational skills to undertake accessibility analysis [22].

Application to Greater London shows that networks of different transport modes can be analyzed over large spatial scales, and reveals considerable spatial variability in transport costs and employment accessibility. Employment accessibility is shown to be a complex function of transport mode, network structure, location of jobs and journey origin. Globally, road-based modes provide the greatest accessibility, suggesting that investment may be needed in lower-carbon forms of transport in order to reduce the dominance of the private car in transport journeys (currently at 44% modal share). The Congestion Charge provides a substantial disincentive to commuting into central London by private transport and thus encouraging the use of other, less carbon-intensive modes of travel to access the jobs and services in this area. However, rail-based modes are still more beneficial on certain routes, highlighting the importance of considering spatial form in this type of analysis.

These accessibility functions provide decision-makers with depictions of the relative accessibility of key facilities (in this case employment locations) from a given location within an urban area. It is quickly demonstrated that certain modes of travel provide access to such locations for a lower generalized cost than other modes. It is possible, with the accessibility tool presented here, to examine the attractiveness of low-carbon modes of transport against less sustainable means of travel. Such functions also allow the testing of possible future infrastructure investment options, to quickly examine the impact of such investments on accessibility and to determine the effectiveness of options to increase the attractiveness of sustainable travel modes.

Rapid analysis of accessibility patterns helps explore a wide range of possible transport options to balance factors such as maintaining accessibility to employment, ensuring connectivity of residential areas, but also wider sustainability goals such as reduction in long-distance commuting, promoting low-carbon transport, and reducing travel emissions. Local accessibility changes (e.g., access to stations, effects of infrastructure improvements on short journeys) can be viewed in the context of city scale effects (e.g., global changes of accessibility) in order to ensure balanced portfolios of measures. Moreover, a rapid assessment tool allows inclusion of transport analysis in multi-sectoral analyses of urban sustainability studies, so that investment in transport infrastructure can be assessed alongside other investments in order to build a balanced portfolio of options which can assist in the development of sustainable urban areas (see Hall et al. [29]; Walsh et al. [30]; Dawson et al., [63]; and Echenique et al. [16] for studies of this kind).

In this study, we have considered only employment accessibility, but other measures such as access to green space, shops and other services could be readily included subject to data availability at the relevant spatial geography. We also considered each transport mode in isolation, and calculated access to each network independently of other possible travel modes. This enabled the relative accessibility of each transport mode to be compared directly, and for results of the analysis to be more readily interpreted and explained. However, we recognize that this may not always offer the least cost result for travel between two zones. A more realistic representation would allow for the computation of a walking, driving network path to a station—or indeed use of multiple modes of transport for the journey—which is the subject of ongoing work. Inevitably, broad scale models of this nature necessitate a number of assumptions that limit its potential for detailed transport infrastructure design and analysis. However, these limitations are justified in this case as the aim was to create an accessibility analysis tool that can be rapidly and widely-applied without the need for large, or bespoke, data collection exercises. Another advantage of such a simple approach is the ability to add more network layers, services or represent additional processes within the generalized cost calculation, if needed.

The value of such a tool, as opposed to a full macro transport model or microsimulation of traffic, is that many alternative scenarios of transport infrastructure and associated policies can be tested and compared quickly and easily. Constructing the modeling tools in a GIS framework facilitates parameterisation and preparation of these scenarios and enables immediate visual exploration and interpretation of the results. The tool has been made available for others to use (http://www.ncl.ac.uk/ceser/researchprogramme/software/). In addition, utilizing widely-available datasets, such as OpenStreetMap, demonstrates the transferability of the tool, and thus its future applicability to other cities in the world, particularly in the developing world where such transformations to sustainable transport are vital and where open and crowd-sourced data sources are becoming more prevalent.

REFERENCES

5. Pooler, J.A. A family of relaxed spatial interaction models. Prof. Geogr. 1994, 46, 210–217.
6. Liu, S.; Zhu, X. An integrated GIS approach to accessibility analysis. Trans. GIS 2004, 8, 45–62.
7. Geurs, K.T.; van Wee, B. Accessibility evaluation of land-use and transport strategies: Review and research directions. J. Transp. Geogr. 2004, 12, 127–114.
8. Bristow, G.; Farrington, J.; Shaw, J.; Richardson, T. Developing an evaluation framework for crosscutting policy goals: The Accessibility Policy Assessment Tool. Environ. Plan. A 2009, 41, 48–62.
9. Hull, A.; Silva, C.; Bertolini, L. Accessibility Instruments for Planning Practice, COST, Brussels; COST: Brussels, Belgium, 2012. Available online: http://www.accessibilityplanning.eu/wp-content/uploads/2012/10/COST-Report-1-FINAL.pdf (accessed on 15 December 2013).
10. Van Wee, B. Evaluating the impact of land use on travel behaviour: The environment versus accessibility. J. Transp. Geogr. 2011, 19, 1530–1533.
11. Grengs, J. Job accessibility and the modal mismatch in Detroit. J. Transp. Geogr. 2010, 18, 42–54.
12. Foth, N.; Manaugh, K.M.; El-Geneidy, A. Towards equitable transit: Examining transit accessibility and social need in Toronto, Canada, 1996–2006. J. Transp. Geogr. 2013, 29, 1–10.
13. Rydin, Y. Spatial planning for sustainable urban development. In Governing for Sustainable Urban Development; Earthscan: London, UK, 2010; pp. 107–110.
14. Hansen, W. How accessibility shapes land use. J. Am. Inst. Plan. 1959, 25, 73–76.
15. Lowry, I.S. A Model of Metropolis RM-4035-RC; The Rand Corporation: Santa Monica, CA, USA, 1964.
16. Levinson, D.M. Accessibility and the journey to work. J. Transp. Geogr. 1998, 6, 11–21.
17. Forrester, J.W. Urban Dynamics; MIT Press: Cambridge, MA, USA, 1969.

18. Wegener, M. Applied models of urban land use, transport and environment: State of the art and future developments. In Network Infrastructure and the Urban Environment. Advances in Spatial Systems Modelling; Lundqvist, L., Mattsson, L.-G., Kim, T.J., Eds.; Springer Verlag: Berlin/Heidelberg, Germany, 1998; pp. 245–267.
19. Waddell, P. UrbanSim: Modeling urban development for land use, transportation and environmental planning. J. Am. Plan. Assoc. 2002, 68, 297–314.
20. Echenique, M.H.; Hargreaves, A.J.; Mitchell, G.; Namdeo, A. Growing cities sustainably: Does urban form really matter? J. Am. Plan. Assoc. 2012, 78, 121–137.
21. Clarke, K.C. A decade of cellular urban modeling with SLEUTH: Unresolved issues and problems. In Planning Support Systems for Cities and Regions; Brail, R.K., Ed.; Lincoln Institute of Land Policy: Cambridge, MA, USA, 2008; pp. 47–60.
22. Batty, M. Building a Science of Cities; UCL Working Papers Series No. 170; University College London: London, UK, November 2012.
23. Hunt, J.D.; Kriger, D.S.; Miller, E.J. Current operational urban land-use-transport modelling frameworks: A review. Transp. Rev. 2005, 25, 329–376.
24. Cooper, S.; Wright, P.; Ball, R. Measuring the accessibility of opportunities and services in dense urban environments: Experiences from London. In Proceedings of the European Transport Conference 2009, Noordwijkerhout, The Netherlands, 5 October 2009.
25. Brown, M.; Wood, T. Accession—Accessibility analysis for local transport planning. In Proceedings of the European Transport Conference 2004, Strasbourg, France, 4 October 2004.
26. Te Brömmelstroet, M.; Silva, C.; Bertolini, L. COST Action—Assessing Usability of Accessibility Instruments; COST Office: Brussels, Belgium, 2014.
27. O'Sullivan, D.; Morrison, A.; Shearer, J. Using desktop GIS for the investigation of accessibility by public transport: An isochrone approach. Int. J. Geogr. Inf. Sci. 2000, 14, 85–104.
28. Lei, T.L.; Church, R.L. Mapping transit-based access: Integrating GIS, routes and schedules. Int. J. Geogr. Inf. Sci. 2010, 24, 283–304.
29. Benenson, I.; Martens, K.; Rofé, Y.; Kwartler, A. Public transport versus private car GIS-based estimation of accessibility applied to the Tel Aviv metropolitan area. Ann. Reg. Sci. 2010, 47, 499–515.
30. Mavoa, S.; Witten, K.; McCreanor, T.; O'Sullivan, D. GIS based destination accessibility via public transit and walking in Auckland. J. Transp. Geogr. 2012, 20, 15–22.
31. Chen, S.; Claramunt, C.; Ray, C. A spatio-temporal modelling approach for the study of the connectivity and accessibility of the Guangzhou metropolitan network. J. Transp. Geogr. 2014, 36, 12–25.
32. Curtis, C.; Scheurer, J. Planning for sustainable accessibility: Developing tools to aid discussion and decision making. Prog. Plan. 2010, 74, 53–106.
33. Hall, J.W.; Dawson, R.J.; Walsh, C.L.; Barker, T.; Barr, S.L.; Batty, M.; Bristow, A.L.; Burton, A.; Carney, S.; Dagoumas, A.; et al. Engineering Cities: How Can Cities Grow Whilst Reducing Emissions and Vulnerability?; The Tyndall Centre for Climate Change Research: Newcastle, UK, 2009.
34. Walsh, C.L.; Dawson, R.J.; Hall, J.W.; Barr, S.L.; Batty, M.; Bristow, A.L.; Carney, S.; Dagoumas, A.; Ford, A.; Tight, M.R.; et al. Assessment of climate change mitigation and adaptation in cities. Proc. ICE: Urban Des. Plan. 2011, 164, 75–84.

35. Wachs, M.; Kumagi, T.G. Physical accessibility as a social indicator. Soc.-Econ. Plan. Sci. 1973, 7, 437–456.
36. Allen, W.B.; Liu, D.; Singer, S. Accessibility measures of U.S. metropolitan areas. Transp. Res. B 1992, 27, 439–449.
37. Handy, S.L.; Niemeier, D.A. Measuring accessibility: An exploration of issues and alternatives. Environ. Plan. A 1997, 29, 1175–1194.
38. Koenig, J.G. Indicators of urban accessibility: Theory and application. Transportation 1980, 9, 145–172.
39. Miller, H.J. Measuring space-time accessibility benefits within transportation networks: Basic theory and computational methods. Geogr. Anal. 1999, 31, 87–212.
40. Makri, M.C.; Folkesson, C. Accessibility Measures for Analyses of Land Use and Travelling with Geographical Information Systems; 1999. 1999. 1999.
41. Hillman, R.; Pool, G. GIS-based innovations for modeling public transport accessibility. Traffic Eng. Control 1997, 38, 554–559.
42. De Ortuzar, J.D.; Willumsen, L.G. Modelling Transport; John Wiley: New York, NY, USA, 2011.
43. Grey, A. The generalized cost dilemma. Transportation 1978, 7, 261–280.
44. Bruzelius, N.A. Microeconomic theory and generalised cost. Transportation 1981, 10, 233–245.
45. WebTAG Transport Analysis Guidance. Available online: http://www.dft.gov.uk/webtag/ (accessed on 30 September 2009).
46. Mackie, P.J.; Wadman, M.; Fowkes, A.S.; Whelan, G.; Nellthorp, J.; Bates, J. Values of Travel Time Savings in the UK. Available online: http://www.dft.gov.uk/pgr/economics/rdg/valueoftraveltimesavingsinth3130 (accessed on 16 May 2009).
47. Nichols, A.J. Standard Generalised Cost Parameters for Modelling Inter-Urban Traffic and Evaluating Inter-Urban Road Schemes, Note 255; Department of the Environment, Mathematical Advisory Unit: London, UK, 1975.
48. Hopkinson, P.; Wardman, M. Evaluating the demand for new cycle facilities. Transp. Policy 1996, 3, 241–249.
49. Noland, R.B.; Kunreuther, H. Short-run and long-run policies for increasing bicycle transportation for daily commuter trips. Transp. Policy 1995, 2, 67–79.
50. Rodrıguez, D.A.; Joo, J. The relationship between non-motorized mode choice and the local physical environment. Transp. Res. Part D: Transp. Environ. 2004, 9, 151–173.
51. Dijkstra, E.W. A note on two problems in connexion with graphs. Numer. Math. 1959, 1, 269–271.
52. ESRI. Network Analyst Documentation. Available online: http://www.esri.com/software/arcgis/extensions/networkanalyst/index.html (accessed on 16 May 2009).
53. Feldman, O.; Simmonds, D.; Zachariadis, V.; Mackett, R.; Bosredon, M.; Richmond, E.; Nicoll, J. SIMDELTA—A microsimulation approach to household location modelling. In Proceedings of the World Conference on Transport Research 2007, University of California, Berkeley, CA, USA, 24–28 June 2007.
54. National Highway Traffic Safety Administration and the Bureau of Transportation Statistics National Survey of Pedestrian and Bicyclist Attitudes and Behaviors. Available online: http://www.nhtsa.gov/DOT/NHTSA/Traffic%20Injury%20Control/Articles/Associated%20Files/810971.pdf (accessed on 2 January 2014).

55. TFL. Transport 2025—Transport Vision for a Growing World City; TfL Group Transport and Planning Policy: London, UK, 2006.
56. TFL. London Travel Report; Transport for London: London, UK, 2006; p. 40.
57. TFL. London Congestion Charge Information; Transport for London: London, UK. Available online: http://www.tfl.gov.uk/roadusers/congestioncharging/6709.aspx (accessed on 12 December 2013).
58. Cyclestreets Journey Planner, How It Works. Cyclestreets.net, Available online: http://www.cyclestreets.net/journey/help/howitworks/ (accessed on 25 November 2014).
59. Crossrail Route Maps. Available online: http://www.crossrail.co.uk/route/maps/route-map (accessed on 29 October 2014).
60. TFL. Citizen Hub Consultation on East-West Cycle Superhighway. Available online: https://consultations.tfl.gov.uk/cycling/eastwest (accessed on 22 October 2014).
61. Travel in London, Report 6. p. 124. Available online: https://www.tfl.gov.uk/cdn/static/cms/documents/travel-in-london-report-6.pdf (accessed on 20 October 2014).
62. GLA. The London Plan: Spatial Development Strategy for Greater London; Greater London Authority: London, UK, 2011.
63. ONS Nomis Annual Business Inquiry Dataset. Available online: http://www.nomisweb.co.uk (accessed on 14 January 2010).
64. DfT. National Travel Survey 2012: Statistical Release; UK Department for Transport: London, UK. Available online: https://www.gov.uk/government/uploads/system/uploads/attachment_data/file/243957/nts2012-01.pdf (accessed on 10 October 2014).
65. ONS National Projections: UK Population to Exceed 65 m by 2018. Office for National Statistics; Available online: http://www.ons.gov.uk/ons/dcp171780_229187.pdf (accessed on 5 December 2013).
66. Farrington, J.; Farrington, C. Rural accessibility, social inclusion and social justice: Towards conceptualisation. J. Transp. Geogr. 2005, 13, 1–12.
67. Dawson, R.J.; Ball, T.; Werritty, J.; Werritty, A.; Hall, J.W.; Roche, N. Assessing the effectiveness of non-structural flood management measures in the Thames Estuary under conditions of socio-economic and environmental change. Glob. Environ. Chang. 2011, 21, 628–646.

Table 2 is not available in this version of the article. To view this additional information, please use the citation on the first page of this chapter.

PART III

SUSTAINABLE HOUSING

CHAPTER 9

Greenhouse Gas Implications of Urban Sprawl in the Helsinki Metropolitan Area

SANNA ALA-MANTILA, JUKKA HEINONEN, AND SEPPO JUNNILA

9.1 INTRODUCTION

Warming of the global climate system is unequivocal [1]. Even though there is a scientific consensus on the existence of climate change and the human role in it [2], the role of cities and urban activities in the global warming has been hotly debated in recent years [3,4]. The general opinion is that sprawling urban development is detrimental to the environment and that the denser the structures are the better. When urban structures grow outwards, the people living in loosely populated surrounding areas are more car-dependent and tend to live in less eco-efficient apartments. Consequently, in the literature the studies on the effects of sprawl have mostly investigated emissions from driving and home heating [5,6,7].

However, urbanization and its supporting infrastructure per se do not alone cause climate change: it is also the people–living, moving, and consuming in cities. Thus, it is not only the emissions from heating, electricity and driving that need to be taken into account but also the emissions from all other consumption must be kept in mind. Only with this kind of holistic assessment, a sufficient level of understanding about the greenhouse gas consequences of sprawl can be reached. Similar consumption-based assessments are gaining more and more foothold in the literature, and it is acknowledged that they suit particularly well for urban environments and should at least act as complements to the traditional production-based methods [8,9].

The proliferation of semi-detached and detached houses at the urban edge is one of the best-known characteristics of sprawl [10,11]. Combined with looser suburban structures and less accessible public transportation, sometimes not even very far from the urban core, this less dense low-rise living represents the sprawl in an intra-urban context. Thus, in this paper, we distinguish between different types of houses and analyze the differences in typically inner-city high-rise and typical suburban low-rise lifestyles and their GHG effects. Furthermore, it has been argued that one reason for low-rise living are policies that encourage home owning and implicitly encourage people to move away from higher density living [11]. The analysis was done within a single metropolis, the Helsinki Metropolitan Area [HMA], in Finland. The HMA is the capital region of Finland that consists of four cities with more than one million inhabitants. This type of sprawl is a very interesting phenomenon to analyze as within a single metropolis the residents still maintain rather close proximity to all the consumption opportunities that the city or metropolitan area offers but rely predominantly on private driving and have more living space to be heated, furnished, and filled with domestic appliances.

In this study, we assess households' GHG impacts by calculating their consumption-based carbon footprints, i.e., direct and indirect life-cycle greenhouse gas emissions either home or elsewhere, which are ultimately caused by consumption of products and services [39]. Following the literature, the footprints are calculated by combining the input data from a Finnish household budget survey with the environmentally-extended input-output [EE I-O] model ENVIMAT based on the Finnish economy [13,14]. The carbon footprints are further elucidated with a multivariate regression analysis. What we add to the previous discussions on sprawl is the perspective of how emissions from consumption in its broad sense, beyond those from solely driving and housing energy, affect the GHG consequences and their policy implications within a single metropolitan area. The study sheds light on the effects of proximity on consumption, suggesting that, in denser agglomerations, indirect emissions from consumption of goods and services grow, as has been argued to happen on a country level [12]. Consumption-based approach is expected to facilitate the understanding of urban lifestyles that are related to urban sprawl. We demonstrate that the GHG impacts of consumption vary by the type of dwelling, but the overall differences between low- and high-rise dwellers remain rather minor and much more equivocal than previously assumed. We believe that an understanding of both the differences and similarities in lifestyles offers important insights for sustainable policy-design and urban planning.

The structure of our paper is as follows: first, the research design, i.e., consumption data, both the methods and research process, is presented in Section 2. The results are presented in Section 3, and a discussion follows in Section 4. We finish with conclusions in Section 5.

9.2 RESEARCH DESIGN

9.2.1 THE CASE AREA AND INPUT DATA

The selected case area, the HMA, with a population of over one million, is the capital region of Finland and consists of four cities: Helsinki, Vantaa, Espoo, and Kauniainen. It is known that the Helsinki region has spread out widely, and the spreading out still continues [15]. The area shares many traits with the globe's biggest metropolises, the results thus giving indication of more general patterns. Furthermore, within such a clearly defined and geographically restricted area, some of the most uncertain assumptions of input-output technique, such as that of homogenous prices, are closer to reality. In addition, the HMA is interconnected by an extensive public transportation network, and possibilities to choose between private and public modes of transport are quite diverse, with free parking made available for those who need private cars to reach public transport hubs. In line with the sprawl hypothesis, the division of housing types within the area is clear: the city core is dominated by apartment houses, and semi-detached and detached houses are the main types of houses at the edge. Some key characteristics of the four areas can be found from the Table 1.

TABLE 1: Key characteristics of Helsinki Metropolitan Area (HMA) areas [36,46].

	Helsinki	Espoo	Vantaa	Kauniainen
Population size (31.12.2012)	603,968	256,824	205,312	8,910
Density (inhabitants per km^2)	2,826	823	861	1,513
Apartment building of housing stock (%)	86	58	62	46

The input data consists of the latest Finnish Household Budget Survey data from 2006 [16]. The dataset is cross-sectional in nature, and alongside the actual detailed consumption expenditure data, arranged according to the international COICOP (Classification of Individual Consumption by Purpose) classification, the data contain a wide array of background and income information for each household. Budget Survey is a sample survey study that employs a one-stage stratified cluster sampling design. The final sample size of the survey was 4,007 households. However, the original sample was double the size, and the magnitude of response was only 47.7%—a situation that can be described as under coverage. However, non-response bias can be significantly reduced using weight coefficients, and systematic biases can be avoided. In order to allow generalization of our results, the weight coefficients are utilized throughout this paper.

In this paper, the sample of households is restricted to the HMA. Even if, with its approximately one million inhabitants, the Helsinki Metropolitan Area is relatively small in size when compared to the world's largest megacities, it has many characteristics of a metropolis: a little more than a third of Finland's GDP is produced there, both the levels of education and wages are higher than the average, and there is an extensive public transport network. According to Alanen et al., the challenges of metropolises are often different than in the rest of Finland, but similar to the other metropolises [17].

In order to support our choice to restrict the sample to the Helsinki Metropolitan Area, it can be briefly compared to the rest of the country. In short, the households in the metropolitan area are richer and smaller: the average household size in the HMA is 1.93, whereas it is 2.16 in the rest of country. Likewise, the average disposable income per household is €42,533 in the HMA, whereas it is €33,634 for the rest of country. Furthermore, for households living in the metropolitan area, the average amount of cars per household is more than a one-third lower than in the rest of the country and the share of carless households in the HMA is 20 percentage points higher, illustrating the availability of public transportation there. In our data, 59% of households living in Helsinki metropolitan area are homeowners. In the rest of the country the figure is 70%. Furthermore, in the metropolitan area, the share of households living in low-rise dwellings (i.e., in detached or semi-detached houses) is relatively low and is 38 percentage points lower than in the rest of the country.

Table 2 describes our sample data. The house types are divided to low-rise and high-rise categories. The low-rise category consists of households living in either detached (n = 97) or semi-detached houses (n = 73), and the high-rise one refers to households living in apartment houses (n = 398).

TABLE 2: Descriptive statistics for high-rise and low-rise sub-samples.

	Low-Rise (n = 170)	High-Rise (n = 389)
Household characteristics:		
Disposable income (€)	62 719	35 410
Consumption expenditure (€)	46 141	29 710
Average household size	2.45	1.75
Number of cars per household	1.00	0.51
Share of carless households	0.26	0.53
Share of households with children	0.37	0.19
Building type characteristics:		
Average living area (square meters)	118.31	61.28
Per-capita living area (square meters)	48.28	35.05
Number of rooms per person*	1.93	1.53
Share of owner-occupied dwellings	0.81	0.51

**rooms [no kitchen] per a household member.*

In household characteristics, there are some interesting differences between sub-samples. As expected, households living in semi-detached or detached houses are the wealthiest and own the most cars, since the hypothesis was that low-rise lifestyle is a lifestyle leading to car-dependency. Accordingly, more than half of the households in the apartment houses are carless, the precise figure being 53%, which indicates that the apartment houses in the Helsinki Metropolitan Area are, on average, located close to public transportation facilities. Furthermore, families with children are likely to live in the areas of sprawl, i.e., in detached or semi-detached houses, the share of families with children in apartment houses being less than one-fifth. Interestingly, 42% of these high-rise households with children are single-parent families.

In addition, housing type differences are rather expected. Average living space in low-rise houses is 118.3 m^2, almost twice the living area of apartment houses. When compared on per-capita level, the living area in low-rise houses is 1.4 times bigger than in high-rise ones. However, the actual difference is lower due to the fact that the apartment dwellers use shared spaces, such as hallways and storage facilities, which are not included in these self-reported figures. Nevertheless, they also create GHG emissions. These common spaces in apartment buildings are taken into account in this paper by allocating the emissions from them for their residents. This allocation is based on official statistics on the finance of housing companies [18]. There are also significant differences in the share of owner-occupied dwellings between the sub-samples. As expected, the majority (81%) of people living in low-rise buildings are home-owners and, thus, urban sprawl seems to be linked with the proliferation of owner-occupied houses at the expense of rented houses. On the other hand, the division between tenants and homeowners in apartment houses is approximately half and half.

9.2.2 ENVIRONMENTALLY EXTENDED INPUT-OUTPUT MODEL

In this paper, we apply an environmentally extended input-output (EE I-O) model, called ENVIMAT, which has been recently developed for the Finnish economy [13,14]. The environmentally extended input-output analysis is one of the possible approaches of life-cycle analyses (LCA) that measure the direct and indirect environmental impacts, i.e., the impacts from cradle to grave, of a functional unit under consideration [19,20]. Besides the environmentally extended input-output analysis, LCAs can be done with a process-based life-cycle analysis, and these two when combined are referred to as hybrid-analysis [19].

The environmentally extended input-output method is based on the input-output tables for national economies and related environmental impact categories. The EE I-O method is often referred to as a "top-down" method, and it measures the product flows in monetary units [20,21]. Usually the benefits of the EE I-O method are said to be its capability to give an overview of the life-cycle effects of production and consumption of an economy, lack of problems related to system boundaries and truncation errors, and the easiness and repeatability of calculations once the model is developed [19,22].

The EE I-O model applied in this article, ENVIMAT, is developed by the Finnish Environmental Institute, University of Oulu's Thule Institute and MTT Agrifood Research Center. It measures the direct and indirect life-cycle environmental impacts caused by the Finnish economy. We utilized the 2005 consumer-price version of the model that has 52 commodity groups, classified according to international COICOP classification (Classification of Individual Consumption by Purpose), and related GHG emission intensity for each sector (emission per monetary amount used to sector). In the consumer-price version of the model the intensity quantifies the amount of life-cycle emissions, both direct and indirect, caused by one euro consumed to a product or service. Contrary to producer-price based version these intensities also take into account the GHGs caused by retail and transportation. All the intensities can be found from Seppälä et al. [13] The expenditure data is also organized according to the COICOP classification, so the model-data fit is excellent. In addition, one of the advantages of ENVIMAT is that it relaxes the often-problematic domestic technology assumption [DTA-assumption] and uses a hybrid approach that distinguishes between domestic and foreign production technologies [13,14]. However, EE I-O models have some built-in weaknesses and uncertainties that are discussed later in the paper.

9.2.3 MULTIVARIATE REGRESSION ANALYSIS

This section discusses our empirical approach, which is based on multivariate regression analysis where the dependent variable is per-capita greenhouse gas emissions, e.g., the carbon footprint. The regression analysis is employed to further analyze the calculated footprints and to find out which factors actually affect them. As our starting point, we use the non-linear exponential relationship between environmental impact and households' expenditures [23,24,25,26]:

$$Y_i = A^{\beta 0} \times E_i^{\beta 1} \times f_1(D_1) \times f_2(D_2) \times \varepsilon_i \qquad (1)$$

The dependent variable Y is per-capita carbon footprint, A is constant, E is per-capita expenditure, and dummy variable D_1 refers to housing type and D_2 refers to household size. For dummy variable $f_i = \exp(\eta D_i)$, and for multiplicative error term $\varepsilon_i = \exp(u_i)$.

The non-linear relationship of (1) is linearized with natural logarithm transformation. The transformed model satisfies the assumptions of a general linear model, and thus the parameters can be conveniently estimated using the highly developed theory of linear relationships. We follow the literature and obtain the following equation:

$$\ln Y_i = \beta_0 + \beta_1 \ln E_i + D_1 + D_2 + u_i \qquad (2)$$

For the log-transformed equation (2), an estimate for slope β and partial regression coefficient for explanatory dummy variable (D_i) can be efficiently estimated with weighted least squares (WLS). One of the benefits of the log-log model (2) is that the estimate β for continuous explanatory variable is elasticity, which in our case tells us the value of the so-called expenditure elasticity of carbon, i.e., it describes how much a relative change in expenditures will affect the relative demand for the dependent variable. The reported standard errors are heteroskedasticity robust.

9.2.4 RESEARCH PROCESS

First, we calculate the carbon footprints of the Helsinki Metropolitan dwellers by combining greenhouse gas intensities, derived from the ENVIMAT-model, with household budget survey data. This is done by aggregating expenditure data's categories to match the 52 COICOP categories of ENVIMAT and then multiplying expenditures with the corresponding ENVIMAT sector's value of greenhouse gas intensity (CO_2 equivalents per euro). It has been broadly acknowledged that combination of the two allows the assessment of the amount of greenhouse gases that consumption choices cause both directly and indirectly [23,24,25,26].

However, we made certain modifications to the straightforward input-output method. Firstly, we multiplied the living area reported by each household in the sample by the average rent in the HMA [27] to erase the bias resulting from variations in the housing price levels within the area. Secondly, the households not living in owner-occupied detached houses pay a significant share of their electricity, heating, and building maintenance with their housing management fees or rents [47]. By utilizing living space information reported by each household and the average expenditure per square meter per month information on what residents are paying for with their management fees from statistics of housing companies [18] we estimated additional heating, electricity, and maintenance expenditures for those living in semi-detached or apartment houses. However, there remains a certain level of uncertainty related to the fact that the heating methods differ, as, naturally, different forms of energy have different CO_2-profiles. For example, relatively CO_2-efficient district heating is the prevailing mode for apartment houses but less widespread among low-rise houses. However, the share of district heating is as high as 57% even in the low-rise area, and the rest of the area is mainly heated by electricity.

Thus, we believe that the uncertainty remains at tolerable levels and do not take these differences into account in our analysis.

We also made one modification to the E-E I-O model itself. We noticed that when using the original model significantly high amounts of emissions were assigned to some households and we traced down these to the fact that household had high expenditure on hot water, district heating, or natural gas. The related ENVIMAT intensity of 34.6 kgCO_2-eq/euro is clearly higher than the intensities of other groups whose average is 0.5 kgCO_2-eq/euro. This turned out to be a model error and we corrected sector's intensity to 4.3, according to the information received from the model developer [50].

In this paper, we associate owner-occupied detached-house living with sprawl. Furthermore, instead of looking only at average footprints, we analyze footprints' direct and indirect shares based on building type. With the direct share we refer to energy demand for home, second home, summer house, and gasoline for private cars. The indirect share consists of all the rest, i.e., greenhouse gases caused by consumption of products and intangible services. In the earlier literature on sprawl, heating and private driving have been widely discussed, whereas the aforementioned indirect consequences of lifestyles have been predominantly ignored, and thus this paper is a step forward. However, a similar direct-indirect distinction has been used by e.g., [28] and [29]. With this setting, we can analyze whether the lifestyles in the sprawl areas differ from the rest of the metropolitan area, in terms of both direct and indirect emissions. Furthermore, conclusions from the overall GHG impacts of sprawl in the HMA can be drawn.

More precisely, the direct categories are energy and fuels for private driving. Energy demand refers to electricity, gas, liquid, and solid fuels, and heat energy requirements. The indirect categories are housing, tangibles, food and beverages, and intangibles. The housing category consists of all non-energy expenditures related to housing, i.e., rentals, maintenance and repair of the dwelling, and miscellaneous services relating to the dwelling. Tangibles include all the products not included in the food and beverage category. Intangibles are services of all kinds. When the results are presented in the next chapter, the direct GHG categories are indicated with letter D in brackets, and letter I in brackets denotes indirect categories.

We then further analyze these footprints with a multivariate regression technique in order to study how expenditure and household size affect the footprints of different types of dwellers. This tradition dates back to the 1970s when the first study of direct and indirect energy consumption of U.S. households was done [30]. In addition, more recent examples of such regression can be found from the literature [22,23,24,25,31,32]. However, our analysis is the first such analysis focusing on the within metropolitan area differences, with a special attention to house types.

TABLE 3: Statistics for sub-samples according to the housing type.

	Low-Rise (N = 170)		High-Rise (n = 397)	
	Mean	SE	Mean	SE
Total carbon footprint	14.83	1.07	12.98	0.46
Energy (D)	3.66	0.44	3.10	0.13
Fuels (D)	1.14	0.14	0.86	0.08
Housing (I)	3.58	0.24	3.12	0.12
Intangibles (I)	2.31	0.29	2.36	0.16
Food and beverages (I)	2.13	0.13	1.86	0.06
Tangibles (I)	2.01	0.22	1.68	0.10

9.3 RESULTS

According to our assessment, the average of annual per-capita carbon footprint of a Helsinki Metropolitan dweller is 13.5 t CO_2-eq, with 95% confidence interval, ranging from 12.6 t CO_2-eq to 14.3 t CO_2-eq. Despite the remarkable differences in disposable income levels for low-rise dwellers, the average footprint of low-rise dwellers is only 1.8 t CO_2-eq–or 14%–larger than that of high-rise dwellers. The average for low-risers is 14.8 t CO_2-eq and 13.0 t CO_2-eq for high-risers.

Besides looking at the average figure, we decompose the footprints to six sub-categories, of which two are direct and the remaining four indirect, as explained earlier. Following the overall result, the partial footprints of low-rise

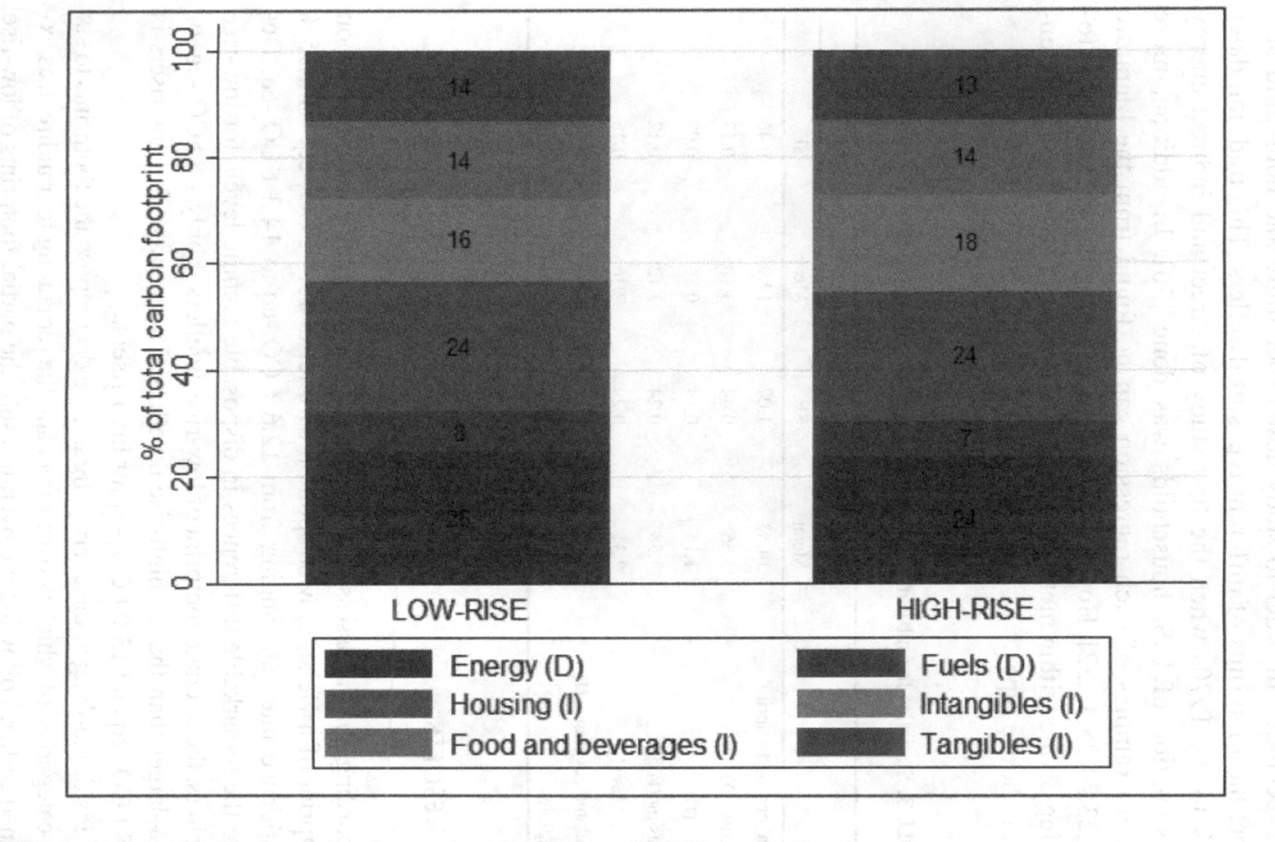

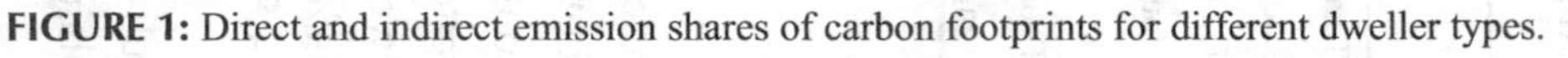
FIGURE 1: Direct and indirect emission shares of carbon footprints for different dweller types.

dwellers in all the direct sub-categories exceed those of high-rise dwellers. Here, the differences between sub-samples are the biggest in the direct energy category. The picture of dweller-type differences is rather similar when looking at the indirect categories, with an exception of the sub-category of intangibles where the footprint of high-rise dwellers slightly exceeds the footprint caused by service demand by those living in the low-rise houses. The averages of these sub-category GHG emissions are presented in Table 3.

TABLE 4: Results from regression models 1–3.

	Regression 1			Regression 2			Regression 3		
Variable	Estimate	SE	% effect	Estimate	SE	% effect	Estimate	SE	% effect
ln (E/capita)	0.82***	0.02		0.82***	0.02		0.78***		
Dwelling (low-rise)									
high-rise				0.01	0.03	1.0	−0.03	0.03	−3.3
Dwellers (1 person)									
2 persons							−0.04	0.03	−3.9
more than 2							−0.16***	0.03	−15.2
N	568	568	568						
R^2	0.7954	0.7955	0.8122						

*** *Statistically significant at 0.001 level.*

One of the most interesting observations about Table 3 is that the absolute amount of greenhouse gases traceable to indirect products and services is relatively stable for both dweller types. Even though the mean disposable income of high-rise dwellers is clearly lower than that of low-rise dwellers, the differences in the amounts of emissions embodied in consumed products and services are fairly moderate. An equally interesting matter is the difference in GHG's from fuel combustion for private driving: the difference between low-rise dwellers and those living in apartment houses is only 0.3 t despite the significant difference in the number of cars possessed on average. In order to further demonstrate the differences between dwellers living in different types of houses, we plot the percentage shares of home energy use, fuel, and emissions embodied in consumption of products and services. The results are presented in Figure 1.

From Figure 1, it can be concluded that there are no substantial differences in the relative sources of greenhouse gas emissions. Actually, all the sub-category shares are rather constant, and the differences are two percentage points at their largest. Figure 1 also tells us that the emissions from indirect sources constitute approximately two-thirds of the total carbon footprints, whatever the type of dweller or dwelling. Energy and housing are the two most important sub-categories, alone covering half of the carbon footprint.

The next steps of our analysis are the regression analyses based on equation (2). When dummy variables are included, the chosen categories are compared to the reference group in parenthesis and the percentage difference in carbon footprints is approximated with $p = 100(\exp\beta - 1)$ [38]. First, we perform a univariate regression with logarithmic per-capita expenditure (regression 1). This is followed by multivariate regressions where dwelling type (regression 2) and household size (regression 3) are added to explanatory variables. The results are presented in Table 4 below.

In the regression model 1 the average expenditure elasticity of an HMA resident is 0.82, meaning that a 10% increase in expenditure is related, on average, to an 8.2% increase in consumption-based greenhouse gas emissions. Thus, carbon footprints, even if due to indirect demand, can be defined to be due to normal goods with expenditure elasticity very close to unity. In regression 2, a dwelling-type dummy is added to the model. When compared to the reference group, that is low-rise dwellers, the estimate for high-rise gets a positive sign, meaning that, when controlling the amount of expenditure, the carbon footprints of high-rise dwellers are slightly larger than those of low-rise dwellers. Furthermore, when a household size variable is added to the model, the result turns around and the explanatory power of the model increases, as presented in regression model 3. Now, the footprints of high-rise dwellers are 3% lower than those of high-risers. The role of household size has a dominant role in the results and seems to have a clear effect on the carbon footprint of different dwelling types. In the literature, it has been suggested that there are so-called economies of scale in GHG's, implying that the coefficients on bigger household sizes should be negative. Likewise, in our model the coefficients on household size get negative values, but the immediate impact is rather small, as the per-capita carbon footprints are only 4% smaller in a two-person household when compared to a single-person household. However, the coefficient is not statistically significant. The main efficiencies related to the size of a household become apparent in households with more than two members, their total per-capita carbon footprint being 15% smaller than that of single dwellers. The difference is statistically significant.

We also control the wealth level with income per capita. The results are presented in Table 5 below. These income elasticity values are smaller than those for expenditure, but in general the results are in line with the results presented in Table 4. The average income elasticity is 0.60, meaning that a 10% rise in per-capita disposable income is related to a 6% increase in carbon footprint. This is related to the role of saved income. It might also be possible that time restricts the growth in purchases when the disposable income increases above a certain level. Therefore, not enough free time is left for consuming the increasing income (see further e.g., [33]). When income, instead of expenditure, is controlled, the effect of household size grows slightly. The effect of dwelling-type remains rather minor and statistically insignificant.

TABLE 5: Results from regression models 4–6.

	Regression 4			Regression 5			Regression 6		
Variable	Estimate	SE	% effect	Estimate	SE	% effect	Estimate	SE	% effect
ln (I/capita)	0.60***	0.02		0.60***	0.03		0.56***	0.03	
Dwelling (low-rise)									
high-rise				0.02	0.03	2.5	−0.03	0.03	−2.9
Dwellers (1 person)									
2 persons							−0.06	0.03	−5.5
more than 2							−0.19***	0.04	−17.6
N	568	568	568						
R^2	0.6107	0.6112	0.6338						

****Statistically significant at 0.001 level.*

Finally, we analyze how direct and indirect emission shares behave when analyzed at the disaggregated level. We use the disaggregated per-capita emission and corresponding expenditure shares on the direct and indirect consumption categories instead of looking at the aggregated per-capita carbon footprints and expenditure levels. The models are of type (2), where the dependant variable is either the direct share of the carbon footprint (regression 7) or the indirect share of it (regression 8). Likewise, E is either direct or indirect per-capita expenditure, and dwelling-type and number of dwellers are used as dummy explanatory variables. The results are presented in Table 6 below.

TABLE 6: Results from regression models 7–8.

	Regression 7			Regression 8		
Variable	Estimate	SE	% effect	Estimate	SE	% effect
ln(Edirect/capita)	0.76***	0.02				
ln(Eindirect/capita)				1.00***	0.01	
Dwelling (low-rise)						
high-rise	0.01	0.03	0.9	0.04***	0.01	3.8
Dwellers (1 person)						
2 persons	–0.14***	0.02	–12.8	0.01	0.01	0.9
more than 3	–0.28***	0.02	–24.6	0.01	0.01	0.7
N	568.00	568.00				
R^2	0.8930	0.9590				

****Statistically significant at 0.001 level.*

Indirect emissions grow more steeply with expenditure level than do the direct ones. The direct expenditure elasticity is 0.76, meaning that a 10% rise in expenditure in housing energy and private driving is related to a growth of 7.6% in direct emissions. The corresponding figure for indirect emissions is 1, meaning that indirect GHGs grow linearly with expenditure on products and services. The dwelling type variable tells us that high-risers have on average 1% higher direct emissions when the amount of expenditure on direct categories is constant. The results is surprising as one could, based on earlier studies hypothesize lower direct emissions for high-rise dwellers. The tendency is similar for indirect GHGs, but the effect is more pronounced and statistically significant, which seems logical as the growing consumption could be expected to be directed more towards indirect consumption. Living in high-rise buildings leads here, on average, to 3.8% more of indirect emissions. Further, for direct emissions the number of dwellers plays a greater role than the type of dwelling, which, based on the earlier results above, also seems to be logical. Following the economies of scale hypothesis, those living in a two-person household are responsible, on average, for 13% less emissions per capita than those living alone. Noticeably, these economies of scale are not present when indirect emissions are explained in regression 8, which could be expected since the economies of scale typically have more direct influence on fixed costs (close to direct emissions here) than variable costs (close to indirect emissions here). Coefficients on household size carry positive signs but lack statistical significance. To sum up, it seems that there are factors related to high-rise lifestyle factors that can lead to higher indirect emissions. Direct emissions are more or less related to household size alone, and dwelling type's role remains insignificant when the influence of household size is separated.

9.4 DISCUSSION

9.4.1 THE GHG IMPACTS OF URBAN SPRAWL IN HMA

The purpose of this paper was to explore how the phenomenon of urban sprawl is reflected in the GHG emissions of the residents of the HMA in Finland. For this purpose, the HMA was divided according to the housing type. We followed a premise that living in detached and semi-detached-houses can be typified as living in less dense sub-urban areas where the proximity to services is lower and public transportation networks are less efficient. This mode of living thus represents the features of sprawl that are considered the most negative in an intraurban context. Consumption-based GHG's were assessed with the EE I-O LCA-model and further elaborated with a multivariate regression analysis.

Our results suggest that there are differences in the characteristics and lifestyles of metropolitan households by type of dwelling but some of these are not as explicitly evident when the actual GHG consequences are assessed. Our case area, the HMA, is a rather typical metropolis with a sprawling urban structure that, at the same time, has multiple positive features typical to metropolitan areas, the most important being its efficient public transportation network. Our results also highlight that, in order to make policies aimed to reduce GHG emissions, the structures of carbon footprints have to be kept in mind. The indirect part accounts for two-thirds of the footprint of a metropolitan dweller, and thus it should not be overlooked.

Looking at the averages, our results indicate that low-rise dwellers have higher carbon footprints than high-rise dwellers. The low-rise residents tend to be households with higher income, more children, and living in owner-occupied houses. However, the differences in the average amount of GHGs, both in absolute and relative terms, between the low-rise and high-rise dwellers are rather moderate compared with the figures from earlier literature. Many authors suggest that less-dense suburban living is approximately two times more energy- or GHG-intensive than inner-city living [3,4,10,34]. VandeWeghe and Kennedy [5], for example, come up with a yearly difference of 1.3 t CO_2-eq and smaller for those living in inner Toronto. Their result is rather close to our estimation of 1.8 t CO_2-eq.

To some extent, our results also support earlier results stating that both the distances driven and the use of private vehicles increase in sprawling areas [5,6,7]. However, according to our assessment these differences in emissions due to private transportation are remarkably small, indicating that private driving-related gains from higher density are moderate at best. This may also suggest that the better public transportation and better possibilities for walking and bicycling available for inner-city dwellers are not utilized up to their fullest potential. For example Kyttä et al. [35] suggest that Helsinki metropolitan dwellers rank smoothness of walking and bicycling very high, and indeed there has been attempts to make Helsinki more biker-friendly in recent years.

Our main results and key findings, however, were revealed when controlling for expenditure or income levels. Firstly, low-rise living is not unambiguously related to more GHG emissions than high-rise living. Actually, our regression models indicate that when expenditure or income levels are kept constant the high-rise dweller might produce equal or even slightly higher amounts of CHGs than the low-rise dweller. However, there was only one specification (regression 8) where the estimate for dwelling type had statistical significance. That told us that when the amount of expenditure on the indirect categories is kept constant the indirect footprints of high-rise dwellers are 4% higher than of those living in low-rise buildings. Secondly, household size, which in our case is usually larger in the low-rise living subset, seems to have a major effect on GHGs that are due to home energy and private driving. Economies of scale in household size are most unambiguous for households with at least three persons. Their per-capita carbon footprints are at least 15% smaller than those of people living alone, depending on the specification. The benefits of larger household size are most clearly seen when looking at the direct emissions, and the often-stated sharing of resources is not apparent when consumption of products and services alone is investigated. Thirdly, compared to household size, dwelling type is of minor importance.

Our analysis demonstrates that there are differences between low-rise and high-rise dwellers also within the metropolitan area. It is, however, worth asking if the differences between the two are to some extent overemphasized, at least in the public debate? Our results suggest that the lifestyles within a metropolis and especially their GHG consequences do not vary as much as the background variables would suggest. On the one hand, it seems that those living in city centers make the most of consumption possibilities and less of the low-carbon possibilities available to them, and e.g., the potential for household-size scale benefits such as resource-sharing is not made use of. On the other hand, those living in sprawl areas are responsible for higher GHG emissions from housing energy and private driving, as is often stated.

However, even if we believe that our paper brings out important insights on how consumption patterns and their environmental consequences vary with building type, the disparities within each category cannot be neglected. Precisely, the less wealthy, whatever their housing conditions were, are not likely to generate great amounts of greenhouse gas emissions.

9.4.2 LIMITATIONS

One of the main methodological limitations of the paper is related to I-O assumption of homogenous prices. Even when a relatively limited area is considered, like the HMA in this paper, it is clear that the monetary quantity does not indicate the quality and even less the GHG consequences. The assumption probably overestimates greenhouse gas implications of the occupants of the wealthiest households, most of whom live in detached houses, and are likely to buy more expensive goods. The EE I-O method assumes that an item that costs 10 times more also causes 10 times more emissions. However, since the area of the study is relatively limited, it can be concluded that prices in sectors such as transportation and services are likely to be rather homogenous. For example, regional wage differences are not present.

There are also other assumptions than that of price homogeneity that are well-known weaknesses of EE I-O models [21]. However, for example the bias related to high level of sector aggregation can be argued to be at a tolerable level. Su et al. [40] suggest that reliable estimates can be produced when the number of I-O sectors is at least 40. The EE I-O model utilized in this paper, therefore, meets this criterion with its 52 sectors. In addition, it has been pointed out that the errors tend to be at least partially negated when the final results are presented at a higher level of aggregation [41].

Lacking exact spatial data on how households on our data are located, the results of this paper are based on a strong assumption about how the different building types are, on average, located within the Helsinki Metropolitan Area. The obvious problem is that there are high-rise buildings that are located at urban edge and vice versa. Furthermore, the distances to centers from low-rise areas vary, and low-rise areas located at farthest corners of HMA, like those in Kauniainen, represent sprawl more self-evidently than those located at some of the Helsinki's suburbs. For example, for those living further afield from the city cores, the distances driven and the resulting emissions are likely to increase. However, we believe that our results give indications of general patterns since in the HMA apartment houses are almost invariably located in the centers of the HMA cities (Helsinki, Espoo, Vantaa, and Kauniainen), the share of apartment houses being highest in the capital (86%) [36]. Furthermore, according to the most recent National Travel Survey, for the apartment dwellers in the HMA the average distance to the nearest public transport stop is 0.29 km, the corresponding figure for low-riser dwellers being 0.55 km [37]. This indicates that our hypothesis is rather accurate.

It is worth noting that this paper restricts itself to a static analysis of the lifestyles with their GHG consequences. Our calculations are limited to consumption-based greenhouse gas emissions, referred as carbon footprints, and do not take into account e.g., the environmental pressure from increased land use and following changes in carbon stock [48]. Often, the term urban sprawl refers to green field development, meaning new residential development taking place at the urban edge. Where sprawling areas replace former open or agricultural lands multiple environmental consequences arise (see e.g., [49]). Furthermore, the expanding urban areas replace former green areas and also the CO_2 storage potential of these areas is lost or decreased. In addition, the carbon footprint comparisons in this study do not include the "carbon spike" related to construction of new infrastructure and buildings, be they located either in city centers or sprawling areas, even if this carbon spike is estimated to be quite substantial [42,43].

We use a rather limited amount of explanatory variables to explain carbon footprints. However, the literature includes a wide array of variables: e.g., education level, car ownership, and age [23,25]. However, none of these have been found to be of a similar importance as the level of expenditure and household size. Furthermore, analyses are often complicated by issues of multicollinearity and endogeneity.

Finally, we would like to highlight that our analysis is static and is not meant to be an analysis of change. Our results indicate an existence of an empirical relationship with given methodology in given point of time–it is unknown whether the results would recur e.g., in other countries. It must be kept in mind that the households who live in different environments are likely to differ in other ways as well. For example, it is impossible to assess to which extent the differences in carbon footprints are related to housing-types and not to households themselves. Living in the city core is also a lifestyle choice. For example, in the United States the liberal and environmentally-conscious prefer higher density areas with good public transport connections [44] that can be considered to be an example of self-selection. This demonstrates the fact that people in green communities, for example, can have lower carbon footprints due to either selection or treatment effects. Here, treatment effect refers to a situation where the environment a person lives in has, for one reason or another, an effect on his or her behavior.

9.5 CONCLUSIONS

According to the study, the phenomenon of urban sprawl is, to some extent, revealed in the increased carbon footprints of suburban dwellers. However, our regression models indicate that when expenditure or income levels are controlled the suburban dweller might actually produce equal or even slightly lower amounts of GHGs than an inner-city dweller. More importantly, household size and the resulting economies of scale effects seem to have by far the greatest effect on carbon footprints.

Notwithstanding certain deficiencies that cause uncertainties in the calculations of carbon footprint and keeping in mind that our calculations do not take into account environmental consequences from increased land use and following changes in carbon stocks, some important policy implications arise. It would seem that the emissions from private driving decrease surprisingly moderately in the dense areas within the HMA, while living in an apartment house is related to lower emissions from housing energy consumption. Thus, controlling buildings' energy efficiency would be of primary importance in preventing the negative effects of sprawl. On the other hand, either the available public transportation facilities may not be utilized up to their fullest potential or there still is room for improvement in the supply side. Using a private vehicle should be unnecessary in the densest areas. In addition, according to this study, the differences in indirect emissions from consumption of products and services are very small despite the large differences in disposable incomes. This suggests that proximity or other lifestyle-related factors may increase consumption of products and services and their indirect GHG consequences especially in city cores. All in all, consumption habits with their GHG consequences are surprisingly uniform across the HMA. Thus, instead of discussing only the differences between high-rise and low-rise areas, it would be essential to address why metropolitan areas have not yet fulfilled the great climate-change mitigation expectations imposed on them. The aim of the paper is not to deny the benefits of high-rise high-density policies but to discuss whether low-rise and not-as-high-density policies could and should act as complements to them. Our results suggests that if families that feel that they benefit from living in the suburban areas (there are also families that feel the opposite) moved to apartment houses, the final outcome, ceteris paribus, would remain almost the same at least when the viewpoint of consumption-based greenhouse gas emissions is taken. That is to say, an ideal metropolitan area would be an area where those living at the city core would not need a car and the benefits of closeness would stem from growing communality not from growing consumption. At the same time, those living in the suburban areas would live energy-efficiently and take an advantage of household size scale benefits as well as positive health effects of the proximity of green areas [45]. Finally, this leads to a conclusion that actually more detailed information about different lifestyles and about the connections between the urban form and the lifestyle choices of households is needed in order to understand and mitigate the GHG consequences of urban sprawl. Sprawl is a complex phenomenon that is too often over-simplified to a single factor such as private driving.

REFERENCES

1. IPCC Climate Change 2007: Synthesis Report. Available online: http://www.ipcc.ch/publications_and_data/publications_ipcc_fourth_assessment_report_synthesis_report.htm (accessed on 5 October 2012).
2. Cook, J.; Nuccitelli, D.; Green, S.A.; Richardson, M.; Winkler, B.; Painting, R.; Way, R.; Jacobs, P.; Skuce, A. Quantifying the consensus on anthropogenic global warming in the scientific literature. Environ. Res. Lett. 2013, 8, 1–7.
3. Brown, M.A.; Southworth, F.; Sarzynski, A. The geography of metropolitan carbon footprints. Policy Soc. 2009, 27, 285–304.
4. Hoornweg, D.; Sugar, L.; Gómez, C. Cities and greenhouse gas emissions: moving forward. Environ. Urban. 2011, 23, 207–227.
5. VandeWeghe, J.R.; Kennedy, C. A Spatial Analysis of Residential Greenhouse Gas Emissions in the Toronto Census Metropolitan Area. J. Ind. Ecol. 2007, 11, 133–144.
6. Glaeser, E.L; Kahn, M.E. The greenness of cities: carbon dioxide emissions and urban development. J. Urban. Econ. 2010, 67, 404–418.

7. Ewing, R.; Cervero, R. Travel and the Built Environment. J. Am. Plann. Assoc. 2010, 76, 265–294.
8. Ramaswami, A.; Hillman, T.; Janson, B.; Reiner, M.; Thomas, G. Demand-Centered, Hybrid Life-Cycle Methodology for City-Scale Greenhouse Gas Inventories. Environ. Sci. Tech. 2008, 42, 6455–6461.
9. Baynes, T.M.; Wiedmann, T. General approaches for assessing urban environmental sustainability. Curr. Opin. Env. Sust. 2012, 4, 458–464.
10. Ewing, R.; Rong, F. The impact of urban form on U.S. residential energy use. Hous. Pol. Debate 2008, 19, 1–30.
11. Glaeser, E. Triumph of the city: How Our Greatest Invention Makes Us Richer, Smarter, Greener, Healthier, and Happier, 1st ed.; The Penguin Press: New York, NY, USA, 2011; pp. 1–400.
12. Heinonen, J.; Junnila, S. Implications of urban structure on carbon consumption in metropolitan areas. Environ. Res. Lett. 2011, 6, 1–9.
13. Seppälä, J.; Mäenpää, I.; Koskela, S.; Mattila, T.; Nissinen, A.; Katajajuuri, J.-M.; Härmä, T.; Korhonen, M.-R.; Saarinen, M.; Virtanen, Y. Environmental impacts of material flows caused by the Finnish economy–ENVIMAT. Suomen. Ympäristö. 2009, 20, 1–134. (in Finnish, abstract in English).
14. Seppälä, J.; Mäenpää, I.; Koskela, S.; Mattila, T.; Nissinen, A.; Katajajuuri, J.-M.; Härmä, T.; Korhone, M.-R.; Saarinen, M.; Virtane, Y. An assessment of greenhouse gas emissions and material flows caused by the Finnish economy using the ENVIMAT model. J. Clean. Prod. 2011, 19, 1833–184.
15. Loikkanen, H.A. Kaupunkialueiden maankäyttö ja taloudellinen kehitys–maapolitiikan vaikutuksista tuottavuuteen sekä työ- ja asuntomarkkinoiden toimivuuteen. (in Finnish). Available online: http://www.vatt.fi/file/vatt_publication_pdf/v17.pdf (accessed on 1 October 2013).
16. Official Statistics of Finland (OSF). Households' consumption of 2006; Helsinki: Statistics Finland, Finland, 2009.
17. Alanen, O.; Hautamäki, A.; Kaskinen, T.; Kuittinen, O.; Laitio, T.; Mokka, R.; Neuvonen, A.; Oksanen, K.; Onnela, S.; Rissanen, M.; et al. Welfare of the Metropolis, 1st ed.; Erweko Painotuote Oy: Helsinki, Finland, 2010; pp. 1–105. (Metropolin hyvinvointi in Finnish).
18. Finance of housing companies. Available online: http://www.stat.fi/til/asyta/index_en.html (accessed 13 June 2013).
19. Suh, S.; Lenzen, M.; Treloar, G.J.; Hondo, H.; Horvath, A.; Huppes, G.; Jolliet, O.; Klann, U.; Krewitt, W.; Moriguchi, Y. System boundary selection in life-cycle inventories using hybrid approaches. Environ. Sci. Technol. 2004, 38, 657–664.
20. Crawford, R. Life Cycle Assessment in the Built Environment, 1st ed.; Spon Press: London, UK, 2011.
21. Wiedmann, T. Editorial: Carbon Footprint and Input–Output Analysis—An Introduction. Econ. Syst. Res. 2009, 21, 175–186.
22. Tukker, A.; Jansen, B. Environmental Impacts of Products: A Detailed Review of Studies. J. Ind. Ecol. 2006, 10, 159–182.
23. Lenzen, M.; Wier, M.; Cohen, C.; Hayami, H.; Pachauri, S.; Schaeffer, R. A comparative multivariate analysis of household energy requirements in Australia, Brazil, Denmark, India and Japan. Energy 2006, 31, 181–207.

24. Roca, J.; Serrano, M. Income growth and atmospheric pollution in Spain: an input–output approach. Ecol. Econ. 2007, 63, 230–242.
25. Kerkhof, A.C.; Nonhebel, S.; Moll, H.C. Relating the environmental impact of consumption to household expenditures: An input–output analysis. Ecol. Econ. 2009, 68, 1160–1170.
26. Shammin, R.; Herendeen, R.A.; Hanson, M.J.; Wilson, E.J.H. A multivariate analysis of the energy intensity of sprawl versus compact living in the U.S. for 2003. Ecol. Econ. 2010, 69, 2363–2373.
27. Rents of dwellings. Available online: http://www.stat.fi/til/asvu/index_en.html (accessed 5 October 2013).
28. Bin, S.; Dowlatabadi, H. Consumer lifestyle approach to US energy use and the related CO2 emissions. Energy Policy 2005, 33, 197–208.
29. Druckman, A.; Jackson, T. The carbon footprint of UK households 1990–2004: A socio-economically disaggregated, quasi-multi-regional input-output model. Ecol. Econ. 2009, 68, 2066–2077.
30. Herendeen, R.; Tanaka, J. Energy cost of living. Energy 1976, 1, 165–178.
31. Vringer, K.; Blok, K. The direct and indirect energy requirements of households in the Netherlands. Energy Policy 1995, 23, 893–910.
32. Weber, C.L.; Matthews, H.S. Quantifying the global and distributional aspects of American household carbon footprint. Ecol. Econ. 2008, 66, 379–391.
33. Heinonen, J.; Jalas, M.; Juntunen, J.K.; Ala-Mantila, S.; Junnila, S. Situated lifestyles: I. How lifestyles change along with the level of urbanization and what the greenhouse gas implications are—a study of Finland. Environ. Res. Lett. 2013, 8, 1–13.
34. Norman, J.; MacLean, H.L.; Kennedy, C.A. Comparing High and Low Residential Density: Life-Cycle Analysis of Energy Use and Greenhouse Gas Emissions. J. Urban. Plann. Dev. 2006, 132, 10–21.
35. Kyttä, M.; Broberg, A.; Tzoulas, T.; Snabb, K. Towards contextually sensitive urban densification: Location-based softGIS knowledge revealing perceived residential environmental quality. Landsc. Urban. Plann. 2013, 113, 30–46.
36. Dwellings and housing conditions. Available online: http://stat.fi/til/asas/index_en.html (accessed 13 June 2013).
37. The Finnish Transport Agency. The National Travel Survey 2010–2011. The Finnish Transport Agency: Helsinki, Finland, 2012.
38. Hardy, M.A. Regression with Dummy Variables (Quantitative Applications in the Social Sciences), 1st ed.; SAGE Publications Inc.: Pennsylvania, PA, USA, 1993; pp. 1–96.
39. Ramaswami, A.; Chavez, A. What metrics best reflect the energy and carbon intensity of cities? Insights from theory and modeling of 20 US cities. Environ. Res. Lett. 2013, 8, 1–11.
40. Su, B.; Huang, B.C.; Ang, B.W.; Zhou, P. Input–output analysis of CO2 emissions embodied in trade: The effects of sector aggregation. Energ. Econ. 2010, 32, 166–175.
41. Wiedmann, T.; Wood, R.; Minx, J.C.; Lenzen, M.; Guan, D.; Harris, R. A carbon footprint time series of the UK–results from a multi–region input-output model. Econ. Syst. Res. 2010, 22, 19–42.

42. Säynäjoki, A.; Heinonen, J.; Junnila, S. A scenario analysis of the life cycle greenhouse gas emissions of a new residential area. Environ. Res. Lett. 2012, 7, 1–11.
43. Stephan, A.; Crawford, R.H.; de Myttenaere, K. Multi-scalc life cycle energy analysis of a low density suburban neighbourhood in Melbourne, Australia. Build. Environ. 2013, 68, 35–49.
44. Kahn, M.E.; Morris, E. Walking the Walk: The Association Between Environmentalism and Green Transit Behavior. J. Am. Plann. Assoc. 2009, 75, 389–405.
45. Maas, J.; Verheij, R.A.; Groenewegen, P.P.; de Vries, S.; Spreeuwenberg, P. Evidence based public health policy and practice: Green space, urbanity, and health: how strong is the relation? J. Epidemiol. Community 2006, 60, 587–592.
46. Population structure. Available online: http://www.stat.fi/til/vaerak/index_en.html (accessed 10September 2013).
47. Kyrö, R.; Heinonen, J.; Junnila, S. Occupants have little influence on the overall energy consumption in district heated apartment buildings. Energ. Buildings 2011, 43, 3484–3490.
48. WRI. The Land Use, Land-Use Change, and Forestry Guidance for GHG Project Accounting; World Resources Institute: Washington, DC USA, 2006. Available online: http://pdf.wri.org/lulucf_guidance.pdf (accessed 1 October 2013).
49. Nuissl, H.; Haase, D.; Lanzendorf, M.; Wittmer, H. Environmental impact assessment of urban land use transitions—A context-sensitive approach. Land Use Pol. 2009, 26, 414–424.
50. Heinonen, J.; Postdoctoral Researcher Aalto University; Mattila, T.; Senior Research Scientist Finnish Environment Institute. Personal communication, 2012.

CHAPTER 10

Renewables in Residential Development: An Integrated GIS-Based Multicriteria Approach for Decentralized Micro-Renewable Energy Production in New Settlement Development: A Case Study of the Eastern Metropolitan Area Of Cagliari, Sardinia, Italy

CLAUDIA PALMAS, EMANUELA ABIS, CHRISTINA VON HAAREN, AND ANDREW LOVETT

10.1 BACKGROUND

A sustainable future for cities significantly depends upon the integration of energy efficiency in regional and urban planning. About 40% of the final energy demand is needed to heat and power homes. This represents a major source of greenhouse gas emissions, making energy savings in the field

---

of residential development a key element of the European climate change strategy [1]. In this context, Europe is also faced with the challenge of implementing growing amounts of intermittent power sources such as microsolar and wind sources in the electricity grid. The generation of renewable energy is characterized by intermittency; therefore, it is imperative that a mix of sources should be selected and used along with the suitable energy storage mechanisms in order to best utilize the available renewable energy resources and ensure the continuity of supply [2].

Two European projects exemplify the state of the art in the energy-efficient residential development: the BedZED development in the south of London and the Vauban development in Freiburg, Germany. These two eco-districts are globally recognized to be models for sustainable environmentally oriented planning using solar energy (photovoltaics and solar thermal collectors). In the BedZED project, the use of solar energy is maximized through the integration of solar cells into the vertical south-facing facades and also through a large installation on the south-facing roofs [3].

In Freiburg, the principles of energy savings and solar optimization are early considered in the planning phase of housing development, e.g., by defining the orientation and position of buildings or by obligatory low-energy construction requirements [4]. These urban multi-residential housing developments are not only models for energy saving, but they also take into account social and economical aspects.

However, the geographical distribution of the renewable energy potential is rarely considered or estimated in the planning of new residential areas [5,6]. Also, in selecting the location, environmental criteria in combination with micro-renewable potentials are still neglected.

Therefore, the state of the art in the field of energy efficiency [7,8] may be advanced by combining the different energy sources in new housing developments and planning their location by estimating the energy potential available for the whole area under consideration. Renewable energy supply is site-specific and variable [9]. A restriction on secure supplies from single renewable energy sources is their output variability. Supply insecurity can be increased by demand variability, especially where this correlates with times of high energy output by renewables, better predictability of their generation output, and the complementarities of different power sources. Also, the combination of different renewable energy sources can increase the supply security.

In recent years, several renewable energy potential mapping methodologies have been developed (e.g., solar irradiation and wind estimation, geothermal and biomass energy) [10-12]. These methods can be used for complementing the urban planning approaches. However, the methodologies have been developed for very small scales and cannot be applied unmodified for selecting new housing locations [13]. Therefore, it is necessary to either adapt the existing methodologies or develop new ones.

Energy efficiency should be integrated from the start of the land use planning process in order to guide the future development to support the sites with the best potential for using renewable micro-generation. These potentials can be developed in a sustainable way by using multicriteria evaluation methods in a Geographical Information System (GIS) to help optimize new settlements in terms of multi-functionality. There is a history of research using such multicriteria evaluation techniques to support collaborative decision-making processes by providing a framework where stakeholder groups can explore, understand, and redefine decision problems with respect to housing location [14,15].

The development and testing of a methodology for an integrated approach to energy-efficient residential development planning is the main objective of the research presented here. Both optimizing the location of new housing development with regard to energy supply and other sustainability criteria as well as optimizing the mix of micro-renewables need to be facilitated. Consequently, the main research questions addressed are

- How to calculate the geographic distribution of energy potentials? How to produce energy potential maps? How to identify the best energy mix combinations? Which criteria and algorithms are needed for identifying the theoretical energy potential in terms of the different energy sources?
- Which environmental and landscape criteria are considered most relevant for the assessment of new housing development with micro-renewable technologies?
- How to support decision makers or planning in the challenge of including multiple criteria in housing development decisions?

Accordingly, this paper describes

- Methodologies (existing, adapted, or newly developed) to estimate the micro-renewable energy potentials in a spatially explicit manner.
- Methodologies for identifying the suitable areas for new sustainable settlements using micro-renewable technologies which enable to support decision makers in planning.

The results of testing these methods are presented for the region of Cagliari in Sardinia.

10.2 METHODS

10.2.1 GENERAL METHODOLOGICAL APPROACH

Part of the approach is not only based on the existing, originally small scale, but also on methods for energy potential assessment, which were pre-tested in the Hannover region (by Master students in cooperation with the State Office for Mining, Energy and Geology (LBEG) [16]. In a second step, the methods were adapted to local/regional scale planning. As there was no suitable method for assessing the bioenergy potential, a new method was developed. This resulted in assessments of the theoretical (potential) supply. Because of the existing technical, ecological, economic, and social restrictions, such theoretical amounts can only be exploited up to a certain percentage [17].

Expert preferences were used to weight multiple assessment criteria for housing developments. These preferences were obtained through a survey conducted with students, academic planners, regional planners, and public authorities in Italy, Germany, and the UK. This expert-based approach was chosen because in most European countries, no clear-cut standards exist about the suitability of micro-energy generation in residential areas (in contrast, e.g., to emission standards). Expert opinions were a simple way of priority setting in such complex decisions. In addition, such a method allows the results of different preferences to be modeled and local or regional stakeholder opinions and interests to be included.

The energy potentials and expert preferences were ultimately combined in a GIS-based analysis to identify the most appropriate housing sites on a regional scale. This analysis made use of multicriteria evaluation (MCE) techniques which are one of the most common GIS-based tools. They have been used to support decision making on complex problems such as site selection, land suitability, resource evaluation, and land allocation [18-22]. Over the last two decades, several MCE methods have been implemented in a GIS environment, including weighted linear combination (WLC) and its variants [23-27] and ideal point methods [28,29]. Among these procedures, WLC and Boolean overlay operations, such as intersection (AND) and union (OR), are the most widely used [15,21] and were adopted in this research.

TABLE 1: Input data for the energy potential estimation

Data	Scale/unit	Data origin
Digital elevation model (DEM 90)	90 × 90 m	CGIAR Consortium for Spatial Information
Wind speeds at 25 m	m/s	Aeolic Italian Atlas
Geological map	1:200,000	Earth Science Department (Cagliari University)
Land use	1:25,000	Region of Sardinia
Irrigation map	1:25,000	Region of Sardinia

10.2.2 DATA

The eastern metropolitan area of Cagliari (Figure 1) covers 591 km2 in the south of Sardinia and has a population of 322,392 inhabitants [30]. Cagliari is the capital of Sardinia, situated at the southern shore of the island and has 157,222 inhabitants [30].

The region is characterized by rural areas around the cities with a large amount of agricultural land (around 46.72%). Other uses, such as residential, commercial, and industrial areas, cover about 40% (land use data, region of Sardinia).

Table 1 lists the main geographical data sources used for the regional assessment. These were supplemented by shape files from the Regional Landscape Plan of Sardinia, scale 1:10,000.

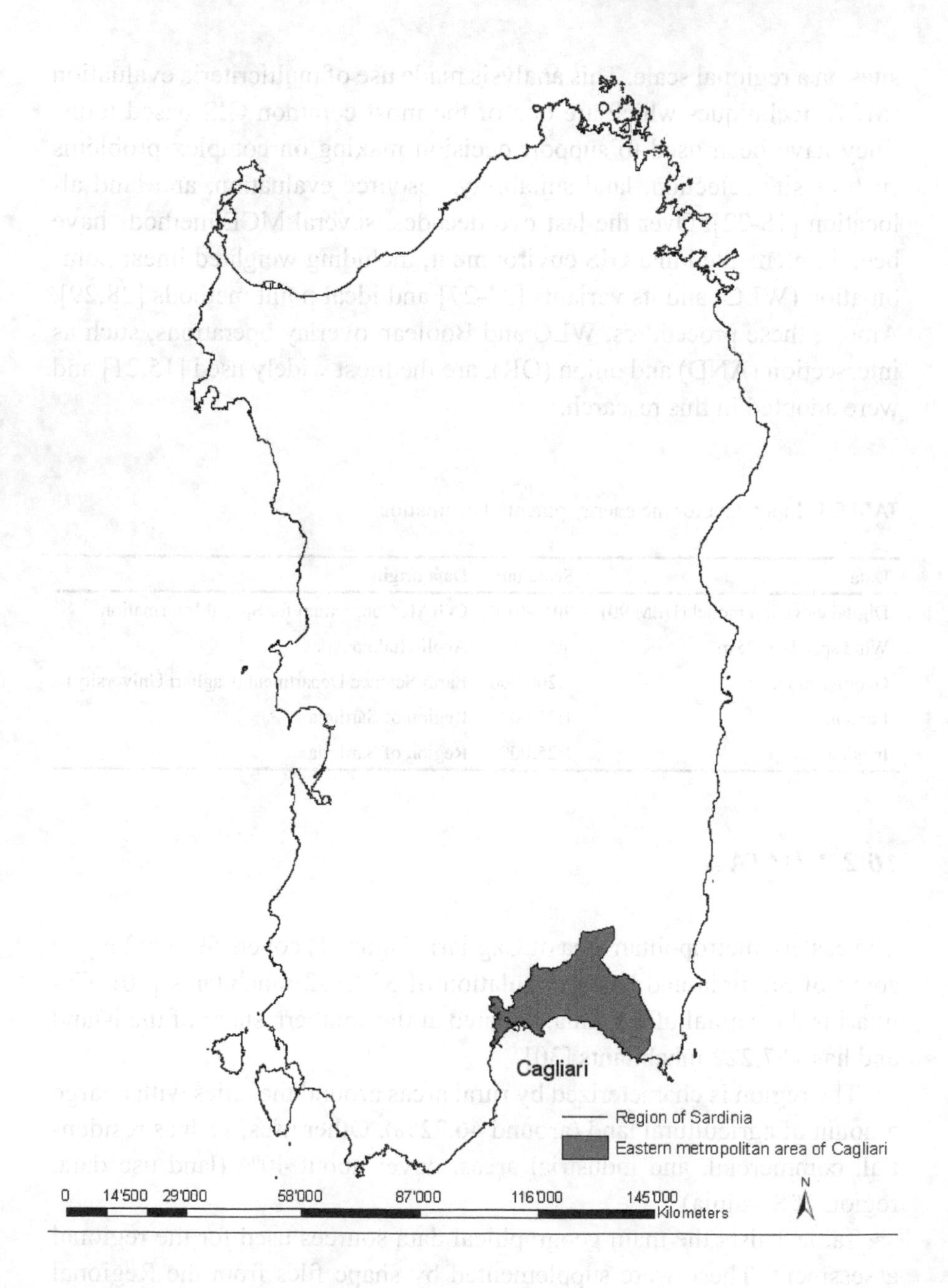

FIGURE 1: Eastern metropolitan area of Cagliari.

10.2.3 ADAPTATION OF AN EXISTING METHODOLOGY FOR IDENTIFYING THE SPATIAL SOLAR ENERGY POTENTIAL UNDER LOCAL-SCALE CONDITIONS

The solar potential raster maps were calculated from the *r.sun* model. The pvgis database, derived from the Photovoltaic Geographical Information System–Interactive Maps [31], was used to validate the data. The algorithm used to calculate the solar irradiation was implemented in the open-source GIS software GRASS, where the beam irradiance normal to the solar beam B_{0c} (in watts per square meter) is attenuated by cloudiness atmosphere and calculated in the model as in Equation 1 [31]:

$$B_{0c} = G_0 \exp\{-0.8662\ T_{LK}\ m\ d_R(m)\} \quad (1)$$

where G_0 is the extraterrestrial irradiance normal to the solar beam (in watts per square meter), −0.8662 TLK is the Linke atmospheric turbidity factor, m is the 'optical air mass,' and dR(m) is the 'Rayleigh optical thickness at air mass m.'

The *r.sun* model operates in two modes. In *mode 1*, the model calculates the instant time (in seconds) of raster maps of chosen components (beam, diffuse, and reflected) of solar irradiance (in watts per square meter) and the solar incident angle (in degrees). In *mode 2*, the raster maps of the daily sum of solar irradiation (in watt hours per square meter per day) are computed as an integration of irradiance values that are calculated within a set day. In this study, *mode 2* was used because we needed to calculate raster maps representing the annual average of daily sums of global irradiation for horizontal surfaces. To compute the irradiation raster maps, *r.sun* requires only a few mandatory input parameters - digital terrain model (elevation, slope, aspect - *elevinslopeinaspin*), *day* number (for mode 2), and additionally, a local solar time (for mode 1). The other input parameters are either internally computed (solar declination) or the values can be set to fit the specific user needs: the Linke atmospheric turbidity, ground albedo, beam, and diffuse components of clear-sky index, and time step are used for the calculation of all-day irradiation [32].

10.2.4 ADAPTATION OF AN EXISTING METHODOLOGY FOR IDENTIFYING THE SPATIAL WIND ENERGY POTENTIAL

The wind speeds were calculated in accordance with the following Equation 2 [[33,34]]:

$$v = v_{ref}\left(\frac{z}{z_{ref}}\right)^{\alpha} \tag{2}$$

where v=wind speed at height z above the ground level; v_{ref}=reference speed, i.e., a wind speed we already know at height z_{ref}; z=height above the ground level for the desired velocity, v; and z_{ref}=reference height, i.e., the height where the wind speed is measured $v_{ref.}$

The exponent α is an empirically derived coefficient that varies according to the stability of the atmosphere. For neutral stability conditions, α is approximately 0.143.

10.2.5 ADAPTATION OF AN EXISTING METHODOLOGY FOR IDENTIFYING THE SPATIAL GEOTHERMAL ENERGY POTENTIAL

The geothermal energy potential maps were generated by considering the physical rock properties for the estimation of the specific heat extraction values, where those for vertical loops followed Kaltschmitt et al. [35]. Here, the geological stratification of rocks to 100 m is derived from a regional geological map and the specific heat extraction is obtained from the following equation:

$$P_{EWS} = (13 \cdot \lambda) + 10 \tag{3}$$

where PEWS = specific heat extraction capacity, and λ = heat conductivity of the rock.

To obtain the specific heat extraction values, the geology was divided in two homogeneous layers: unconsolidated and solid rocks. Further information on the geological stratification for vertical loops and soil characteristics for horizontal loops was obtained from Dott. Geol. Fausto Pani, freelance, and Prof. Giovanni Barrocu, Cagliari University.

10.2.6 DEVELOPING A METHODOLOGY FOR IDENTIFYING THE SPATIAL BIOMASS ENERGY POTENTIAL

Given the focus on housing development, not every type of biomass is relevant. Attention was focused on wooden biomass which is suitable for producing heat and electricity in residential areas with the installation of a cogeneration system. Important criteria for identifying the potential include the distance of the source of wood from the settlement and the capacity of the forest in terms of the wood reservoir. From an economic perspective, the energy-efficient use of biomass can be defined as a use within a radius of 30 km around a potential biomass facility [36]. According to the sustainability principles, e.g., the needs of localizing new settlements near the biomass source, we assume that the energy biomass efficiency is related to a use within a radius of 15 km around a potential biomass facility, as shown in Equation 4. We assume that the biomass energy potential P_i is defined as the theoretical qualitative potential for a hypothetical settlement location or users V_i.

$$P_i = \sum_j \left[\left[\frac{A_j}{A} \cdot \frac{(15 - d_{ij})}{15} \right] - T_j \right] * i = 1,2,K,N; j \neq 1 \tag{14}$$

where P_i = biomass energy potential, V_i = potential settlement, A_j is the forest cell area, A is the total forest cell area, d_{ij} is the distance between the centers of the cell of settlement potential location and of the cell of the forest areas $d_{ij} \leq 15$ km, and T_j is the factor depending on transport and wood extraction cost.

To differentiate between the areas of varying potential, a Monte Carlo method was introduced. Broadly speaking, Monte Carlo integration methods are algorithms for the approximate calculation of the numerical value of a definite integral, usually multidimensional ones, in our case the sum of forest areas (cf.[37]). The usual algorithms evaluate the integrand at a regular grid. Monte Carlo methods, however, use random samplings to approximate probability distributions. This is performed by selecting some numbers of random points over the desired interval and summing the function evaluations at these points [38].

10.2.7 BEST ENERGY MIX COMBINATIONS

The suitability maps of the theoretical energy potentials were integrated into combined layers showing the best locations for the new settlement development according to the most appropriate energy mix for the area under consideration. Maps were normalized and two versions were produced depending on whether geothermal vertical or horizontal loops were included.

10.2.8 SURVEY OF EXPERT PREFERENCES

Decisions about the spatial resource allocation require prioritizing multiple criteria. Different criteria were selected for (1) assessing housing development in general as well as for (2) settlements with micro-renewables. The selection of criteria took into account the possible environmental and landscape impacts as well as the availability of relevant geodata in order to transform the preferences into spatially explicit representations. The criteria used in this research were proximity to existing urban areas, proximity to major roads and train lines, distance from environmentally valuable and vulnerable areas or from protected areas, proximity to water (lakes and rivers), and the slope gradient (see Table 2).

Other factors such as the location, size, and accessibility of a site and its proximity to amenities and services are also important for future housing developments. These could be considered on a broader scale.

TABLE 2: Criteria for new settlement development

Factor/criteria	Type
Proximity to existing urban areas	Planning factor (compact development)
Proximity to major roads and train lines	Transport factor
Distance from environmentally valuable and vulnerable areas or from protected areas	Environmental factor
Proximity to water (lakes and rivers)	Attractiveness factor
Slope gradient	Physical factor

TABLE 3: Criteria for settlement development using micro-renewable technologies

Micro-technology	Distance from flooding areas
Solar panel and thermal collectors (S)	Distance from landscape-protected areas and other beauty areas
	Distance from historic/cultural facilities (historical centre, areas of historical and cultural interests, archeological sites)
Wind turbines (W)	Distance from historic/cultural facilities
	Distance from Special Protection Areas (SPA) and others avifaunistic important areas
	Distance from landscape-protected areas or other beauty areas
Biomass power plants (B)	Distance from historic/cultural features
	Distance from landscape-protected areas or other beauty areas
Geothermal vertical loops (GVL)	Distance from historic/cultural features
	Distance from drinking water or aquifers
Geothermal horizontal loops (GHL)	Distance from historic/cultural features
	Distance from flooding areas

The criteria were divided into continuous suitability factors and constraints (binary yes/no restrictions). The constraints were built-up areas, water (lakes and rivers), and areas characterized by hydrogeological instability.

The criteria for the survey of the settlements with micro-renewables shown in Table 3 focused on landscape and environmental impacts, because the technical factors were included in the potential maps.

Both parts of the survey were conducted in Italy, Germany, and the UK and sought to gain insights into perceptions about new energy-efficient settlement development. This required the participation of people who had expert knowledge regarding landscape and environmental planning and/or renewable energy, so the survey focused on students and academic planners, regional planners, and public authorities. The questionnaire was distributed in person and by email, with participants returning the completed surveys in the same ways.

The expert preferences were converted into values using pairwise comparison methods, a procedure in the Analytical Hierarchy Process (AHP) [39]. As an input, the method takes the pairwise comparisons of the different criteria and produces their relative weights as an output. According to the relative importance, the weights, which were assigned to the different criteria, were calculated using MathCAD, an engineering calculation software. Consistency ratios were also calculated to assess the reliability of the pairwise comparisons [39].

The output maps were generated using a Boolean approach and a WLC method [21]. The Boolean approach is based on a reclassification operation and specified cutoffs. WLC was used to produce suitability raster maps for housing development and micro-renewable preferences with respect to environmental and landscape impacts. The suitability maps were generated as shown in Equation 5:

$$\text{Suitabilitymap} = \Sigma[\text{factormap}(cn) * \text{weight}(wn) * \text{constraint}(b0/1)] \quad (5)$$

where cn = standardized raster cell, wn = weight derived from AHP pairwise comparison, b0/1 = Boolean map with values 0 or 1, and n = number of raster cell

To identify the optimal sites for new residential areas by using micro-renewables, the three GIS layers (energy potentials, suitability for new settlement development, suitability for new settlements with micro-generators) were overlaid. This integration was conducted using Spatial Analyst functions available in ArcGIS 9.x [40].

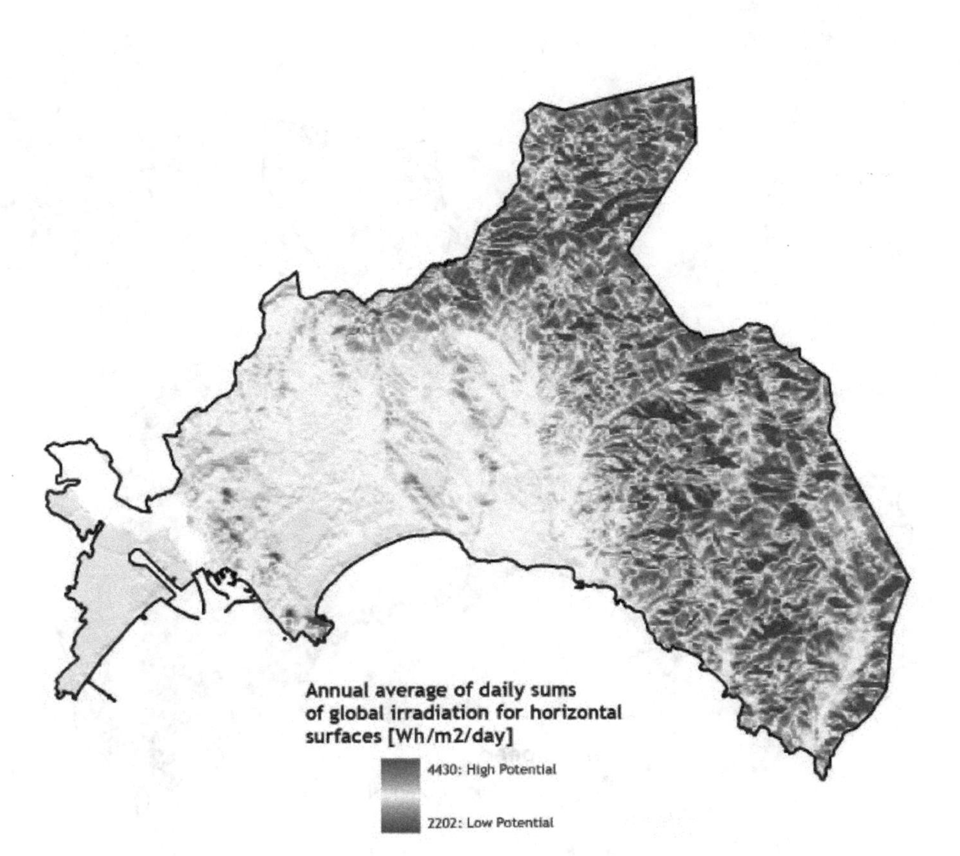

FIGURE 2: Annual average of the daily sums of global irradiation for horizontal surfaces (Wh/m^2/day).

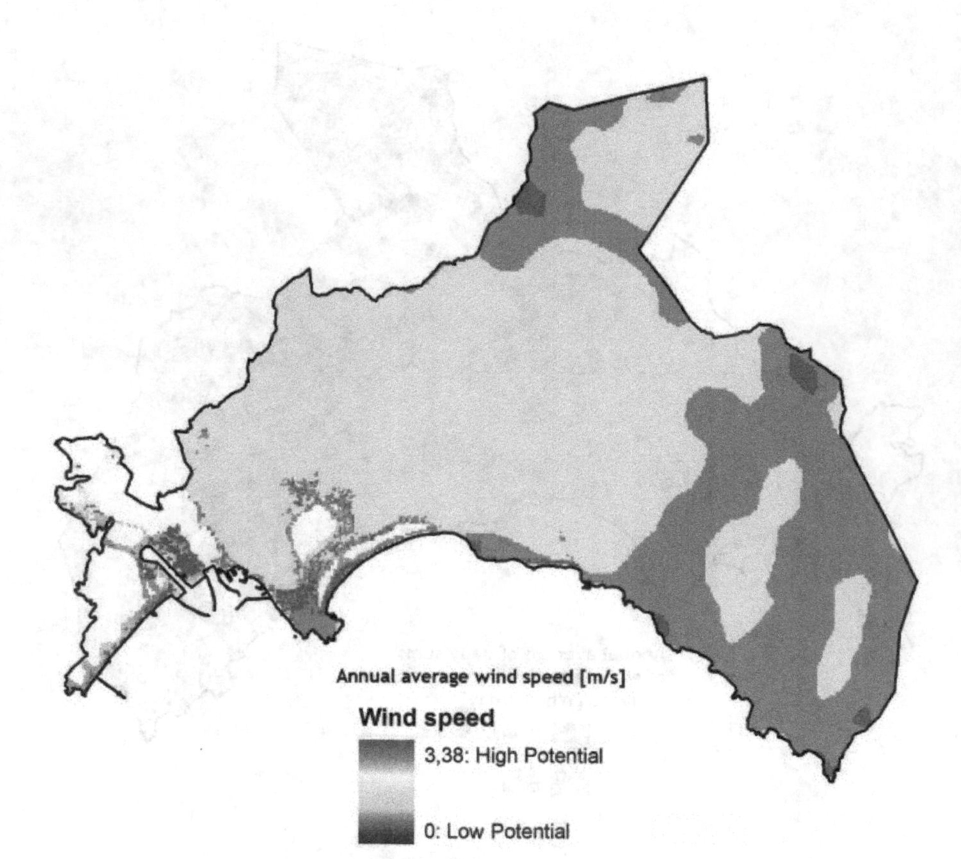

FIGURE 3: Wind energy potential.

10.3 RESULTS

10.3.1 SOLAR ENERGY POTENTIAL

Parameters such as the albedo factor (0.2) and the Linke turbidity (3.0) were assumed constant across the region as a first approximation. The clear-sky indices were not available. The influence of terrain shadowing was taken into account by setting the -s flag. After validation of the data, the output raster map showed the annual average of the daily sums of global irradiation for horizontal surfaces (in watt hours per square meter per day) (see Figure 2).

10.3.2 WIND ENERGY POTENTIAL

To create the wind energy potential maps, speeds at 25 m above the ground with 1-km resolution were used. The data were derived from the Italian Atlas Wind Energy (Atlante eolico italiano) developed by the Genoa University and the CESI research center [41]. Equation 2 was applied to obtain a final average wind speed raster map (at 10 m above the ground) with a resolution of 90 m (see Figure 3).

10.3.3 GEOTHERMAL ENERGY POTENTIAL

For the unconsolidated rocks, there were some data limitations. Therefore, the thickness was sometimes only a rough estimate. The data for solid rocks were more accurate. The information regarding the groundwater flow component was not considered according to the VDI 3640 German directive.

10.3.4 GEOTHERMAL VERTICAL LOOPS

The geological map of the region of Sardinia, scale 1:200,000, was consulted to evaluate the specific heat extraction capacities, which were combined by values from the literature with regard to the specific heat conductivity (cf.[42]). The resulting map was classified into three categories (see Figure 4). The unsuitable areas are not suitable for economic reasons.

10.3.5 GEOTHERMAL HORIZONTAL LOOPS

The geological map, the map for irrigation, and the land use map (scale 1:25,000) were considered to select the suitable and unsuitable areas for the installation of horizontal loops. Given the variety of soil conditions (e.g., evapotranspiration) and characteristics (e.g., presence of aquifers), soil types, and the absence of quantitative data regarding all these factors [43], it was only possible to give a qualitative potential estimation for the use of horizontal loops on this scale, as is shown in Figure 5.

10.3.6 BIOMASS ENERGY POTENTIAL

The study did not take into account the factor T_j in Equation 4 which depends on road types and conditions as well as variable factors such as fuel prices for wood transportation and extraction costs, because the necessary data were not available. These factors can be better considered in a more detailed local view.

A grid with a 250-m spacing was overlaid over a larger section of the eastern metropolitan area of Cagliari. A total of 5,000 random points were used, giving an average density of around 1 point per 250 m^2. After the application of the Monte Carlo integration to the existing data, a biomass potential map was generated (Figure 6).

10.3.7 BEST ENERGY MIX POTENTIAL

We assumed that all the micro-energy potential maps are of the same weight. The maps obtained show the best locations for the integration of solar, wind, and biomass micro-generators with geothermal vertical loops (Figure 7) or horizontal loops (Figure 8).

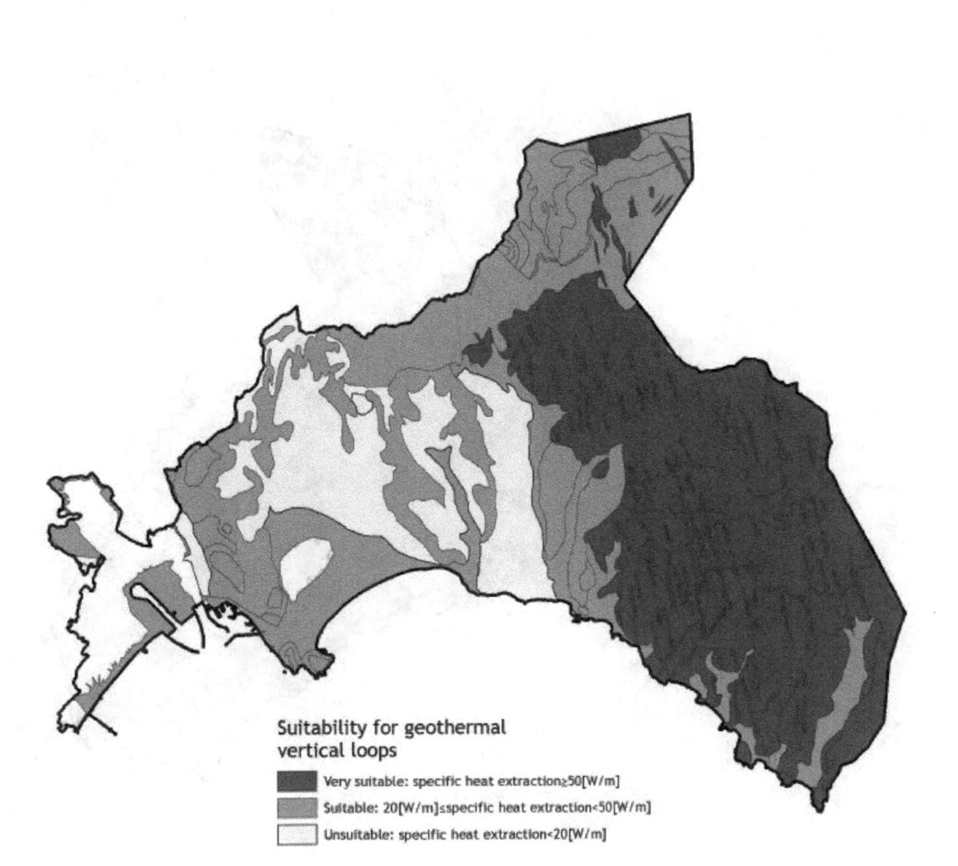

FIGURE 4: Geothermal energy potential for geothermal vertical loop

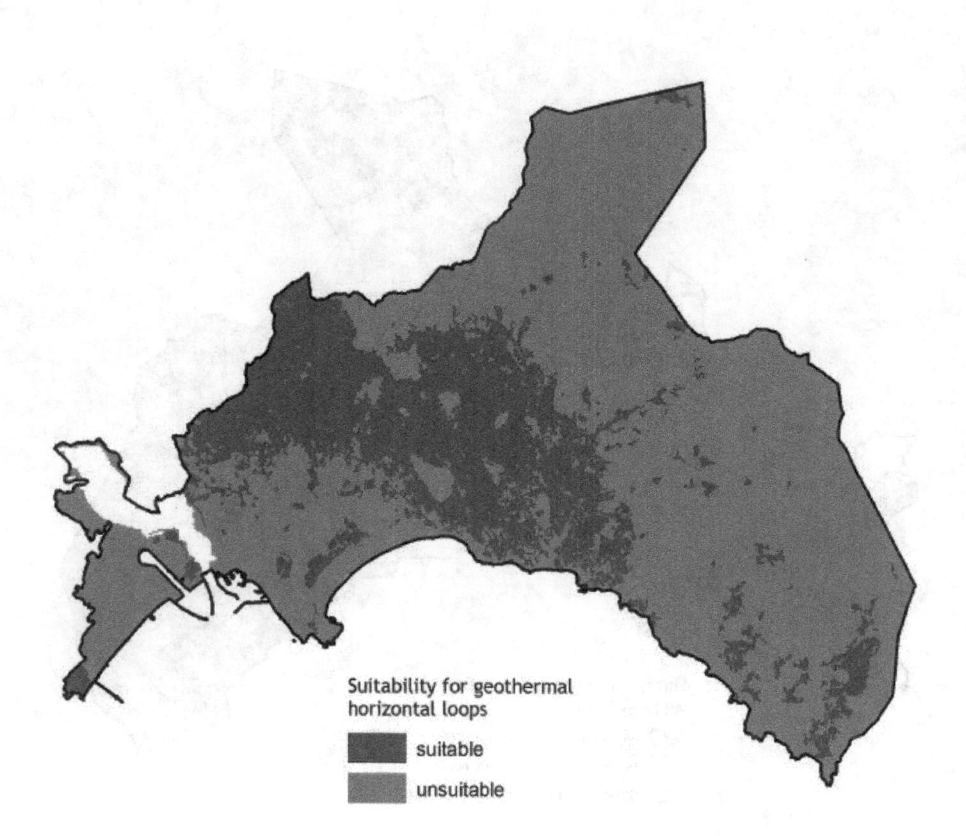

FIGURE 5: Geothermal energy potential for geothermal horizontal loops.

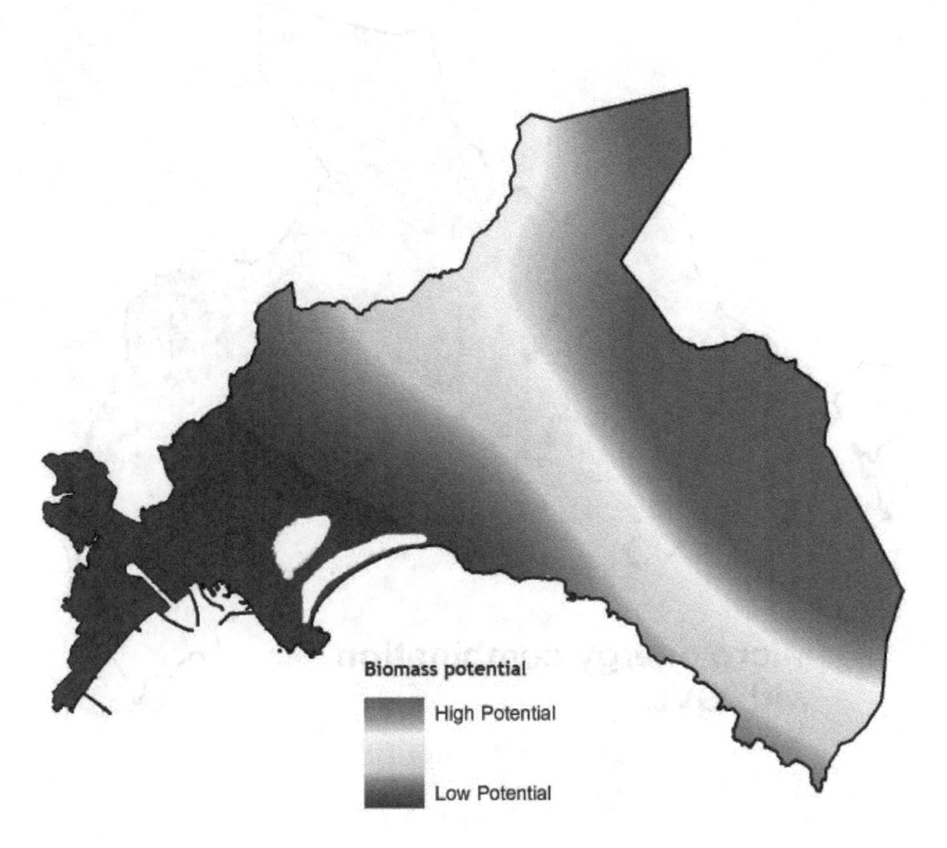

FIGURE 6: Biomass energy potential.

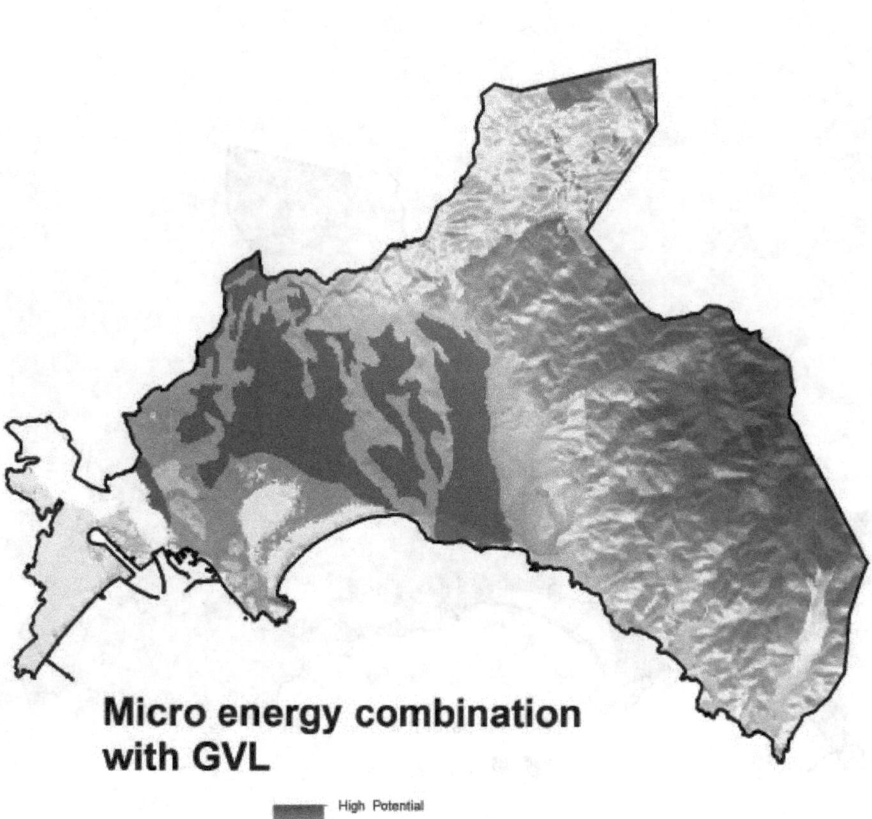

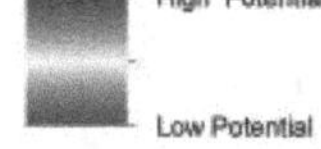

FIGURE 7: Best energy mix potential for solar, wind, and biomass micro-generators using geothermal vertical loops.

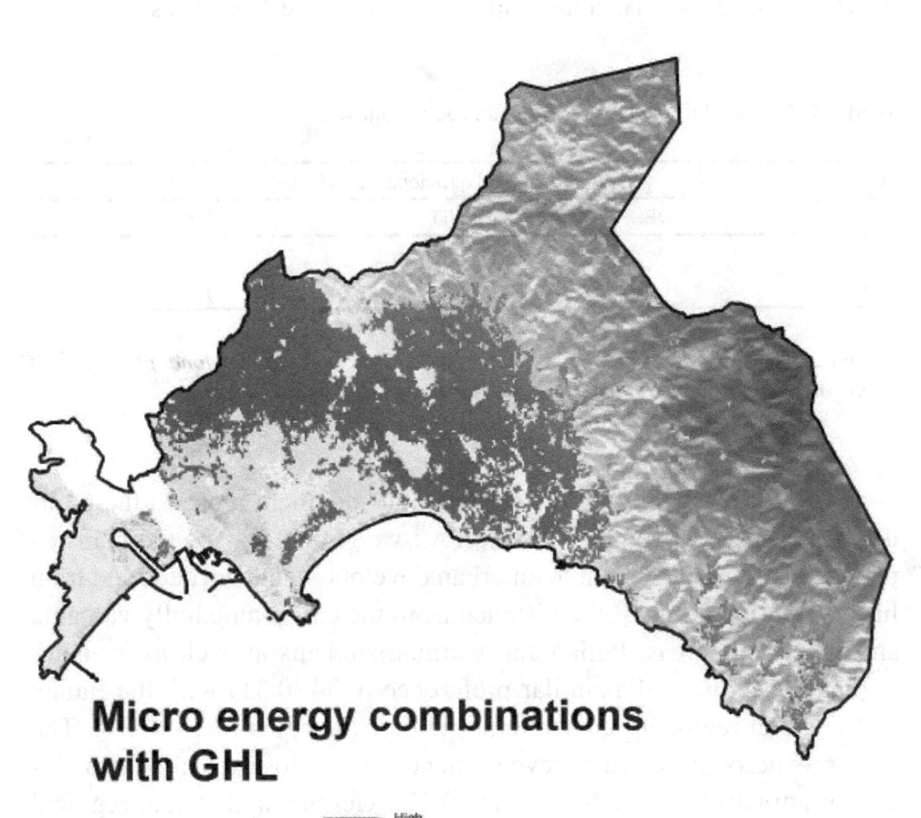

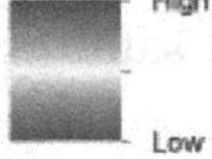

FIGURE 8: Best energy mix potential for solar, wind, and biomass micro-generators using geothermal horizontal loops.

10.3.8 SURVEY RESULTS

A total of 120 questionnaires were completed, but only 108 were considered further (consistency ratio<0.1). As shown in Table 4, for each category of experts, we had a minimum of 15 valid questionnaires.

TABLE 4: Total evaluated questionnaires for each country

Experts	Total evaluated questionnaires with consistency ratio<0.1		
	DE	IT	UK
S.&AP.	19	16	15
P.&PA.	15	28	15

Obtained from students and academic planners (S.&AP.) as well as regional planners and public authorities (P.&PA.). DE, Germany; IT, Italy; UK, United Kingdom.

English experts preferred a compact development close to the built-up areas (S.&AP.: weight 0.29; P.&PA.: weight 0.35). The German experts gave the same weight to an urban development near roads and train lines (0.22). In terms of the distance from the environmentally valuable and vulnerable areas, Italian and German students as well as academic planners expressed a similar preference (0.34; 0.31) with the Italian and German regional planners and public authorities (0.23; 0.24). The Italian experts preferred a development that is close to lakes and rivers for attractiveness reasons (0.15; 0.17). German and Italian regional planners and public authorities gave the same consideration to the slope gradient (0.15).

Table 5 presents the weights for each criterion regarding the new housing development obtained from students and academic planners (S.&AP.) as well as regional planners and public authorities (P.&PA.) from each nationality. The weights sum to 1 with a higher value corresponding to more emphasis on the relevant criteria.

Table 6 shows the averages of the standard deviations expressed as a percentage of the means for the different expert groups and nationalities.

These results show some variation in weighting but do not exceed 100% so the variations are not too high.

TABLE 5: Weighting for settlement development

Criteria/factors	Experts	Weights		
		DE	IT	UK
Proximity to existing urban areas	S.&AP.	0.20	0.26	0.29
P.&PA.	0.26	0.25	0.35	
Proximity to major roads and train lines	S.&AP.	0.22	0.16	0.26
P.&PA.	0.22	0.20	0.22	
Distance from environmentally valuable areas	S.&AP.	0.31	0.34	0.21
P.&PA.	0.24	0.23	0.19	
Proximity to water	S.&AP.	0.14	0.15	0.11
P.&PA.	0.14	0.17	0.12	
Slope gradient	S.&AP.	0.13	0.10	0.12
P.&PA.	0.15	0.15	0.13	

Obtained from students and academic planners (S.&AP.) as well as regional planners and public authorities (P.&PA.). DE, Germany; IT, Italy; UK, United Kingdom.

TABLE 6: Averages of standard deviations in percentages of the means for housing development

Experts	Average of the SD in percentage of the mean		
	DE	IT	UK
S.&AP.	65.74	79.62	49.78
P.&PA.	48.02	66.83	40.50

S.&AP., students and academic planners; P.&PA., regional planners and public authorities; SD, standard deviation; DE, Germany; IT, Italy; UK, United Kingdom.

The final suitability maps were identified through the expert survey localizing the new settlement with renewable energy (see Table 7). The German students and academic planners (weight 0.54) as well as regional planners and public authorities (weight 0.60) gave more consideration to the visual impact caused by solar panels and solar thermal collectors on the cultural heritage. Italian academic (0.58) and environmental planners (0.54) and English academic (0.54) and environmental planners (0.54), by contrast, more intrusively considered the solar power plants near landscape-protected areas and other beautiful areas.

German experts paid more attention to the environmental impact represented by the distance of wind turbines from important avifaunistic areas (respectively, weights: S.&AP., 0.50 and RP.&PA., 0.44). On the contrary, Italian experts expressed their preferences to the visual impact near historical and cultural facilities (S.&AP., 0.35; RP.&PA., 0.40), while English experts assigned almost equal weights to all three criteria, including the visual impact to landscape evaluable areas.

All experts, in particular the Italian regional planners and public authorities (0.71) as well as the German students and academic planners (0.70), assigned the highest weight to the criteria 'Distance from drinking water or aquifers' for geothermal vertical loops. Similarly, the experts, except for the English regional planners and public authorities (0.32), were in agreement regarding the importance of 'Distance from flooding areas' (average 0.63).

However, Italian experts assigned a similar weight regarding the visual impact of an additional chimney for a single power plant or a central biomass power plant near cultural/historical areas (0.48; 0.47) and landscape areas (0.52; 0.53).

The results of the survey showed similarities and differences between the stakeholder group preferences from the three countries. This outcome stems from national contrasts in planning systems and in attitudes towards micro-renewables. Transferring these preferences into a spatial representation resulted in an environmental suitability map which was overlaid with the energy potential for each micro-renewable technology. Figure 9 presents the three layers for solar energy. The combined result in Figure 10 shows the optimum sites for a new settlement development according to the energy potentials and expert preferences.

TABLE 7: Weighting for housing development using micro-renewable technologies

Criteria/factors	Experts	Weights		
		DE	IT	UK
Distance from landscape-protected areas and other beauty areas (S)	S.&AP.	0.46	0.58	0.54
	P.&PA.	0.40	0.54	0.54
Distance from historic/cultural facilities (S)	S.&AP.	0.54	0.43	0.46
	P.&PA.	0.60	0.46	0.46
Distance from historic/cultural facilities (W)	S.&AP.	0.26	0.35	0.30
	P.&PA.	0.31	0.40	0.35
Distance from Special Protection Areas (SPA) and avifaunistic important areas (W)	S.&AP.	0.50	0.45	0.39
	P.&PA.	0.44	0.30	0.38
Distance from landscape-protected areas or other beauty areas (W)	S.&AP.	0.25	0.19	0.31
	P.&PA.	0.25	0.30	0.28
Distance from historic/cultural features (B)	S.&AP.	0.30	0.38	0.34
	P.&PA.	0.35	0.47	0.39
Distance from landscape-protected areas or other beauty areas (B)	S.&AP.	0.70	0.62	0.66
	P.&PA.	0.65	0.53	0.61
Distance from historic/cultural features (GVL)	S.&AP.	0.31	0.45	0.36
	P.&PA.	0.37	0.29	0.68
Proximity to drinking water or aquifers (GVL)	S.&AP.	0.69	0.55	0.64
	P.&PA.	0.63	0.71	0.32
Distance from historic/cultural features (GHL)	S.&AP.	0.43	0.48	0.39
	P.&PA.	0.42	0.54	0.53
Distance from flooding areas (GHL)	S.&AP.	0.57	0.52	0.61
	P.&PA.	0.38	0.46	0.47

GVL, geothermal vertical loops; GHL, geothermal horizontal loops; S.&AP., students and academic planners; P.&PA., regional planners and public authorities; SD, standard deviation; DE, Germany; IT, Italy; UK, United Kingdom.

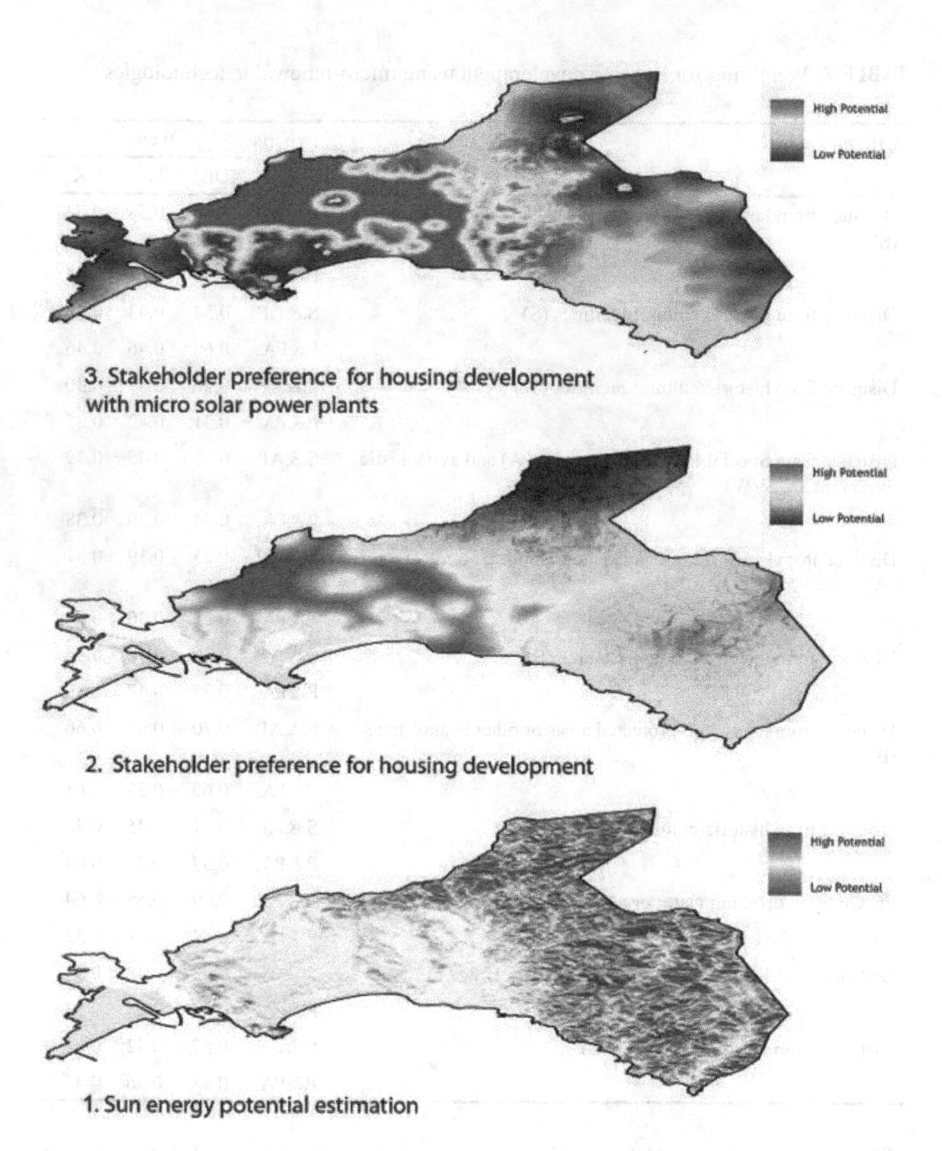

FIGURE 9: Overlaying of the three GIS layers (energy potential, settlement development, and settlement development using micro-generators).

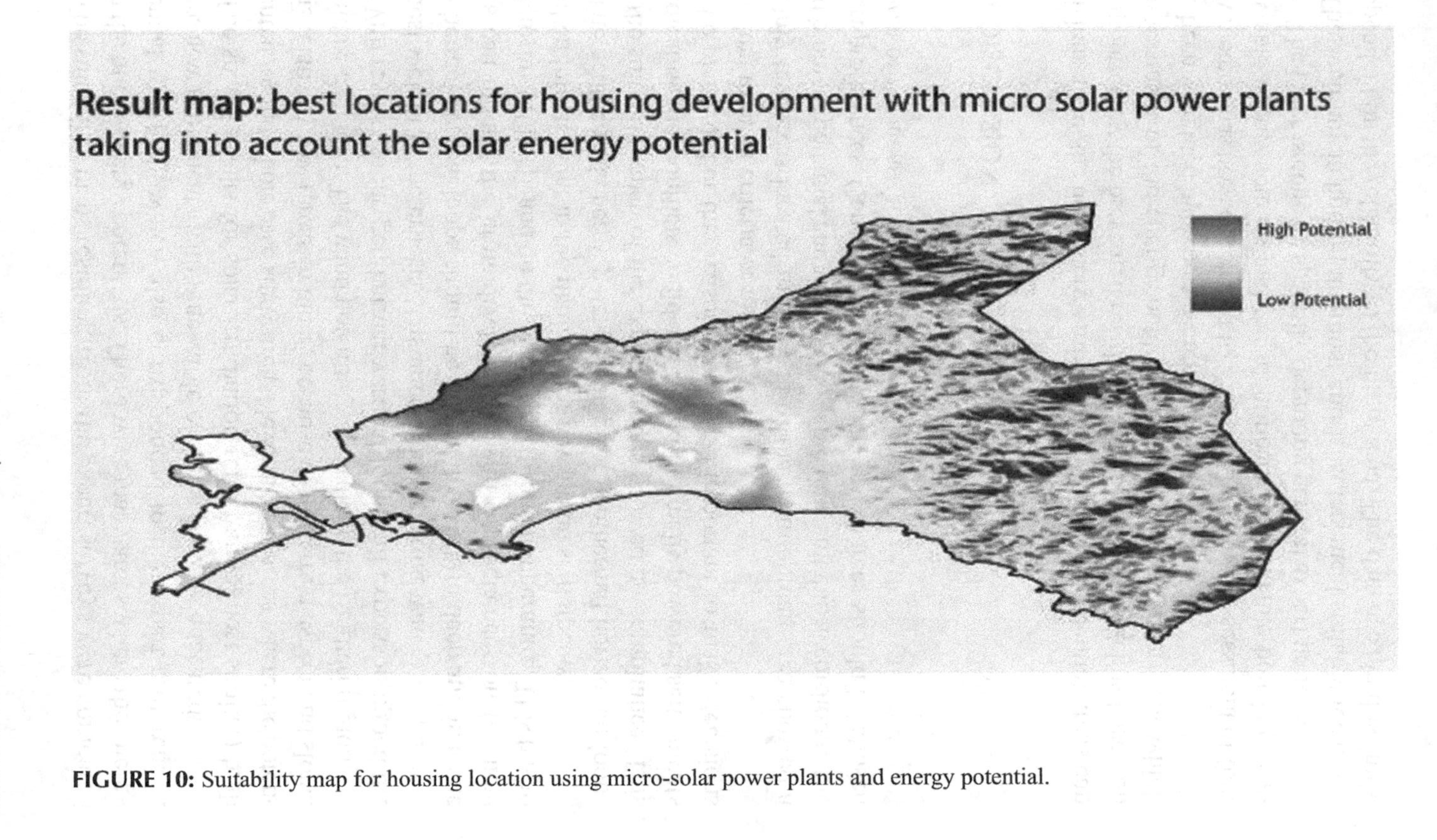
Result map: best locations for housing development with micro solar power plants taking into account the solar energy potential
High Potential
Low Potential

FIGURE 10: Suitability map for housing location using micro-solar power plants and energy potential.

The suitability maps resulting from the expert surveys were compared with the micro-energy potentials. There were many areas where the energy potential was high, which also corresponded to the expert preferences, where new settlements with renewable energies should be located.

The spatial results for solar irradiation reflect an (almost similar) high potential for the whole case study area. Nevertheless, areas where the potential is relatively low (areas in blue), because of the terrain aspect and slope, should be excluded. The wind potential varies along the Cagliari region.

With respect to the geothermal vertical loops, all experts assigned the highest weight to the criteria 'Distance from drinking water or aquifers.' Geothermal vertical loops should be buried up to 100 m deep and in some cases can modify the groundwater flow with consequences on the new settlements (cf.[44]), and on the water quality and temperature. This has to be taken into account in the planning of new urban settlements.

The geothermal energy potential map for geothermal horizontal loops and the suitability map of the expert groups showed no compliance. This is indicating that conflicts could arise if geothermally powered settlements will be planned. For this reason, it may be important to make decisions according to other criteria or needs.

Only few areas located in the east of the metropolitan area showed a good biomass potential and were at the same time suitable according to the expert preferences. Consequently, other energy sources should be chosen in most of the areas far away from any forests.

10.4 DISCUSSION

Decisions about the best energy mix for the different residential areas can be supported by the presented results. The proposed approach is based on an assessment of the energy potential and other relevant criteria which have been weighted by expert preferences.

A method has been developed for both the integrated assessment of four renewable energy sources and the identification of suitable housing locations. The latter is done by spatially depicting general expert preferences.

The accuracy in finding the most and least technical suitable locations is dependent on the reliability of the input data. The data used in this study originated from different sources and therefore also showed different levels of accuracy. For that reason, the resulting maps are less accurate than the least accurate layer used in their composition. As the methods have been successfully tested under German as well as Italian data conditions, it may be assumed that they can be applied in many European countries.

Since the calculation of the solar energy potential estimation depends on the application of the *r.sun* model and on the *pvgis* data, it can be applied in every region. However, the accuracy depends on input data (DEM) and on *pvgis* data availability.

Data of wind speeds are also available for every country. The only difference between the German and the Italian wind speeds was that the Italian data are calculated at a height of 25 m and the German ones at a height of 10 m, a height which better suits the requirements of microgeneration.

Wind speeds can deliver a good approximation to the wind energy potential, but for the planning of new settlements, a simulation of the wind flow would be more useful.

The accuracy of geothermal energy estimation is dependent on the data availability (e.g., the profiles) of the rock layers under the ground. This study demonstrated that even if no data on stratification and soil characteristics have been available, the information needed can be generated by the assistance of geologists with local knowledge. Nevertheless, it should be restricted to the creation of suitability maps for the use of horizontal loops. For more precision, it will be necessary to conduct further specific studies. It will also be important to have more detailed data regarding the groundwater level and movement in order to estimate the geothermal energy potential using the groundwater flow.

The biomass potential estimates can be calculated in every region. Only the data about forested areas which are available for biomass use or short rotation coppice are required. The next step will be to estimate the wood extraction capacity and transportation costs.

The best energy mix, calculated after normalization and overlay, could be derived more precisely if other criteria were included, e.g., costs, local characteristics, as well as electricity and heat needs. Such results would better support the practical planning. Methodologically, multicriteria evaluation methods in a GIS could be used for this optimization of settlement allocation with respect to sustainability.

Expert weighting of criteria about the location of an energy-efficient residential development in combination with the use of GIS and multicriteria analysis were useful for supporting the complex planning process. Various experts independently came to a considerable degree of agreement about their general preferences. The proposed method offers some advantages over the classical site suitability analysis techniques: First, it provides a structured approach to derive the suitability by 'decomposing' a complex problem into three levels (energy potentials, expert preferences for housing development, expert preferences for housing development with micro-generation technologies). This allows planners and public authorities to focus on a systematic analysis of the factors for each level. A disadvantage is that the criteria are less differentiated than in a conventional environmental impact or suitability assessment. Also, supplementing with new criteria needs considerable effort. Second, this method allows for incorporating criteria, which differ in nature. Furthermore, the method is a suitable way to weight the different criteria if no democratic legalized standards are available as a basis for weighting and decision making. Third, the approach provides an opportunity for decision makers to incorporate their own judgments. However, for a transparent practical application, the general expert preferences, which substitute the legal valuation and assessment standards, have to be presented separately from the preferences of local politicians and stakeholders. Fourth, the general preference and not a special site-specific individual interest is relevant, which may help at the same time to support rational decisions, in particular in local development and achieve a good acceptance of the results. Fifth, if regional/local stakeholder preferences are taken as a basis, the methods can be used in order to model the probable future expansion of housing development according to local interests. If mandatory zoning is weak or non-existent, land use planning can use this information for strategy building.

In the future, more concrete legal standards and priorities for decisions about energy-efficient housing and the environment may more strongly confine the importance of the expert preferences. In that case, both more predefined priorities should be included in the method and their combination with conventional impact assessment should be recommended.

Environmental planners and public authorities often make complex decisions within a short period of time when they must take into account sustainable development and participation. A set of land-use suitability maps (e.g., as part of a landscape plan) would be very useful for supporting fast decisions. Once the maps are available, land planners can analyze any new project by using simple operations such as map overlay or statistical analysis of a given area.

Clearly, the criteria selected for housing development and for micro-renewable preferences need to be combined with other siting factors which are relevant on broader scales. In this context, it will be interesting to compare the landscape plan of Sardinia and the land use plans of the municipalities with the results obtained in this work to identify the benefits and limitations. For further research, we will also integrate a few territorial and landscape-geographical aspects on a larger scale (e.g., a local scale) in order to compare them with those of the energy potentials. The aim is to give a more complete assessment to support planning decisions by integrating relevant territorial, environmental, and landscape criteria for assessing the new housing development.

10.5 CONCLUSIONS

The need to reduce the oil consumption and to produce renewable energy favors the integration of micro-renewable energy generation into housing development. Urban and regional planning can optimize this integration by selecting the best suited areas with the highest energy potential and the least environmental impacts as well as by choosing the best mix of renewables for each individual residential site. This paper suggests a methodology for finding the best locations for new housing developments which use micro-renewable technologies. The results obtained are of direct relevance for practical planning in different European countries. The methodology proposed can be an effective tool for planners in Europe to simulate new residential areas and evaluate their energy potentials for tracking changes and identifying the best solutions.

REFERENCES

1. EU (2010) Renewable energy White Paper laying down a Community strategy and action plan. http://europa.eu/legislation_summaries/other/l27023_en.htm . Accessed 21Sep 2010
2. Macleod A (2008) Using the microclimate to optimise renewable energy installation. Renewable Energy 33(8):1804-1813
3. Zedfactory (2010) Projects. http://www.zedfactory.com/about.html . Accessed 21 September 2010
4. Deleske A (2010) Stadtteil Vauban, Freiburg. http://www.vauban.de/ . Accessed 21 September 2010
5. Jenks M, Dempsey N (2005) Future forms and design for sustainable cities. Elsevier/Architectural Press, Oxford.
6. Droege P (2007) The renewable city: a comprehensive guide to an urban revolution. Wiley, Chichester.
7. Brookes L (2004) Energy efficiency fallacies—a postscript. Energy Policy 32:945-947
8. Linden A-L, Carlsson-Kanyama A, Eriksson B (2006) Efficient and inefficient aspects of residential energy behaviour: what are the policy instruments for change? Energy Policy 34(14):1918-1927
9. International Energy Agency (IEA): Key world energy statistics. 2007.. http://www.iea.org/textbase/nppdf/free/2007/key_stats_2007.pdf . Accessed 21 September 2010.
10. Maxwell EL, Renne DS (1994) Measures of renewable energy resources. NREL/MP-463-6254. National Renewable Energy Laboratory, Golden, Colorado.
11. Ivanov P, St L, Trifonova L, Renne D, Ohi J (1996) An investigation of renewable resources and renewable technology applications in Bulgaria. Environ Manage 20(Suppl 1):583-593
12. Schneidera DR, Neven D, Zeljko B (2006) Mapping the potential for decentralized energy generation based on renewable energy sources in the Republic of Croatia. Energy 32(9):1731-1744
13. Vettorato D, Zambelli P (2009) Estimation of energy sustainability at local scale. In: 45th ISOCARP Congress 2009. University of Porto, Porto.
14. Jankowski P, Nyerges T (2001) Geographic information systems for group decision making. Taylor & Francis, London.
15. Malczewski J (2006) GIS-based multicriteria decision analysis: a survey of the literature. Int J Geogr Inf Sci 20(7):703-726
16. Bredemeier B, Dandova M, Jähnchen I, König T, Liang X, Palmas C, Peschel I, Stegemann J (2009) Mikrogeneration erneuerbarer Energien. Energiepotenziale in der Region Hannover (Micro generation from renewables. Energy potential in the Region of Hannover). Master project. Department for Environmental Planning, Faculty of Architecture, Hannover University, Hannover, Alemanya.
17. Rode M, Schneider K, Ketelhake G, Reisshauer D (2005) Naturschutzverträgliche Erzeugung und Nutzung von Biomasse zur Wärme- und Stromgewinnung. Ergeb-

nisse aus dem F+E-Vorhaben 80283040 des Bundesamtes für Naturschutz. Bundesamt für Naturschutz, Bonn.

18. Voogd H (1983) Multicriteria evaluation for urban and regional planning. Pion, London.
19. Nijkamp P, Voogd H (1986) A survey of qualitative multiple criteria choice models. In: Nijkamp P, Leitner H, Wringley N (eds) Measuring the unmeasurable, Kluwer/Nijhoff, Dordrecht.
20. Eastman JR, Kyem PAK, Toledano J (1993) GIS and decision making. UNITAR, Geneva.
21. Malczewski J (1999) GIS and multicriteria decision analysis. Wiley, New York.
22. Geneletti D (2005) Formalising expert's opinion through multi-attribute value functions: an application in landscape ecology. J Environ Manage 76:255-262
23. Janssen R, Rietveld P (1990) Multicriteria analysis and geographical information systems: an application to agricultural land-use in the Netherlands. In: Scholten HJ, Stillwell JCH (eds) Geographical Information System for urban and regional planning, Kluwer Academic Publishers, Dordrecht.
24. Eastman JR, Jiang H, Toledano J (1998) Multi-criteria and multi-objective decision making for land allocation using GIS. In: Beinat E, Nijkamp P (eds) Multicriteria analysis for land-use management, Kluwer Academic Publishers, Dordrecht.
25. Banai R (1993) Fuzziness in geographic information systems: contributions from the analytic hierarchy process. Int J Geogr Inf Syst 7(4):315-329
26. Carver SJ (1991) Integrating multicriteria evaluation with geographical information systems. Int J Geogr Inf Syst 5:321-339
27. Joerin F, Theriault M, Musy A (2001) Using GIS and outranking multicriteria analysis for landuse suitability assessment. Int J Geogr Inf Sci 15(2):153-174
28. Pereira JMC, Duckstein L (1993) A multiple criteria decision-making approach to GIS based land suitability evaluation. Int J Geogr Inf Syst 7(5):407-424
29. Jankowski P (1995) Integrating geographical information systems and multiple criteria decision making methods. Int J Geogr Inf Syst 9(3):251-273
30. ISTAT (2010) Dati demografici (demographic data). http://www.istat.it . Accessed 21 September 2010
31. EU Joint Research Centre (JRC) (2010) Solar model *r.sun*. http://re.jrc.ec.europa.eu/pvgis/solres/solmod3.htm. Accessed 21 Sep 2010
32. Suri M, Hofierka J (2004) A new GIS-based solar radiation model and its application for photovoltaic assessments. Trans GIS 8(2):175-190
33. Touma JS (1977) Dependence of the wind profile power law on stability for various locations. Air Pollution Control Association 27:863-866
34. Counihan J (1975) Adiabatic atmospheric boundary layers: a review and analysis of data from the period 1880–1972. Atmospheric Environment 79:871-905
35. Kaltschmitt M, Huenges E (1999) Energie aus Erdwärme. Deutscher Verlag für Grundstoffindustrie, Stuttgart.
36. Gerlinger E, Gerlinger E (2008) oral. Biokompakt. www.biokompakt.com. Accessed 21 Sep 2010
37. Ueberhuber CW (1997) Monte Carlo techniques. In: Numerical computation 2: methods, software, and analysis. Springer, Berlin.

38. Cheney W, Kincaid D (2004) Numerical mathematics and computing. Thomson Learning, Belmont.
39. Saaty TL (1980) The analytic hierarchy process. McGraw-Hill, New York.
40. (2010) ESRI. http://www.esri.com/ Accessed 16 Jun 2012
41. Genoa University CESI Research Center (2002) Italian Wind Atlas (Atlante Eolico d'Italia). http://atlanteeolico.rse-web.it/viewer.htm. Accessed 13 Jun 2012
42. Verein Deutscher Ingenieure (VDI) (2010) Thermische Nutzung des Untergrundes.– Grundlagen, Genehmigungen, Umweltaspekte. Blatt 1, Düsseldorf, Richtlinie 4640.
43. Aru A, Baldaccini P, Vacca A (1991) Nota illustrativa alla carta dei suoli della Sardegna (Explanatory note to the soil map of Sardinia). Regione Autonoma della Sardegna, Università degli Studi di Cagliari. Region of Sardinia, Cagliari University.
44. Sass I, Burbaum U, Petrat L (2009) Staufen im Breisgau: Artesisches Grundwasser, Anhydrit und Karst im Konflikt mit geothermischen Bohrungen.Schwerter R (ed) 17. Tagung für Ingenieurgeologie, Zittau.

CHAPTER 11

Sustainability Appraisal of Residential Energy Demand and Supply: A Life Cycle Approach Including Heating, Electricity, Embodied Energy and Mobility

GERNOT STOEGLEHNER, WOLFGANG BAASKE,
HERMINE MITTER, NORA NIEMETZ, KARL-HEINZ KETTL,
MICHAEL WEISS, BETTINA LANCASTER,
AND GEORG NEUGEBAUER

11.1 BACKGROUND

The Energy Roadmap 2050 of the European Union aims at 80% to 95% greenhouse gas reduction until 2050 [1]. Energy saving, energy efficiency and the shift towards renewable energy supplies have to be jointly applied in order to reduce the environmental overshoot of the current energy systems. This environmental overshoot is due to the high energy intensity of society and the extensive use of fossil and nuclear energy sources [2]-[4]. All sectors of society and economy have to contribute to achieve this shift in energy supplies.

Households are important consumers of energy in the European Union, using up about 27% of the European energy demand within the settlement infrastructure plus a considerable part of the 33% of transport energy. Finally, 24% energy is consumed in the industrial sector, which can also be attributed to the consumption of private households to a large share (all figures based on 2009 data from Eurostat, [5]). Much of this energy demand is related to the dwelling itself, and concerning the mobility of persons, to the site of the residential area.

Planning decisions of settlements and houses have considerable impacts on the energy demand and the technological options for the energy supply of residential areas, and, therefore, on the energy consumption of society. These decisions are not only confined to the energy consumptions of households but affect the transport sector and, via embodied energy in goods and services, many industrial sectors as well. As spatial structures are long lasting, planning decisions determine the energy consumption of society in the long term and include, inter alia, the choice of a site, the determination of the infrastructure, the building densities aspired, the energy standards and the construction materials of buildings as well as the energy sources and energy provision technology.

In order to reach sustainable development, such decisions should be made on the sound assessment of alternatives including the aspects of energy demand and supply and their related environmental and socio-economic effects. Furthermore, the sustainable construction and operation of buildings and settlements will have to become a fundamental part of any sustainability oriented energy strategy.

Therefore, planning tools are required that empower decision makers to recognize the long-term consequences of their actions regarding the environmental and socio-economic impacts of buildings and settlements along their life cycle. Such planning tools have to offer a sound estimation of the energy demand of settlements, and an evaluation of the environmental and socio-economic impacts depending on the energy supply. Various tools that help optimizing buildings, are already available in great number (see e.g. BREEAM [6], baubook [7], WECOBIS [8]). Yet, these tools still have to be complemented by further approaches in order to support a comprehensive, sustainability-oriented quantitative assessment of planning decisions related to the energy demand and supply of housing that includes not only buildings but also technical infrastructure and the mobility of residents. In order to fill this gap, we developed a model to calculate the energy demand of households related to dwelling and location, as well as its overall environmental and socio-economic impacts, based on a life cycle approach: the Energetic Long term Analysis of Settlement structures (ELAS). This model was transferred to a freely available online decision-making tool that allows assessing and optimizing settlements, the ELAS calculator (www.elas-calculator.eu). The aim of this article is to introduce the complex ELAS model and the ELAS calculator and to show how this model can support stakeholders in planning processes to take more sustainable decisions about residential development.

When characterising the dwelling-related energy demand of settlements, the following components of energy consumption have to be taken into account:

- Energy demand and supply for the construction of buildings and public infrastructure like roads, sewage systems, water supply systems, street light etc.;
- Energy demand and supply for the maintenance of public infrastructure;
- Energy demand and supply of buildings for room heating, warm water and electricity; and
- Energy demand for the mobility of residents, which depends on the demographic structure of the population, the supply structure and the location of the settlement.

Based on these categories of energy demand and supply, an overall sustainability assessment is calculated with the following fundamental indicators: (1) environmental indicators: ecological footprint (as sustainable process index, SPI), life cycle CO_2 emissions and; (2) socio-economic indicators[a]: regional economic turnover, revenue, regional imports as well as jobs created. This fundamental assessment has the quality of an 'unsustainability' test [3]: Planning alternatives, that fail this test, should not be followed. If the alternatives pass, more detailed issues have to be covered applying further quantitative as well as qualitative indicators in order to guarantee sustainability in a broad perspective. Therefore, the ELAS model can help to reduce the information load on decision-makers by sorting out alternatives that do not provide the chance to strive for sustainable development from the perspective of climate change mitigation and sustainable energy supplies. ELAS can achieve that task by just using a few indicators that can be easily generated by the end users, even though the models behind this assessment are complex.

The model can be applied to existing settlements as well as to planned projects spanning from (1) new settlements as greenfield developments, (2) renewal and renovation of existing settlements with/without expanding them and (3) tearing down and reconstructing settlements on the same site or on a different site.

11.2 THE ELAS CONCEPT

11.2.1 CONTENT OF THE EVALUATION

As already pointed out in the introduction, activities to provide housing permeate through all economic sectors. Basic decision like the site chosen for a building and the technical standard of a building will have considerable impact on the energy consumption during its life cycle. This has to be reflected in any reliable decision support tool. The life cycle in this case therefore more resembles a 'life cycle-network' with the dwelling in the centre. Figure 1 shows the life cycle network includes the construction, maintenance and operation of buildings and infrastructure like roads, sewage systems, energy provision grids. The respective life cycles consist of the provision, transport demolition and final disposal or recycling of building materials as well as of building materials with the life cycle of infra-structure, and the life cycles of the energy supply of the buildings for electricity, heating and cooling. Furthermore, the life cycle regarding the mobility of residents is taken into account.

The concept realized in the ELAS calculator is to evaluate the impact of a building or settlement as well as any planned changes to a building or a settlement (including radical changes such as demolition and construction at a different site with a different technological standard) on this 'life cycle-network'. That means that all energy demand-and-supply-relevant impacts generated along the life cycles constituting this network are calculated and made transparent, not only in environmental but also in economic and social terms. This applies to current operation as well as to planned changes to the existing structure. As resource consumption already took place in the past, the status quo of buildings and infrastructure are not rated according to their environmental, economic and social impacts. By this approach, the users of the calculator can assess not only the direct impacts of their decisions but may also gain a comprehensive view on the consequences of their action on nature, economy and society.

The ELAS calculator is designed to support informed decisions about design and operation of buildings and settlements by all relevant actors. It therefore addresses different audiences in order to provide actors on different levels of decision taking with coherent information, thus enabling a discourse within as well as across different levels of actors. In particular, the ELAS calculator is designed to be used by the following:

- Single households in order to assess their individual decisions regarding design, operation and maintenance of their houses and flats, also including the impact their behaviour has on construction, operation and maintenance of infrastructure supporting their buildings and mobility induced by their choices;
- Municipal administrators in order to assess the ecological, economic and social impacts of existing settlements; and
- Planners in order to assess consequences of their activity regarding design and changes in settlements (enlargement, change of technical standards, changes regarding supporting infrastructure and measures concerning energy efficiency and energy provision systems, etc.), including impacts generated by necessary supporting infrastructure and induced mobility by residents.

This requires that the necessary data as well as the representation of results must be adapted to the needs of these actor groups. Figure 2 shows the overall architecture of the calculator. The two principal modes, the 'private mode' and the 'municipal mode' refer to the two main user groups, private households and professional users, respectively. Within the 'municipal mode', the separation of assessment of an existing settlement (status quo analysis) and planning of a new settlement (planning mode) represent the realization of the two tasks, analysis of impacts of existing settlements and planning of new settlements. It is however possible to change from the analysis of an existing settlement to the planning mode and thus assess changes for an existing settlement. Using this option, municipal administrators as well as planners can evaluate not only direct impacts of enlarging of settlements and/or upgrading of technical standards of buildings and infrastructure but also impacts of measures of spatial planning and zoning.

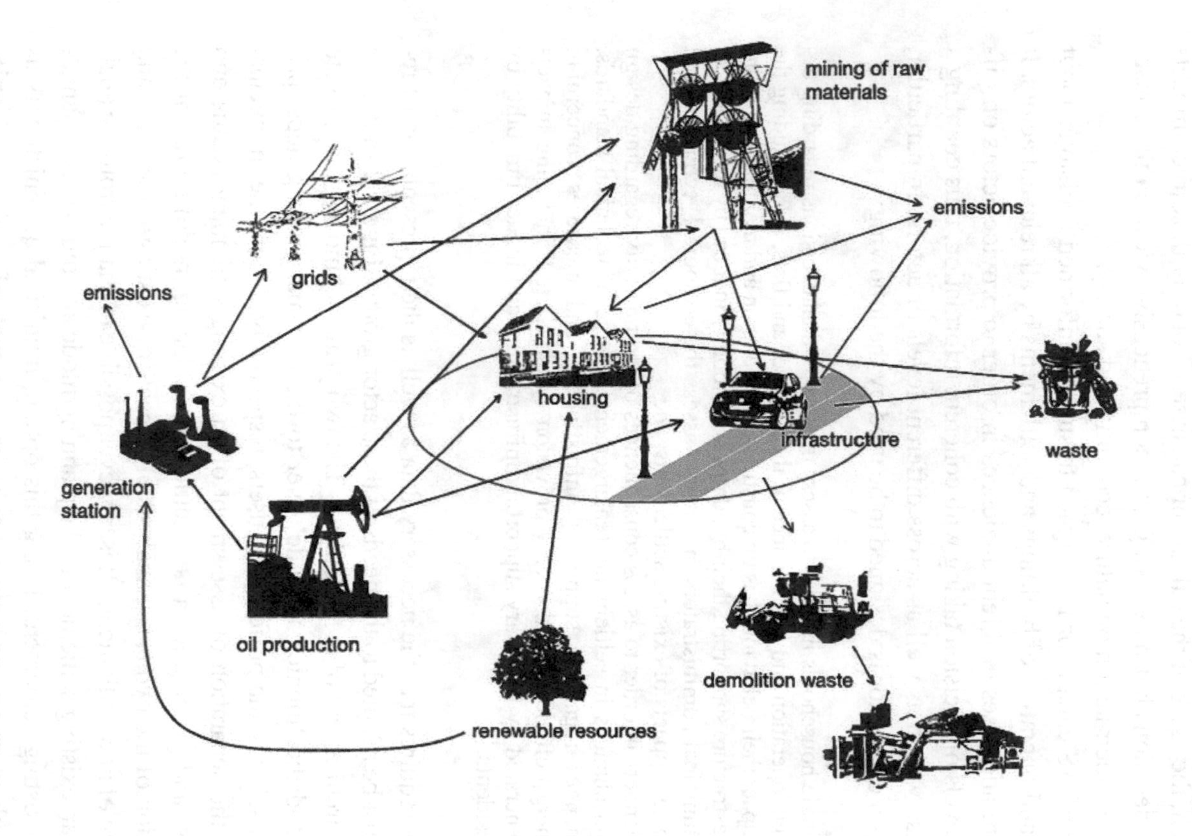

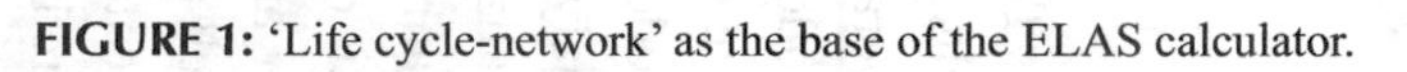

FIGURE 1: 'Life cycle-network' as the base of the ELAS calculator.

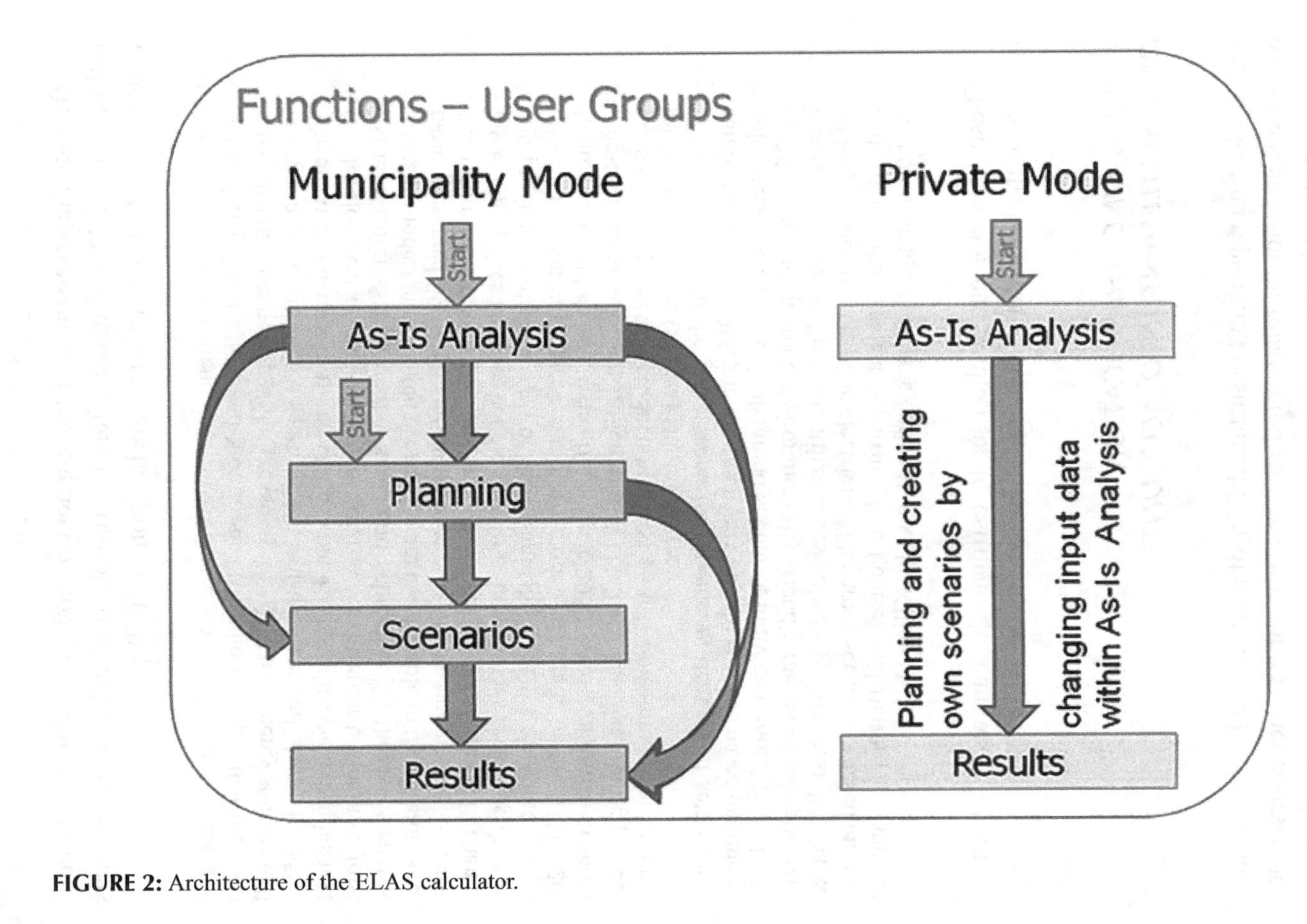

FIGURE 2: Architecture of the ELAS calculator.

Besides the assessment of the impacts of housing under current conditions, the ELAS calculator offers the option to calculate these impacts under future scenarios (with a time horizon of 2050):

- Trend scenario: The trend scenario is based on studies by Kratena and Schleicher [9] and Friedl and Steininger [10] who estimate that the total travel distance per person will increase by 25% and that energy sources will change: biogas will hold a share of 10% of the fleet and electricity 15%. The electricity demand will rise at 2.2% p.a. Both studies yield insights into the development of gross output in Austria for different business sectors. In the 'Baseline-Szenario' of Kratena and Schleicher [9], the rise in energy prices is modest at most, and the results are comparable to the baseline scenario presented by Friedl and Steininger [10]. It must be noted, however, that future economic projections in both models are not the result of changing energy prices as such, but are due to policy choices regarding the use and production of energy [9] or the sustainability of the transport system [10]. Furthermore, projections in both models are made up to 2015/2020. In order to establish long-term effects up to 2050, we use information about long-term scenarios from Bollen et al. [11]. They calculate future projections of energy use and production based on the 'Four futures for Europe' model of the Dutch Central Planning Bureau.
- Green scenario: In the green scenario, the energy demand of settlements can be reduced by 33%, and the electricity supply is 100% based on renewable energy sources. As in the trend scenario, individual mobility will rise by 25%, but the share of environmentally friendly means of transport will increase and the car fleet will utilize renewable energy sources: 70% based on biogas and 30% based on electricity. The green scenario is based on several studies like Arpaia and Turrini [12], Polasek and Berrer [13], and the economic analysis of energy price shocks by Kilian [14].

These scenarios enable the user to estimate the bandwidth of future impacts of the settlement within reasonable boundaries.

11.2.2 TAKING THE LOCATION OF A SETTLEMENT INTO ACCOUNT

Any evaluation regarding the sustainability of a settlement has to consider the spatial situation in relation to locations providing all necessary functions for residents living in it. Construction and maintenance of infrastructure as well as mobility requirements for residents depend critically on the distance between settlement and the provision of functions such as retail stores, educational institutions, health provision, administrative centres, etc. The ELAS calculator meets the challenge of systematizing this fundamental factor for the evaluation of settlements by defining different 'levels of centrality' that define what services are provided in the community the settlement is located in, according to Table 1. Higher levels of centrality refer to communities offering a broader spectrum of functions. The calculator offers a step-by-step help function to support the user in defining this fundamental set of data[b]. Using the distances to the different locations providing basic functions for residents (and using demographic information and modal split information explained later), the ELAS calculator is able to evaluate the part of the impact of settlements and buildings related to their locations.

TABLE 1: Definition of 'levels of centrality'

Level of centrality	Description
1	Community without centrality, no basic function and no local supply available
2	Community without centrality, no basic function but local supply available
3	Small centre
4	Regional centre (e.g. main municipality of a county)
5	Supra-regional centre (e.g. main municipality of a federal state)

11.2.3 SUSTAINABILITY EVALUATION METHODS USED IN THE ELAS CALCULATOR

With the ELAS calculator, a quantitative evaluation of settlements can be carried out from the viewpoint of sustainable development. The ELAS calculator is grounded on a coherent and consistent model and database that allows for an appraisal of environmental and socio-economic aspects. Therefore, the ELAS calculator has the strength to quantify decisions regarding residential development including spatial planning decisions. Yet, the ELAS calculator is not designed to evaluate qualitative issues of residential developments, which are as important as the numerical issues addressed by the ELAS calculator. Such aspects include, e.g. effects on biodiversity, landscape, quality of life, social and inter-generational equity or gender issues.

11.2.4 OVERALL ENVIRONMENTAL APPRAISAL

The overall environmental appraisal allows for a general assessment of the environmental impacts of settlements within the life cycle network according to Figure 1. The ELAS calculator uses the same database in order to provide a comprehensive, multidimensional appraisal of the environmental impacts by using several general indicators. Therefore, the ELAS calculator offers a comprehensive planning and assessment tool that relieves the users from considerations about the conclusiveness of life cycle limits or data provision, as the ELAS calculator secures coherence of the life cycle assessment. Furthermore, information overload of the users by too many indicators has to be avoided, as the ELAS calculator is designed to be used by decision-makers at the municipal level, who are often voluntary politicians and non-experts in urban planning and energy planning. Therefore, only few indicators have to be selected that provide a complex assessment and a clear picture about the environmental effects of a proposed settlement development concerning the building stock and/or new settlements. Therefore, measures were carefully selected. Finally, the following three measures are applied:

- Cumulative energy demand: This measure reveals the energy flows related to the life cycle network of a settlement and includes embodied energy from construction, renovation and infrastructure provision, energy for operation of the buildings and settlement as well as the mobility of residents. The measure expresses the fact that energy use causes an important share of environmental pressures and drafts a clear picture about the impacts of energy efficiency measures.
- Life cycle CO_2 emissions: This measure was chosen to express the effects of settlements on global warming and to assess the impacts of planning decisions regarding settlements on greenhouse gas reduction policies. With this measure, not only energy efficiency but also the contribution of different alternatives of energy sources and energy provision technologies for a settlement can be assessed.

- The sustainable process index (SPI) method as one calculation method for ecological footprints: the SPI method shows environmental pressures from all material and energy uses of the life cycle network by calculating the area of land which is associated with the supply of resources and the dissipation of emissions and wastes. The SPI method is applied to compare the overall environmental impact of a wide range of planning alternatives.

The three methods are described in more detail below.

11.2.5 CUMULATIVE ENERGY DEMAND

Energy is a major factor of the ecological pressure exerted by a settlement. The ELAS calculator accounts for all energy flows generated by the whole life cycle network. This includes the energy to operate the buildings (heating, cooling, electricity demand of appliances), the necessary supporting infrastructure (energy to operate sewage systems, street lighting, road service etc.) and mobility of residents. The calculator however also includes all 'embodied energy' that is necessary to provide the materials of construction for buildings and infrastructure or used in construction, renovation or demolition and disposal (when appropriate) for any planned changes to the current structure. This embodied energy is calculated using the methodology of the 'Kumulierter Energieaufwand - KEA' according to [15]. Embodied energy input will be related to 1 year by taking lifetimes of buildings and infrastructures (66 years) into account.

11.2.6 LIFE CYCLE CO_2 EMISSIONS

The calculation of CO_2 emissions is directly coupled with the calculation of the ecological footprint (see below). All fossil carbon inputs across the whole life cycle network as explained by Figure 1 form the base of the calculation. This includes also CO_2 emissions from synthetic materials used in construction of new buildings and infrastructures as well as in renovation, depreciated over the life time of the building and the renovation interval, respectively.

11.2.7 ECOLOGICAL FOOTPRINT AS SUSTAINABLE PROCESS INDEX

SPI is a method to calculate ecological footprints that takes emissions to air, water and soil besides resource provision into account [16]. This method compares anthropogenic and natural flows according to the following sustainability criteria [17]:

- Principle 1: Anthropogenic mass flows must not alter global material cycles and;
- Principle 2: Anthropogenic mass flows must not alter the quality of local environmental compartments.

The results of the footprint calculations are broken down into partial footprints for direct area consumption, fossil resource consumption, renewable resource consumption and emissions to air, water and soil.

11.2.8 SOCIO-ECONOMIC APPRAISAL

The socio-economic appraisal is based on regional economic input-output analysis [18]: the impacts of specific economic activities on the whole economic system are estimated by modelling the economic interaction of the different economic sectors. The model applies input-output coefficients that connect the different sectors. The sectors of this input-output analysis are based on the NACE systematic of economic sectors [19], dividing the sector 'electricity, gas and water supply' into non-renewable energy, renewable energy and water supply. The economic effects of building, maintaining and operating settlements are attributed to building construction, building operation, public infrastructure construction and operation as well as external effects (primarily related to mobility).

A special challenge is the regionalisation of socio-economic impacts taking the economic structure of the region—the province—into account where the settlement is situated. Applying techniques in accordance with Clijsters et al. [20] and Baaske and Lancaster [21], the regionalisation of the economic input-output coefficients are based on the national data provided by Eurostat [22]. With the techniques applied, problems of common regionalisation techniques, like the overvaluation of transformators and multipliers, can be avoided.

As a result of the socio-economic appraisal, the ELAS users are provided with estimations of the regional and national turnover and revenue, the imports induced as well as the regional jobs created with the construction, renovation, operation and the maintenance of settlements[c]. In this way, the socio-economic appraisal complements the environmental appraisal using indicators at the same degree of abstraction.

11.2.9 APPROPRIATE ROLE OF THE ELAS CALCULATOR IN PLANNING TASKS

Any meaningful evaluation tool must correlate with the requirements of a certain task within the course of taking decisions. The ELAS calculator is designed to support decisions within a planning process for buildings and settlements from the vantage point of sustainable energy systems. Stoeglehner and Narodoslawsky [3] provide a critique of proposed requirements for sustainability evaluation within planning processes and offer a framework for allocating evaluation methods to different assessment tasks, summarized in the indicator pyramid shown in Figure 3. The representation of this indicator pyramid reflects the fact that the information load and solid data increase along the decision pathway in planning. Evaluation should help to distinguish between alternative pathways, eliminating those that will in the end not lead to the desired goal of sustainability. The 'tip' of this pyramid is formed by general indicators that allow eliminating alternatives that clearly contradict the objectives of sustainable development. According to this framework, these general indicators should have a strong environmental bias, screening for alternatives that in the long term harm the natural base for human development. In order to facilitate elimination of unsustainable alternatives, highly aggregated indices (like the ecological footprint) are advantageous at this level.

As fewer alternatives remain in the planning process, information on the performance of each alternative increases, allowing to evaluate environmental impacts in more detail as well as adding other dimensions to the decision support provided by evaluation tools. This level of evaluation rates alternatives within the different dimensions of sustainability but aggregates data only within their particular field. The result of this evaluation then forms the base for the discourse between relevant actors who then weigh the different aspects to finally reach a planning decision for a certain alternative.

The ELAS calculator is intended as a planning tool for the 'pre-assessment level' in the indicator pyramid according to this framework. True to its objective to support planning of buildings and settlements to conform to sustainable energy systems, it has a strong energy and environmental fundament whilst providing numerical evaluation of economic as well as social aspects of projects regarding renovation, extension or greenfield development of buildings and settlements.

11.2.10 DATABASE OF THE ELAS CALCULATOR

The ELAS calculator draws on a large and comprehensive built-in database. Describing this database in detail would exceed the scope of this article by far. The reader is therefore kindly asked to consult the extensive background material provided on the ELAS calculator homepage at www.elas-calculator.eu for specific information regarding data sources and statistical material behind the ELAS calculator. The purpose of this chapter is to explain the general approach for data acquisition followed in the development of this tool.

Evaluating the complex life cycle network linked to operating and changing buildings and settlements requires a comprehensive database that must be adapted to the many specific aspects of individual buildings and settlements. It is the general approach of the ELAS calculator to allow the user as much leeway as possible to individualize his/her data in order to provide reliable decision support whilst at the same time reduce amount of data required from the user in order to increase user-friendliness of the program. Wherever possible, the program will provide sensible default values.

Besides striking a delicate balance between individualization and generalization of data, the ELAS calculator strives for data coherence. This means that life cycle data are taken only from one source [23] whenever possible, SPI-related data where taken only from the SPIonExcel (homepage: http://spionweb.tugraz.at). Statistical data underlying socio-economic evaluation are all taken from the material provided by Eurostat as already mentioned.

11.2.11 SPECIFIC DATA BASE OF THE ELAS CALCULATOR

Many data required to evaluate the impact of the life cycle network underlying the ELAS assessment method are not available in existing statistics. This applies in particular for individual mobility of residents according to the levels of centrality that form the base of characterising the spatial interaction between settlements and services used by residents. As this aspect is a prominent factor for the sustainability of settlements, great care has been applied during the development of the ELAS calculator to come up with reliable and recent data for evaluating mobility of residents.

Mobility associated with a building or settlement is influenced by a complex set of factors. Besides the distance to particular service providers (represented by the different levels of centrality and the distance from the settlement to the nearest cities associated with these levels), the demographic structure of the settlement has to be taken into account as residents within different age brackets show vastly differing mobility requirements and behaviours. The modal split used by residents to travel to service providers is in turn dependent on the centrality level of the settlement as the fraction of public transport increases with a higher level of centrality. Finally, the modal split also depends on the age bracket the individual resident belongs to.

In order to obtain realistic data for evaluation of crucial aspect of the sustainability of settlements as well as to verify statistical data from other sources, a thorough analysis of ten Austrian settlements in seven municipalities, representing all levels of centrality, was undertaken (see Table 2). The settlements ranged in size from 20 to 428 households. Within this analysis, all relevant parameters about buildings and infrastructure used in the ELAS calculator were gathered. In particular, this analysis encompassed a survey of households, inquiring the demographic set up of the household, consumer behaviour and technical building standard.

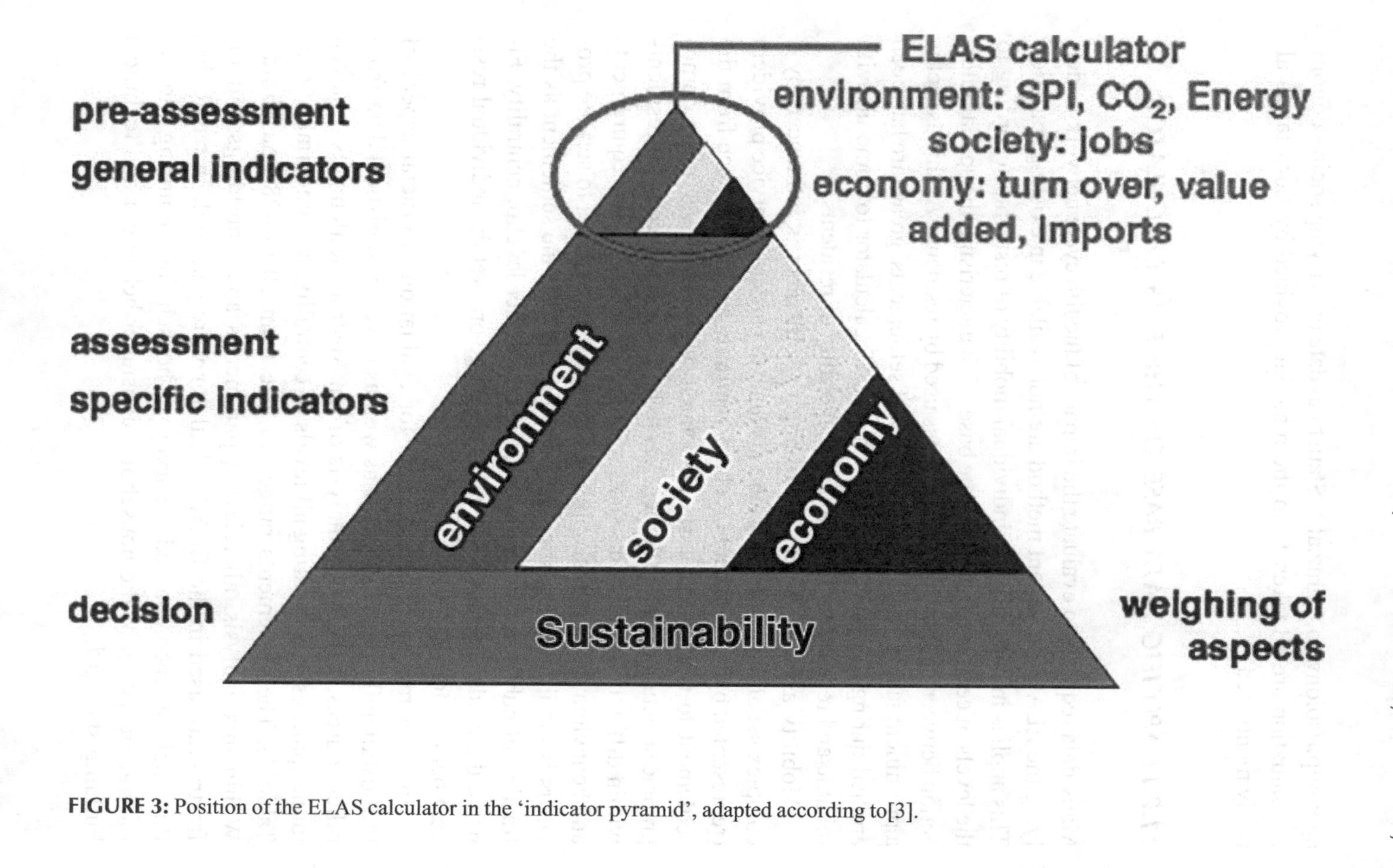

FIGURE 3: Position of the ELAS calculator in the 'indicator pyramid', adapted according to[3].

TABLE 2: Analysed settlements

Number	Settlement/municipality	Level of centrality	Characteristic building type	Number of households
1	Holzstraße/Linz	5	Multi-storey apartment buildings	233
2	Pregartenteich/Freistadt	4	Row houses	39
3	Billingerstraße/Freistadt	4	Multi-storey apartment buildings	109
4	Petringerfeld/Freistadt	4	One family houses	248
5	Row house settlements/Gallneukirchen	3	Row houses	42
6	One family house settlm./Kottingbrunn	3	One family houses	291
7	Whole town/Laab a. W.	2	Miscellaneous	428
8	Centre/Vorderstoder	1	Miscellaneous	25
9	Periphery/Vorderstoder	1	One & two family houses	20
10	Großnondorf/Guntersdorf	1	Miscellaneous	150

Questionnaires inquired the mobility behaviour of individual residents regarding frequency of travels, leisure mobility and modal split for all categories of mobility. This analysis was coupled to a participatory evaluation process in all settlements, involving all residents as well as stakeholders and political representatives, with public auditing events throughout the process. Local actors distributed the questionnaires and helped with additional information. Due to this participatory nature of the analysis, 37% of the 1,585 (i.e., 587) household questionnaires could be recovered, on top of 1,047 individual questionnaires. This statistical material allowed the formulation of 75 different modal splits linked to all 5 levels of centrality and 5 age brackets.

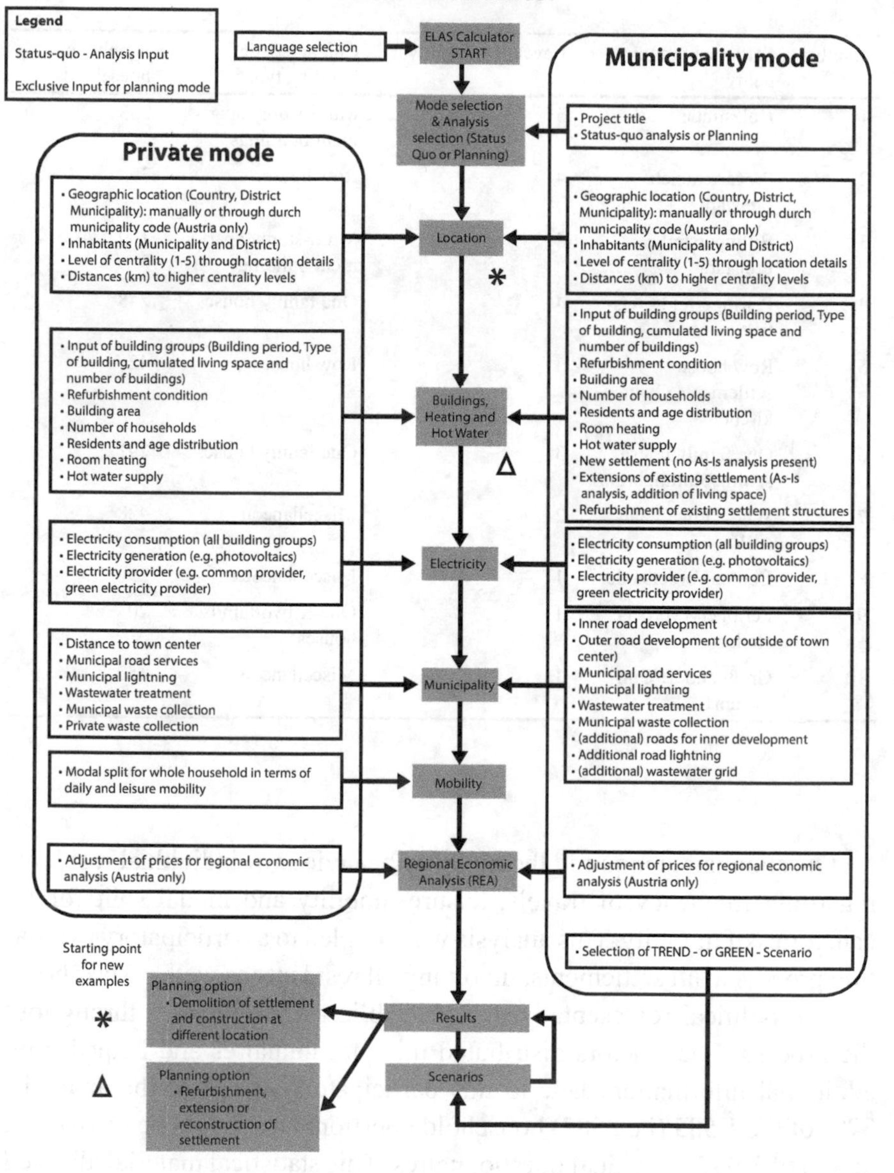

FIGURE 4: User-supplied input to the ELAS calculator (white entries apply to the planning mode).

11.2.12 USER-PROVIDED DATA

The user has to provide all data that define the building/settlement in sufficient detail to allow for reliable sustainability evaluation. The ELAS calculator offers extensive help functions to guide the user through the evaluation exercise as well as realistic default values wherever possible. A thorough discussion of the input data is outside the scope of this paper, the user may however draw extensive support material offered on the web page of the ELAS calculator.

Figure 4 offers a schematic overview over the input that has to be supplied by the user, differentiated into private mode and municipal mode. The calculator is available in German and English language version. In general, the input to the private mode is less complex and geared towards an audience of interested lay persons. Using the municipal mode and in particular the planning mode requires detailed knowledge about the infrastructure of the settlement and/or about planned changes.

Most sections in the private mode are in line with corresponding sections in the municipal mode (although the detail and volume of the required data differ). Starting the calculator requires decision about the language and the mode used, with the municipal mode differentiating further more into status quo analysis and planning mode.

A first set of input data requires the user to identify the site in terms of the level of centrality and distances to cities of higher level of centrality. Following the site definition, the user is asked to define the buildings in terms of age categories, size and technical status, number of households and residents as well as demographic distribution of residents. On top of that, the user has to define the technology used to provide heat and warm water. In this section, buildings with the same age and technical status are subsumed in 'building groups'.

The next section deals with electricity demand, electricity provision technology mix and (if applicable) production of electricity via photovoltaic panels on the building or within the settlement.

In the subsequent input section, the user may define the situation of the building/settlement within the municipality in terms of distance to the town centre as well as municipal services like road service, street lighting, waste and wastewater disposal.

Following that section, users of the private mode may define their mobility behaviour in detail. The last input section allows the user to change prices and flows of goods to adapt the regional economic analysis to the actual situation regarding the settlement to be evaluated.

In the municipal mode, the user may chose the 'trend scenario' and the 'green scenario' to estimate the impact of the building or settlement in the midterm future. Following the presentation of the results, the user in the municipal mode may then switch to the planning mode, adding more groups of building and additional infrastructure or evaluating the impact of any energy efficiency measures. The user may also evaluate the impact of dismantling the whole settlement and rebuilding it at a new site.

11.3 RESULTS FROM THE ELAS CALCULATOR

Ecological evaluation results from the ELAS calculator are represented in terms of total energy consumption, life cycle CO_2 emissions and SPI in graphical as well as tabular form. Using either the private mode or the 'As-Is Analysis' of the municipal mode will provide life cycle wide ecological impacts corresponding to 1 year of the operation of building or settlement. Results of the planning mode will always include the ecological impact of the planned infrastructure, both for buildings, renovation (summarised in the category heating) and municipal services and, if applicable, for demolition of buildings. Infrastructure will also be referred to 1 year, taking the life time into account.

Graphical representation for all valuation methods will be broken down into heating and hot water provision, electricity, municipal services daily mobility and leisure mobility. In case of the SPI valuation, an additional graphic will provide the breakdown of the ecological footprint according to infrastructure, fossil resources, renewable resources, non-renewable resources and emissions to air (excluding CO_2 which is rated in the fossil resources part), water and soil (see Figure 5).

Tabular results provide overall results as well as more detailed information, breaking down results of energy consumption into different energy forms (for heating) and applications (heating/warm water provision), electricity into grid provided and own production via photovoltaic panels, municipal services into waste and wastewater disposal, street lighting and street services. Mobility is differentiated according to the modal split and depicted for daily as well as leisure mobility. Tables relating to the SPI will also provide details about the breakdown of the ecological pressure of the different aspects of the life cycle into impact categories.

Socio-economic analysis shows turnover, value added, import and jobs created, differentiated to the national and regional level in tabular form. Results will also provide a breakdown for economic effects of different aspects such as construction, operation of buildings, municipal services and external effects (in particular mobility) on the categories defined above (see Figure 6). In addition to that, imports induced by the building or settlement will be shown in the overall summary of results. The results have to be interpreted carefully, because the logic of 'more is better' cannot necessarily be applied to the regional economic analysis: 'more' in the energy sector could mean that the regional population might be restricted in other areas of consumption which would be more effective in the regional economy. Therefore, we added the imports in relation to turnover and revenue. If regional renewable resources are used, the money spent on energy aspects of a settlement will be regionally operative.

A multitude of ELAS calculations shows that the impact of spatial planning decisions, such as developing mixed-use structures with sufficient daily supply and social infrastructure, as well as pursuing settlement density has an important effect on the energy demand and the environmental pressures related to the energy supply of residential areas. The efficient supply with renewable energy sources can be supported, and the energy demand for the mobility of the residents can be reduced as people in more central residential areas drive less and are able to cycle, walk and use public transport.

With the optimisation function for existing settlements, we explored that through integrated spatial and energy planning measures enormous positive environmental effects can be achieved. Such measures combine spatial developments based on mixed-use, short-distance supply and (moderate) density with innovative energy technologies as well as energy efficiency measures in buildings and infrastructures. They make it possible to reduce the energy demand by 30% to 50% and the associated negative environmental effects in terms of footprint and life cycle CO2 emissions up to 80%.

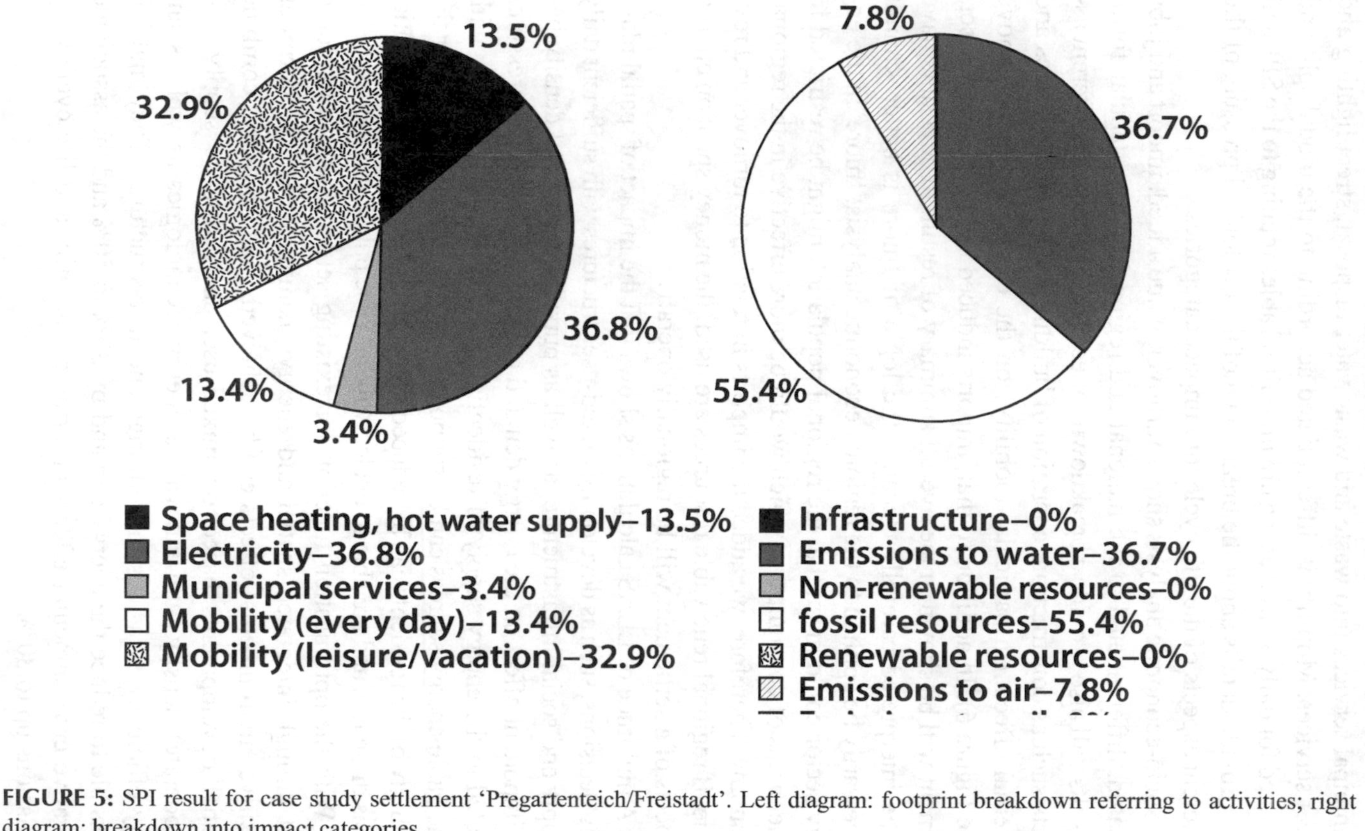

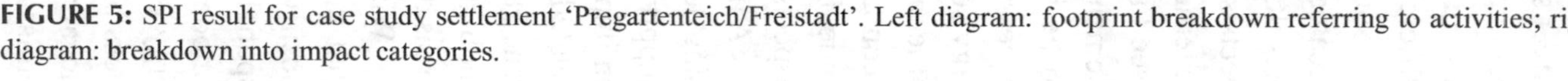

FIGURE 5: SPI result for case study settlement 'Pregartenteich/Freistadt'. Left diagram: footprint breakdown referring to activities; right diagram: breakdown into impact categories.

Value added effects according to initiator - Austria

Category	Turn over	Value added	Imports	Jobs
Living Space, construction	0 €	0 €	0 €	0.0
Living Space, operation	85,137 €	38,274 €	13,731 €	0.3
Municipal Infrastructure, construction and operation	11,738 €	6,541 €	1,217 €	0.1
External Effects (Mobility)	210,229 €	97,010 €	35,049 €	1.4
Total	307,105 €	141,825 €	49,996 €	1.8

Value added effects according to initiator - Oberösterreich

Category	Turn over	Value added	Imports	Jobs
Living Space, construction	0 €	0 €	0 €	0.0
Living Space, operation	76,975 €	32,227 €	19,778 €	0.3
Municipal Infrastructure, construction and operation	9,886 €	5,415 €	2,342 €	0.1
External Effects (Mobility)	168,268 €	67,200 €	64,858 €	0.9
Total	255,130 €	104,842 €	86,979 €	1.3

FIGURE 6: Socio-economic results for case study settlement ‘Pregartenteich/Freistadt’.

In new developments, we found out that in low-dense settlements relying on stand-alone single-family houses, the energy demand for streets, sewage systems and other related technical infrastructures in the settlement can exceed 50% of the total energy demand of a settlement, causing up to 90% of the life cycle CO_2 emissions and up to 80% of the ecological footprint. The construction of houses plays a relatively low role with around 5% share on the total energy demand. By more dense settlement structures like four- to five-floor apartment buildings, the share of infrastructures can be reduced to about 10% to 15% accounting for less than 50% of the life cycle CO_2 emissions and 40% of the ecological footprint. Further determinants of the energy demand of settlements are the degree of centrality of the municipality and the distance of the settlement to the supply centres within the municipality.

11.4 DISCUSSION OF THE ELAS APPLICABILITY

Decisions concerning construction and operation of buildings and settlements are pivotal to achieving sustainability and in particular to arrive at a sustainable energy system of society. This is due on the one hand to the large fraction of energy used in buildings (in particular in Europe) and on the other hand to the considerable influence the way settlements are situated has on other energy uses, in particular by shaping mobility requirements of residents and modal split of travels. As such decisions, due to the longevity of buildings, have long ranging consequences, thorough and reliable decision support is necessary.

Providing such comprehensive decision support for individual house owners as well as for municipal administrations and planners was the goal of the 'energetic long term analysis of settlement structures' study that forms the background to this article. Based on a thorough analysis of ten settlements in different spatial situations and featuring a variety of building types and demographical set-ups of residents, a web-based calculator was developed that evaluates buildings and settlements from the point of view of their sustainability. The ELAS calculator can support decision-making in spatial planning in the following situations:

Comparison of settlements of the same type on different sites;

- Comparison of variants concerning the design of settlement on one site, e.g. concerning the size and density of the settlement, efficiency of buildings, the selected energy sources and technologies and on-site renewable energy production;
- Comparison of renovation of existing buildings, with re-densification and enlargement of settlements; and
- Iterative optimisation of settlement site and settlement design.

Therefore, the calculator can be used in different stages of a planning and design process. In planning processes, decisions are taking place at least at three different levels, which can be explained by the types of alternatives included in the planning process [24]:

- System alternatives: choice of demand, size of projects, density, technological networks etc.;
- Site alternatives: choice of sites for projects; and
- Technical alternatives: choice of technical implementation of a project on a given site.

The ELAS calculator allows to 'jump' between the scales: sometimes, the quality of the available sites or certain site-specific limitations hinder the implementation of system alternatives. Therefore, the ELAS calculator can be used at different stages of planning processes: (1) to evaluate system alternatives with few assumptions about technical aspects of potential energy systems concerning their overall energy demand and supply and the related environmental and regional socio-economic effects; (2) to evaluate and rank site alternatives for given kinds of projects, e.g. multi-storey housing, terrace houses etc.; and (3) to assess technical options, e.g. certain insulation or energy sources and technologies for project implementation. Finally, objectives for site-specific design processes can be set from the perspectives of SPI, life-cycle CO_2 emissions and overall regional socio-economic effects.

By applying the ELAS calculator, users can assess if their decisions are in line with energy policies and climate change policies promoting energy saving, energy efficiency and the use of renewable energy supplies and can optimize planning decisions to reduce energy demand and environmental impact in depending on the chosen energy supply technologies and sources. This is valuable for the following different target groups the ELAS calculator is intended to address [25]:

Legal bodies (municipal mode) are enabled to assess based on case studies, how legal proposals in spatial planning, housing subsidies, building codes etc. might impact the energy demand and supply of settlements and might contribute to achieving international and national energy policy and climate protection targets.

Municipal decision-makers and planners (municipal mode) can assess local spatial planning activities (land use plans, master plans, zoning plans) concerning residential developments with respect to the environmental and socio-economic impacts of energy demand and supply. They are enabled to choose planning alternatives with a high potential to be sustainable for detailed assessments, e.g. in strategic environmental assessments.

Developers (municipal mode) are able to estimate the energy demand as well as environmental and socio-economic impacts of their future dwellers. In doing so, they can assess alternatives of sites and optimize the design of their settlement projects. Furthermore, they might use the results in marketing their products to customers interested in low energy demand and sustainable energy supplies.

Single households (private mode): Individuals can also assess their choices regarding their dwellings taking the structural aspects of the settlement and their individual behaviour into account. Especially interesting decisions situations for the ELAS application might be the comparison of different flats when intending to move, decisions about thermal insulation, change of heating devices or PV-installation, change of mobility patterns etc. so that interested individuals can choose planning options that help them to lead a more sustainable life and contribute to climate protection and the shift of energy supplies towards a renewable resource base. In the private mode, less knowledge and information is needed. Only information has to be entered by the users that is within their decision scope.

11.5 CONCLUSIONS

Within the research carried out, it became obvious that building and infrastructure construction and maintenance as well as site-induced mobility are crucial for the sustainability of settlements. The site of the settlement is a determining factor for both. The ELAS approach addressed this by introducing different levels of centrality, differentiated according to the services available in the town the settlement belongs to. Thorough studies linking these levels of centrality and the demographic set-up of residents with daily and leisure mobility allow the evaluation of these factors within the ELAS calculator without requiring excessive data acquisition from the user. The ELAS calculator offers free accessible evaluation for stakeholders involved in long ranging decisions regarding buildings and settlements. With this tool stakeholders can readily integrate aspects of sustainability in their decisions, both regarding the operation of buildings and settlements as well as in planning renovations, enlargements or even relocation of settlements. As can be seen from the list of target groups and their potential benefits from ELAS applications, the ELAS calculator supports consistent decision-making from the policy level via the regional and local planning levels to the household level.

ENDNOTES

[a]In the current version of the ELAS calculator, the socio-economic evaluation is only available for Austrian cases.

[b]For application of the ELAS calculator to settlements in Austria, the user may only supply the ZIP code to define the distances to cities with other (higher) level of centrality as the calculator will draw on a database based on [25],[26]

[c]In the current version, regionalized input-output tables are available for Austrian Federal States.

REFERENCES

1. (2011) Communication from the Commission to the European Parliament, the Council, the European Economic and Social Committee and the Committee of the Regions: Energy Roadmap 2050, (COM/2011/0885 final).
2. Narodoslawsky M, Stoeglehner G (2010) Planning for local and regional energy strategies with the ecological footprint. J of Environmental Policy & Planning 12(4):363-379
3. Stoeglehner G, Narodoslawsky M (2008) Implementing ecological footprinting in decision-making processes. Land Use Policy 25:421-431
4. Stoeglehner G (2003) Ecological footprint - a tool for assessing sustainable energy supplies. Journal of Cleaner Production 11(3):267-277
5. [http:/ / epp.eurostat.ec.europa.eu/ portal/ page/ portal/ esa95_supply_use_input_tables/ data/ workbooks] Eurostat (2011) ESA 95 supply use input-output tables. European Commission, Accessed December 2012
6. [www.breeam.org] BREEAM International (2014). . Accessed Mar 2014
7. [www.baubook.at] Energieinstitut Vorarlberg, Werkzeuge für ökologische Produktauswahl. . Accessed Mar 2014
8. [www.wecobis.de] BMVBS (n.d.) Federal ministry of transport, building and urban development. Germany. . Accessed Mar 2014
9. Kratena K, Schleicher S (2001) Energieszenarien bis 2020. WIFO, Wien.
10. Friedl B, Steininger KW (2002) Environmentally sustainable transport: definition and long-term economic impacts for Austria. Empirica 29:163-180
11. Bollen J, Manders T, Mulder M (2004) Four futures for energy markets and climate change. No.52, CPB. Netherlands Bureau for Economic Policy Analysis, The Hague.
12. Arpaia A, Turrini A (2008) Government expenditure and economic growth in the EU: long-run tendencies and short-term adjustment. European Commission, Brussels.
13. Polasek W, Berrer H (2005) Regional growth in Central Europe, long-term effects of population and traffic structure. IHS, Wien.
14. Kilian L (2007) The economic effects of energy price shocks. CEPR Discussion Papers 6559, published as: Kilian, L., 2008, The Economic Effects of Energy Price Shocks. Journal of Economic Literature 46(4):871-909
15. (1999) Erarbeitung von Basisdaten zum Energieaufwand und der Umweltbelastung von energieintensiven Produkten und Dienstleistungen für Ökobilanzen und Öko-Audits. i.A. des Umweltbundesamtes, Darmstadt. Freiburg, Berlin.
16. Krotscheck C, Narodoslawsky M (1996) The sustainable process index - a new dimension in ecological Evaluation. Ecological Engineering 6(4):241-258
17. Narodoslawsky M, Krotscheck C (2000) Integrated ecological optimization of processes with the sustainable process index. Waste Management 20:599-603
18. Miller RE, Blair PD (2009) Input-output analysis: foundations and extensions, 2nd edition. Edinburgh, Cambridge

19. (2002) Commission Regulation (EC) No 29/2002 of 19 December 2001 amending Council Regulation (EEC) No 3037/90 on the statistical classification of economic activities in the European Community.
20. Clijsters GMJ, Oude Wansink MJ, Peeters LMK, Baaske WE (2003) Adapting QFD for evaluating employment initiatives. In: Transactions from International Symposium on QFD 2003/15th Symposium on Quality Function Deployment. QFD Institute, Orlando. 12–13 December 2003
21. Baaske WE, Lancaster B (2004) Evaluating local commitment for employment - towards a realisation of the European employment strategy. Trauner, Linz
22. (2007) Energy, transport and environment indicators, Eurostat pocketbooks.
23. [http://www.ecoinvent.org/database/] EcoInvent database (2010) Swiss Centre for Life Cycle Inventories. Zürich. . Accessed December 2012
24. Stoeglehner G (2010) Enhancing SEA effectiveness: lessons learnt from Austrian experiences in spatial planning. Impact Assessment and Project Appraisal 28(3):217-231
25. Stoeglehner G, Narodoslawsky M, Kettl KH (2013) Assessing the energy aspects of settlements. In: Passer A, Höfler K, Maydl P (eds) Sustainable buildings construction products and technologies, Collection of Full Papers. Verlag der Technischen Universität Graz, Graz. pp 1537-1544
26. (2010) Feststellung der Bevölkerungszahl für den Finanzausgleich gemäß § 9 Abs. 9 FAG 2008.

Author Notes

CHAPTER 1

Acknowledgments

This paper draws on the work of my friend and colleague Gordon McGranahan and was much improved by his comments and suggestions.

CHAPTER 2

Funding

Funding was provided by the National Natural Science Foundation of China (41201598), http://www.nsfc.gov.cn; Fujian Provincial department of Science and Technology (2011R0093,2010I0014), http://xmgl.fjkjt.gov.cn/; and China Postdoctoral Science Foundation (20110490614), http://www.chinapostdoctor.org.cn/. The funders had no role in study design, data collection and analysis, decision to publish, or preparation of the manuscript.

Competing Interests

The authors have declared that no competing interests exist.

Author Contributions

Conceived and designed the experiments: TL. Performed the experiments: TL YY LF JW. Analyzed the data: TL LF XB. Contributed reagents/materials/analysis tools: TL XB. Wrote the paper: TL XB JW.

CHAPTER 3

Acknowledgments

Funding for this research was from Australian Research Council Discovery Project DP 0878231, Consuming the Urban Environment: A Study of the Factors That Influence Resource Use in Australian Cities. The authors acknowledge the contributions of Terry Burke and Maryann Wulff to the development of the Living in Melbourne survey that provided the data for this paper.

Conflict of Interest

The authors declare no conflict of interest.

CHAPTER 4

Competing Interests

The author declares that he has no competing interests.

CHAPTER 5

Acknowledgments

The data was collected by researchers in Kigali with cooperation of the local community leaders. The authors would like to thank the researchers Roger Mugisha and Carine Tuyishime. This research is part of the 3K-SAN project, funded by SPLASH, Swiss Agency for Development and Cooperation (SDC), which is investigating how to catalyse self-sustaining sanitation chains in low-income informal settlements in Kigali (Rwanda), Kampala (Uganda) and Kisumu (Kenya). Self-sustaining sanitation chains are defined here as socio-technological systems that provide continued health and environmental improvement, as required to meet the MDGs, without continued external intervention. This definition includes, but is not limited to, construction, maintenance, and management of the waste through pump-out/collection services, transport, treatment and re-use or disposal.

Conflicts of Interest

The authors declare no conflict of interest.

CHAPTER 7

Competing Interests

The authors declare that they have no competing interests.

Author Contributions

MDD, AL, MCG and AA devised the study and wrote the manuscript. MDD and AL performed the data analysis and data-driven simulations. All authors reviewed and approved the complete manuscript.

Acknowledgments

MDD is supported by the European Commission FET-Proactive project PLEXMATH (Grant No. 317614), AA by the MULTIPLEX (grant 317532) and the Generalitat de Catalunya 2009-SGR-838. AA also acknowledges financial support from the ICREA Academia, the James S. McDonnell Foundation, and FIS2012-38266. MCG acknowledges Accenture and the KACST-Center for Complex Engineering Systems.

CHAPTER 8

Acknowledgments

This research was funded by the Engineering & Physical Sciences Research Council ARCADIA: ARCADIA: Adaptation and Resilience in Cities: Analysis and Decision making using Integrated Assessment project (Grant No. EP/G061254/1), the European Community's Seventh Framework Programme under Grant Agreement No. 308497 Project RAMSES—Reconciling Adaptation, Mitigation, and Sustainable Development for Cities, and the Tyndall Center for Climate Change Research Cities programme. Richard Dawson is funded by an EPSRC fellowship (EP/H003630/1) All maps contain Ordnance Survey data © Crown copyright and database right 2014. Light rail data is copyright Transport for London 2009. OpenStreetMap data is open data, licensed under the Open Data Commons Open Database License (ODbL).

Author Contributions

All authors have contributed to the development of the research described in this article. Alistair Ford was the lead researcher in the development of the tools and production of the data, Richard Dawson analyzed the accessibility patterns and produced the resultant curves, whilst Stuart Barr oversaw the research project and reviewed the article during production. Philip James contributed to the review and preparation of the article.

Conflicts of Interest

The authors declare no conflict of interest.

CHAPTER 9

Acknowledgments

The authors thank Innovative City Partnership Programme for financial support.

Conflict of Interest

The authors declare no conflicts of interest.

CHAPTER 10

Competing Interests

The authors declare that they have no competing interests.

Author Contributions

CP conceived and carried out the study and drafted the manuscript. She worked on all parts and in particular on the adaptation and development of methods for the estimation of the micro-energy potentials. EA helped define the general methodology of the study, participated in the design, landscape criteria selection, and planning in the eastern metropolitan area of Cagliari, and supervised the survey in Italy. AL helped develop the structure of the paper and supervised the multicriteria analysis, GIS work, and the survey in UK. CvH helped define the general methodology of the study, participated in the environmental impact selection of micro-renewable technologies, landscape, and regional planning, and supervised the survey in Germany. All authors read and approved the final manuscript.

Author Information

CP received the title of Doctor Europeus from the universities of Cagliari and Hannover in 2011. Her research interests are centered on the integration of micro-renewable technologies in regional and urban planning.

Acknowledgments

This study of the integration of micro-renewable technologies in regional planning was financially supported by the Italian Ministry of Education, University and Research (Ministero dell'Istruzione, dell'Università e della Ricerca (MIUR)).

CHAPTER 11

Competing Interests

The authors declare that they have no competing interests.

Author Contributions

All authors have contributed to the ELAS project, GS as project manager. All authors read and approved the final manuscript.

Acknowledgments

The model was developed by the team of authors in an interdisciplinary research project named 'ELAS - energetic long-term analysis of settlement structures' co-funded by the Austrian Climate and Energy Fund (project number: 818.915), the Provinces of Upper Austria and Lower Austria as well as the Town of Freistadt. Therefore, we will further refer to the model as ELAS model. The ELAS model was transferred to an open-access internet tool, the ELAS calculator, available at www.elas-calculator.eu. This paper reveals the ELAS concept, the main approaches to calculating the outcomes and the way the database for the calculation was created.

Index

C

D

E

Q

R

S

T